Dr. Craig Harper and his students have used scientific rigor over the past 15+ years to research every facet of wildlife food plots and habitat management for land managers and hunters seeking to improve their property for wildlife. Luckily for us, all this research has been compiled and condensed in a single book for the layperson and wildlife professional alike. This book is the most significant and comprehensive work to date on the subject. If you are looking for information on food plots and early successional plants for wildlife management, you need this book.

DR. BRONSON STRICKLAND
St. John Family Professor of Wildlife Management
Extension Wildlife Specialist
Mississippi State University

Dr. Craig Harper is the go-to guy for wildlife habitat issues. He always has a workable solution and this book is full of them. This book should be the cornerstone of your wildlife habitat development program.

NEIL DOUGHERTY
North Country Whitetails

WILDLIFE FOOD PLOTS AND EARLY SUCCESSIONAL PLANTS

Craig A. Harper
Professor of Wildlife Management
Department of Forestry, Wildlife, and Fisheries
University of Tennessee Institute of Agriculture

Produced by NOCSO Publishing
Maryville, TN 37802
http://nocsopublishing.com

Printed by Thomson-Shore

CONTENTS

FOREWORD

Perhaps the first narrative describing the use of an agricultural planting to harvest white-tailed deer was penned in 1854 by Philip Tome in his classic volume Pioneer Life; or Thirty Years a Hunter. Tome describes a visit (circa 1810) to a settlement originally colonized in 1805 by a group along Pine Creek in Pennsylvania: "...a man in the vicinity conceived the idea of attracting [deer] to fields cleared and sowed with wheat... and built a fence around it... The planting had the desired effect and during the season he took in this manner between sixty and seventy deer." Certainly, the scenario of harvesting deer over agricultural fields was repeated on myriad occasions as settlers established farmsteads across the range of the whitetail.

However, it wasn't until more than a century later that the utility of agricultural plantings to aid in managing wildlife populations was recognized by the emerging field of wildlife management. In the 3rd volume of the Journal of Wildlife Management (1939), Aldo Leopold describes a 4-year study that evaluated various agronomic plantings for wildlife food patches. Subsequent to World War II, many state wildlife agencies established programs to enhance and manage habitat for various species of wildlife, including planting food plots. By the 1970's, agronomic plantings were a common practice on many state lands managed for wildlife.

Despite increased understanding of the important role that agronomic plantings can play in wildlife management, food plot plantings on private lands lagged until the late 1980's. The year 1988 saw the launch of the commercial food plot industry, which concentrated initially on plantings for white-tailed deer. Subsequently, university research/extension programs and private industry began conducting studies to evaluate the productivity, palatability, nutritional value, and culture of various food plot plantings.

Regrettably, despite the myriad publications and web sites that have been produced since describing the science of food plot establishment, few have provided a thorough and holistic discussion of all aspects of food plot management. Arguably, notable exceptions would include Quality Food Plots, published by the Quality Deer Management Association in 2006, and Dr. Harper's original 2008 publication A Guide to Successful Wildlife Food Plots. However, Craig's subsequent revisions and especially this volume makes significant strides beyond these earlier publications by not only providing detailed guidelines for managing plantings for wildlife, but also highlighting the importance of combining these planting with the management of native and naturalized plant communities. Perhaps Craig's most important message in the entire book is found in Figure 1.1: "Food plots do not replace other habitat management practices, but complement them" – Amen!

If there is anyone more knowledgeable about "all things food plot," I certainly am unaware. Furthermore, the compilation of information related to the use of herbicides in managing food plots in this volume is unmatched between two covers of any other reference. Without question, Craig's book will help land managers take their management efforts to the next level. Simply put, Craig has made a science out of incorporating food plot management into an overall wildlife habitat management plan. Wildlife Food Plots and Early Successional Plants contains accurate, science-based information for everyone interested in establishing and managing high-quality food plots – from the seasoned biologist to the novice landowner. The most significant addition to this volume is the inclusion of a pictorial guide and description of more than 300 commonly occurring plant species or species groups found in food plots and fallow fields. Important information concerning the wildlife value of these plants, along with management guidelines, will enlighten and prove helpful for any wildlife manager, and is a great botanical accompaniment to Forest Plants of the Southeast.

Easy to read, yet flush with detailed recommendations and supporting data, this book will be an invaluable reference for conscientious wildlife managers across the country.

Karl V. Miller
Wheatley Distinguished Professor of Deer Management
The University of Georgia

December 2018

PREFACE

I began researching food plots and old-field communities in 1994. A lot has been learned since with regard to what we should plant and how we should manage food plots. A tremendous amount also has been learned about the value and management of naturally occurring plants. Without question, more people today are interested in managing early successional plant communities than ever before. With increased knowledge of early successional communities, there is less emphasis and reliance on food plots. That is good! However, make no mistake, planting food plots still is a legitimate and often most appropriate management practice to help landowners achieve their objectives, whether that be attracting wildlife for hunting or to increase the nutritional carrying capacity of an area to support various wildlife species.

My purpose for this book is to provide the latest science-based information related to growing and managing food plots for wildlife. It also is my intention to increase awareness and knowledge of the value of naturally occurring plants. Wherever you grow food plots, plants arising from the seedbank also grow. It only makes sense to recognize the importance of both and to use plants occurring naturally to your advantage when possible.

Agronomic principles established over previous decades provide the foundation for this book. Specific recommendations for planting and management come from research conducted with several of my graduate students as well as from other researchers across the country. Research with my graduate students progressed over time. Initially, back in the late '90s and early 2000s, our main research focus with food plots was to investigate preference of various forages by white-tailed deer. Along the way, we documented changes in forage selectivity with regard to deer density. We noted susceptibility and resistance of various forages to grazing, compatibility of various forages in mixtures, and best management practices for extending use and efficiency of those forages. We worked to come up with planting mixtures and strategies that would allow weed control with various herbicides. We then began to look closely at how the availability and quality of "natural" foods (various forbs and browse) in managed forests, woodlands, and early successional plant communities (such as old-fields) influenced food plot use. This information should be of great interest to serious land managers (as well as crop producers). Along the way, we also evaluated various plantings and management strategies for other species, such as mourning dove, eastern cottontail, northern bobwhite, and waterfowl. Working with both professional and private wildlife managers from across the country through the years has made this effort very rewarding and a lot of fun.

The Plant Identification Guide in the back of the book should be of particular interest to anyone wanting to manage food plots and early successional communities for wildlife. This guide is unique in that it provides pictures and clues for plant identification, as well as information on the value of various plants for wildlife, and information on how to get rid of undesirable plants. Many people ask me how to get rid of the "weeds" in their food plots. However, the plants they want to get rid of often are as valuable as what was planted! Knowledge of these plants begins with identification. I hope the Plant Identification Guide helps you with your management efforts.

Some of the information in this book goes against dogma and recent popular opinion. Various plantings and practices implemented today continue to represent "what we've always done" (especially true among some governmental agencies) as well as the latest fads (driven by marketing and advertisement). Unfortunately, much of what continues to be done today is not based on experimentation or biological evidence, but on what might seem "obvious" or what will sell the most. My objective for this book is to give wildlife enthusiasts and managers detailed information needed to grow and manage high-quality food plots, while also guiding them as to how they might incorporate food plots into a sound wildlife management plan without relying on food plots as their primary habitat management effort, which is not sensible. The recommendations within are science-based, but they are coupled with real-world experience from across the country that I think will help you, as hunters and land managers, when determining objectives and evaluating program success.

In the scientific community, there continue to be many who disdain food plots. In fact, there are some who ignorantly compare food plots to artificial feeding. I understand and agree with some of their contentions — that food plots are not holistic habitat management and other management practices are needed. That is why you will find more information related to the importance and need for other habitat management practices in this book than any other publication concerning food plots you have read. However, the naysayers seemingly don't understand land management, nor do they grasp the importance of private landowners in conserving our natural resources. Few landowners practice ecosystem management. Instead, most who manage their property do so for focal wildlife species, such as white-tailed deer or wild turkey, and they use food plots to help them reach their property management goals. Agricultural plantings, especially when combined with other habitat management practices, can provide nutrition beyond what is available otherwise, yet in a manner that is natural (planted and growing, as opposed to feeders), and enable many wildlife species to respond in a positive manner (increased survival, weights, reproduction, antler growth, etc.).

The interest surrounding food plots and early successional plant communities continues to be immense, and there is a tremendous demand for additional information. To me, that is refreshing because it means more people want to work with the land. And if more people want to work with the land, as opposed to developing it, then there is hope that we can conserve more acres for future generations of wildlife and people. And if food plots can be the impetus to help get more people involved with holistic habitat management, then I'm glad to be a part of it.

Craig A. Harper

Isaiah 48:6-7
Matthew 13:3-9; 18-23

"Education progresses through a series of steps from awareness, to enlightenment, to understanding, and culminates with respect. This epic book will take readers on an educational journey. Travelers will become future stewards armed with the knowledge and respect necessary to ensure sound management of land and its wild inhabitants."

R. Joseph Hamilton
Founder and Senior Advisor,
Quality Deer Management Association

Proverbs 3:13-14

One

Chapter 1 — Introduction

Planting food plots is by far the most popular habitat management practice among landowners wanting to enhance habitat for wildlife. Indeed, planting food plots is an excellent way to increase available nutrition and the nutritional carrying capacity of an area for wildlife. Research has indicated warm-season forage plots can provide at least two to eight times the amount of digestible energy and protein available per acre in recently regenerated forests or within actively managed mature forests. And though forage available in early successional areas during summer may rival that in warm-season food plots, forage availability in cool-season food plots far exceeds that in early successional areas during winter, and grain or seed availability from various plantings far exceeds what is available without planting. Food plots not only provide nutritional benefits, but they also can increase and enhance hunting and wildlife viewing opportunities. Planting and managing food plots also provides recreational activity, and the satisfaction of working with the land often exceeds the value of hunting and wildlife viewing for many people.

Increased available nutrition through food plots can positively influence wildlife in many ways, including weight, reproduction (both timing and recruitment), survival (adults, broods and fawns), hatchability (percentage of eggs that hatch), lactation rate, and antler development. Food plots also may influence daily movements and home range size. Typically, daily movements and home range size for a particular species decrease as habitat quality increases. However, decreased movements and home range size also can be reflective of poor habitat quality, especially when a highly attractive food source is made available in the midst of otherwise poor habitat (see discussion in *Chapter 8*). Understanding habitat quality is an important consideration that should influence management decisions.

Although food plots can provide nutritional and recreational value, it is important to not let the potential benefits of a food plot program overshadow the importance of holistic habitat management. Establishing food plots is only one habitat management practice, and food is only one component of habitat, which is defined as ***the collection of resources (cover, food, and water) in an area of sufficient space required to support and sustain a particular wildlife species.*** Food may *attract* wildlife, but cover *holds* wildlife. **In most cases, food plots should be the last step in a habitat management plan. Much more important than food plots is providing and maintaining the appropriate successional stages and vegetation types in a suitable arrangement for the desired wildlife species across the property and landscape. Until the composition and arrangement of habitat is addressed, the overall impact of a food plot program will be minimal and, in fact, could be detrimental.**

Food plots do not produce magical results. Land managers must adjust their food plot expectations according to the relative productivity of the property they are working on, the surrounding landscape, and the local wildlife populations. Food plots should supplement naturally occurring foods made available through other habitat management practices when increased available nutrition is an objective. In particular, food plots should provide a nutritious food source when the quantity and quality of naturally occurring foods decline during specific times of the year. Other habitat management practices that should be considered for various wildlife species might include managing forests with the appropriate regeneration methods, thinning them when ready, and burning the understory when and where needed. Early successional plant communities (such as old-fields) should be managed and maintained by burning, disking, and using selective herbicides to remove undesirable plants and promote desirable species. Hard- and soft-mast-bearing trees and shrubs may be promoted to provide additional food and cover. Field borders of native forbs and grasses should be established around crop fields to provide increased usable space for various wildlife species that use such areas. Food plots cannot replace the value of practices such as these.

Although planting food plots might be less important than providing the necessary vegetation types, a well-designed food plot program *can* help increase the nutritional carrying capacity of a property and support healthy wildlife populations. Along the way, they also can increase hunting and wildlife viewing enjoyment and stimulate increased landowner involvement in land management. What can be wrong with that?

Fig 1.1 Food plots do not replace other habitat management practices, but complement them. Well-managed forests and fields provide outstanding cover as well as an abundance of forage and seed. This mixed hardwood stand was thinned 15 years previously and has been burned five times since.

Fig 1.2a (above) and 1.2b (below); The importance of early successional communities for various wildlife species cannot be overstated. Fallow fields, old-fields (above), and savannas provide excellent food and cover resources for many species. Woodlands (below) and shrublands also provide excellent food and cover for many species, some of which use food plots and others that do not.

Two

Chapter 2 — Initial considerations

When managing wildlife habitat, it is essential to identify factors limiting growth and health of species whose populations are declining or stagnant. Food is not always a limiting factor. In fact, on many properties, suitable cover is more often a limiting factor than food for several wildlife species. Nonetheless, where food is a limiting factor, attention should be given to making sure adequate nutrition is available throughout the year, not just during the hunting season. And in the case of white-tailed deer, the astute wildlife manager often realizes planting additional food may not be the answer. Population reduction may be what is really needed.

Before planting, there are many factors to consider. One is how much work is required to establish and maintain a high-quality food plot. Food plot management often involves soil amendment, applying herbicides, plowing, subsoiling, disking, mowing, and rotational planting. If you do it right, it is an involved process that requires time and effort. **If you aren't prepared to put in the time, effort, and money, you really shouldn't bother getting started.**

Determine your objectives

The first thing you should do when thinking about planting a food plot is define your objectives. Why are you planting a food plot? Are you merely trying to attract wildlife for hunting or viewing opportunities? Or, do you really intend to improve available nutrition for wildlife and increase the nutritional carrying capacity on your property? Are you targeting only white-tailed deer, or are wild turkeys, northern bobwhite, mourning dove, and/or other game or nongame species also a major interest? Answers to these questions influence not only what should be planted, but also how plots are managed and your overall management strategy.

As I see it, there are essentially 3 broad reasons why you might consider planting a food plot:

1. To provide additional food for a particular wildlife species during a time of year when natural food availability may be limiting the population or limiting your wildlife viewing opportunities.
2. To help boost a focal wildlife species in an area where natural foods are available for the present population, but increased production (density, weight, antler size, etc.) is desirable.
3. To increase or enhance hunting and viewing opportunities and experiences.

There are things to consider with each of these reasons to plant food plots. With reason No. 1, special consideration should be given to other habitat management practices. Why are natural foods limiting? Providing added nutrition through other habitat management practices should be conducted also; however, it is recognized that these opportunities vary greatly among properties. With reason No. 2, it can be irresponsible to promote increased productivity if you cannot balance the population (as it responds) with existing habitat and without harming other species. The classic example is allowing deer density to increase to the point that the forest understory is decimated by overbrowsing, thereby destroying cover required by other species. With reason No. 3, there is no question that food plots can enhance hunting and viewing opportunities, but be aware of the danger of falling into the trap of thinking you must have a food plot in order to be successful when hunting. This line of thinking can actually lead to decreased hunting excitement and satisfaction over time, and it can lead to increased reliance on gimmicks and decreased hunting prowess and skill.

Incorporating food plots into a sound wildlife management plan

Property management generally is directed in one of two ways, either to manage for an ecosystem or to manage specifically for one or more focal wildlife species. **Ecosystem management** focuses on restoring or maintaining the communities that comprise an ecosystem, such as a bottomland hardwood forest ecosystem that extends along a major drainage, a longleaf pine ecosystem and its associated pine savannas and oak hammocks, the eastern deciduous forest ecosystem that may contain open oak woodlands, cove forests, and closed-canopy beech-birch stands, or the shortgrass prairie ecosystem. Management practices to maintain ecosystems usually are focused on historic disturbance regimes, such as frequency and timing of prescribed fire or timing and longevity of flooding, and invasive species control. Managers aim to restore or maintain associated native plant and animal communities within the ecosystem, including rare, threatened, or endangered species. The system is managed for maintenance, and the wildlife community associated with the system responds accordingly, but nothing necessarily is done to specifically manage for a particular wildlife species. **Focal species management** is different from ecosystem management in that a property is managed specifically to meet the habitat requirements of one or more focal species. Focal species management can be refined into **objective-driven management** where the property is managed not only for one or more focal species, but to meet the specific objectives of the landowner. For example, if deer hunting is the primary objective of the landowner, then management not only is implemented to maximize habitat value for white-tailed deer, but also to adjust the arrangement and distribution of food and cover resources across the property to influence predicted movements of deer to facilitate successful hunting. Implementation of food plots in a wildlife management plan differ greatly depending on which approach you take. Indeed, food plots would not be necessary or have a place in some plans or approaches, whereas they may play a major role in others.

Regardless of whether you are interested in ecosystem management or focal species management, if you want to have a positive impact on wildlife populations and increase the nutritional carrying capacity of your property, then it is important to involve a Certified Wildlife Biologist® — that is, a professional who has met the strict guidelines for accreditation by The Wildlife Society — and develop a wildlife management plan for the property. The plan will include an inventory of the property, an assessment of the surrounding landscape, your goals and objectives, realistic opportunities and limitations (including biological and financial), management strategies to reach your objectives, and a timeline for implementation that includes expected outcomes. A Certified Wildlife Biologist® will assess the property and current habitat value for various species of interest and identify if and how food plots might be used in a holistic habitat management strategy. This assessment should be completed by considering the focal species in the plan as well as the potential impact on other species that also occur on the property. If you are serious about land stewardship and want to make sure your management efforts are successful, then do not overlook or underestimate the importance of a wildlife management plan developed by a competent, trained professional.

Cost considerations

Food plots can be expensive, and the associated equipment can be a huge investment. Of course, some plots do not require large equipment. No-till top-sowing can be completed on some sites with nothing more than a backpack sprayer and a hand-held seed sower. Small plots can be planted and managed with all-terrain vehicles and associated implements. However, some operations, such as subsoiling or chisel plowing in heavy clay, or planting large acreage to impact food availability over a large property, requires heavy equipment. Tractors, plows, subsoilers, disc harrows, cultipackers, drills or planters, spreaders, spray rigs, and rotary mowers vary greatly in price, but they may be necessary depending on the situation.

Fig 2.1; Food plots can be expensive and require work, but the benefits often are worth it. Bennie Riddle killed this buck as it walked into a food plot and never thought about the expense or effort. Habitat management was a lifestyle for Bennie.

Throughout much of the U.S., soils require liming and fertilizing to optimize plant growth. It is not unusual for fertilizer costs to exceed $200 per acre. Liming is most economical (often $30 – 40 per ton) when a lime truck is contracted from a local agriculture supply dealer. Most dealers have a minimum order (such as 10 tons) for delivery, so it is economical to lime multiple areas if needed. Bagged lime is much more expensive by volume (often $150 – 200 per ton), but is efficient if you are amending soils in small plots.

Herbicides can be expensive, but they often are necessary for successful food plots. Despite the apparent "sticker shock," herbicide applications can be quite cost-effective (usually $10-20 per acre), especially considering how other costs can be wasted if weeds overtake your planting. A sprayer can be just as important as a seed sower on many sites. Seed cost is variable according to what you are planting, but averages $50 to $80 per acre.

Considering average costs for lime, fertilizer, seed, and herbicide, establishing initial food plots typically averages $300 to $350 per acre, not including costs for equipment, fuel, or labor. Only you can justify your costs when watching a 160-class buck feed in your soybean plot or when killing limits of doves with family and friends in your dove field.

Fig 2.2; Habitat composition and surrounding land-use practices greatly influence how much acreage should be devoted to food plots and what should be planted. Less acreage in food plots may be needed where there is a diversity of vegetation types that include row-crop agriculture. When possible, use agriculture to your advantage. Given the cost and effort involved with planting food plots, paying a farmer to leave a portion of a crop for wildlife could save you money. Here, the producer was paid to leave a 40-foot-wide swath of soybeans unharvested.

What and how much to plant?

Not all wildlife species benefit from all food plot plantings. What you plant and how much you plant should depend upon the wildlife species you want to attract, the seasonal requirements of those species, your area (region/climate), and surrounding habitat quality. Habitat composition and the surrounding land-use practices should be evaluated carefully before deciding what and how much to plant. Species such as white-tailed deer, wild turkey, eastern cottontail, and northern bobwhite thrive in areas with considerable vegetation diversity. Productivity and carrying capacity usually are greatest where several vegetation types and successional stages are well interspersed. If agriculture is prominent on adjacent properties, it may be important that your food plots provide a food source after those crops are harvested. Often, food/nutrition is not limiting in these areas. Instead, cover may be a limiting factor. Evaluation of your property and the surrounding area should not focus only on available food sources, but also on other habitat requirements, including food and water.

Where surrounding properties are primarily forested, food plots can make a dramatic impact on available nutrition. In those areas, considerable acreage in food plots (as well as forest management) may be necessary to meet your objectives. Some properties are too small to provide a variety of vegetation types, successional stages, or food plot plantings. In this situation, it is particularly beneficial to work with adjoining landowners and form a cooperative program that encompasses as large an area as possible. Managers of several small properties working together can make a big impact on available nutrition and overall habitat quality. This principle holds true regardless of focal species, from white-tailed deer to mourning dove to monarch butterfly.

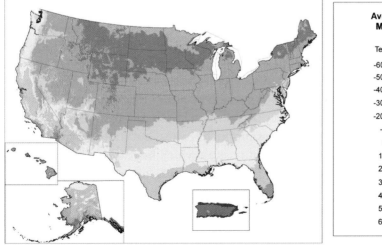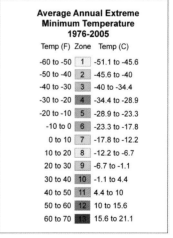

Fig 2.3; Climate is a major consideration when determining what to plant. Obviously, you don't want to plant something in a region where the plant is not adapted. This map, adapted from the U.S. Department of Agriculture, shows plant hardiness zones. Refer to this map when considering various plantings to make sure what you are planting is adapted to the climate in your area.

Another major consideration when evaluating the surrounding landscape is deer density. If deer density is so great that food plots have little chance to establish, you are wasting time and money until deer density is lowered and/or additional habitat management is implemented (see *Chapter 8*).

Where to plant food plots

Where you locate food plots should depend upon several factors. You should identify locations on the property where targeted wildlife species frequent, then look for sites nearby that are suitable for planting. The best sites generally are flat, where there is sufficient moisture, with high nutrient levels, and it is easy to operate equipment. Keep in mind that for several wildlife species — such as deer, quail, and rabbits — food resources are more readily used if located near suitable cover. Sometimes suitable openings (existing fields, woods roads, log decks, utility rights-of-way) are nearby and available for planting. If not, don't overlook the possibility of creating new openings, which may sound extreme, but it isn't. Bulldozer/front-end loader operators typically charge around $100 per hour. If the site is timbered, the cost of road building may be offset by the value of the trees removed. A respectable equipment operator can easily clear a 2-acre opening in a day. At $100 per hour, a 2-acre food plot is created for approximately $800, plus there is easy access into the area and the woods road itself can be planted. Constructing and clearing a woods road for equipment access can save money over time and help make food plots much more productive. With a clear road to get to the plot, equipment access is much easier, which makes it much more likely the plot will be managed and maintained as needed.

Site quality should be considered when planning food plots. Availability of nutrients and moisture are strong determinants of plant growth and yield. Locating food plots on high-quality soils helps ensure maximum plant growth and yield with minimal fertilizer amendment. I recommend using soils maps available free online from the Natural Resources Conservation Service (NRCS) to locate various soil types on your property when considering where to locate food plots. Considering site quality is especially important when you are trying to provide maximum food per acre, such as a large feeding plot for white-tailed deer or a dove field. That is not to say food plots cannot or should not be grown on less productive sites, but more amendments (fertilizers and lime) are needed on poor-quality sites, which increases cost per acre planted. Plant growth and yield are less on poor-quality soils, often substantially less, which reflects how soil quality can influence the amount of food available for wildlife on some properties. However, relative to the amount of food in the surrounding area, a food plot on a poor-quality site can far exceed that occurring naturally without management. Nonetheless, if maximizing plant growth and yield per acre is important, then locating the best sites on the property (with regard to soil nutrients, moisture, slope, etc.) should be a consideration. It also is important to consider how open areas

Fig 2.4; Identifying the best sites for food plots is important. Animal movements, habitat composition, soil fertility, and equipment access should be considered. This corn plot and wheat/clover plot are intended for deer and turkeys and are located in a bottom, along a brushy creek drainage with plenty of early successional cover on both sides. These plots are out of sight of any road and have easy equipment access.

Fig 2.5; Have you ever found just the right spot for a food plot, but there was no opening? Then create one! This linear strip for deer was created with a front-end loader in just a few hours, costing less than $300.

Fig 2.6; This opening is along a ridge with rocky and poor-quality soil. Soil phosphorus and potassium values were very low and soil pH was 5.4. Instead of amending the soil for a food plot, we simply managed the site with naturally occurring plants. Forage value of forbs was outstanding, with protein, phosphorus, and calcium levels exceeding 20%, 0.3%, and 0.5%, respectively, during spring and early summer. The structure is perfect for wild turkey broods, and various songbirds, such as indigo bunting and yellow-breasted chat, as well as eastern box turtles, used the site regularly.

on poor-quality sites still can be managed and made very productive for wildlife without planting. These sites typically are best for management of early successional plant communities. The food and cover value on these sites can be outstanding. Not only may cover be provided for species such as white-tailed deer, wild turkey, northern bobwhite, eastern cottontail, and a host of nongame species, but the food value can be exceptional. Native forbs naturally adapted to poorer soils provide high levels of protein, phosphorus, and calcium, and many of these plants provide seed critical for upland game birds and songbirds (see Appendix 7 and 8). So, as you can see, consideration of soil type can help you develop your management plan with regard to location of various vegetation types as well as your food plots.

When creating new openings, remember to consider not only soil type, but orientation and drainage patterns also. Various plantings have different requirements and some are better adapted to various conditions than others. For example, do not expect ladino clover to do well in shallow soils on ridgetops and southern exposures, especially in the South. North- and east-facing slopes, as well as the lower third of slopes and bottomlands, hold considerably more moisture than south- and west-facing slopes and ridgetops. Not only do south- and west-facing slopes and ridgetops receive more sun during the afternoon and evening, these areas typically receive more wind. Nonetheless, for any given exposure, you can influence soil moisture by orienting plots in certain directions and by considering plot width. For example, if soil moisture might be a limitation, consider relatively narrow plots oriented north-south, or plant the southern side of plots oriented west-east. If the area might be too moist, the northern and eastern sides of plots will be drier because they receive the most sun during the hottest part of the day.

Also, when creating new openings for deer where hunting is an objective, refer to an aerial photo and consider prevailing wind directions and how the terrain influences travel patterns via corridors and funnels. Think about using smaller food plots that lead from bedding areas to larger food plots or other feeding areas. Now is the time to influence deer movements to your advantage.

Do not plant food plots within sight of a road or property boundaries unless you want to advertise your work and encourage poachers and trespassers. Visibility from roads can be limited by planting trees and shrubs adjacent to the road and by allowing naturally occurring vegetation to grow up along a road edge. When planting trees and shrubs along a road, evergreens, such as white pine and eastern redcedar, should be considered; trees and shrubs that might attract wildlife (soft-mast producers) should not be used for this purpose.

Size and shape

According to your objectives and the targeted wildlife species, food plot size and shape may vary considerably. Specific recommendations concerning food plot size and shape (as well as other initial considerations) are provided in subsequent chapters for particular wildlife species. Nonetheless, when planning food plots, it is important to consider equipment operation, especially where large tractors and implements and lime trucks will be operated. Food plots with irregular edges make it difficult to maneuver equipment when preparing the seedbed, planting, fertilizing, liming, spraying, and cultipacking the plot. Managing such a plot with an ATV is less of a problem than when using an 80-horsepower tractor. However, the meandering edge of a half-acre food plot planted with an ATV is meaningless relative to food plot use (see discussion in *Chapters 8* and *10*). If additional early successional or brushy cover is needed, thin the woods adjacent to food plots by killing or removing undesirable trees. You also can establish field borders around relatively large fields. These practices do not require "wavy" edges of the planted area. Again, consider the overall habitat arrangement and the size of the plot. Where larger plots are warranted and distance to cover from the middle of the plots is a concern, plots more rectangular in shape may be appropriate. Relatively narrow strips may be planted for some species — such as quail, rabbits, and songbirds — and firebreaks may be seeded as well (see *Chapter 11*).

Figs 2.7a and b (below); Planting logging roads and power line rights-of-ways is a good way to provide more forage in forested areas where forage availability may be relatively low. Rectangular-shaped plots can provide an advantage by keeping distance to cover relatively short, and size can be increased as needed with added length. Photo below by Ryan Basinger.

Measuring plot size

Measuring plot size is an often overlooked step that is important in order to spread the correct amount of lime and fertilizer, spray the correct amount of herbicide, and plant the correct amount of seed. **Do not estimate plot size by guessing!** Measuring plot size is easily accomplished with a GPS, a smartphone, a rangefinder, a 300-foot tape, or by pacing. Plot area can be determined by measuring length × width in feet and dividing by 43,560 (the number of square feet in an acre). If the plot is irregularly shaped, it may be necessary to mark off sections and measure them separately. Or, numerous length and width measurements may be taken, then averaged before calculating area.

Distribution of food plots

In some cases, according to habitat composition, focal wildlife species, and your objectives, food plots should be distributed across the property to make them available to as many animals as possible and lessen foraging pressure in any one plot. Special consideration should be given to areas on the property where food availability may be limiting. For example, if white-tailed deer is the focal species and half of the property contains considerable early successional cover and agriculture, and the other half is wooded, additional food plot acreage may be needed in the forested half (depending on property size). Planting woods roads and log landings is a good way to increase food plot acreage and influence movements in forested areas (see *Chapter 14*). These "linear wildlife openings" can provide forage throughout an area, potentially reaching more animal home ranges (per acre planted) than a single opening of similar size. Similarly, power line rights-of-ways can provide additional opportunities to "partner with the neighbors" and influence a larger area. Where deer hunting is important, planting woods roads, log landings, and rights-of-ways also helps ensure more hunters have access to food plots across the property, which can be an important aspect in hunter management and satisfaction. Linear wildlife strips also can be integrated with other habitat management practices. Firebreaks around old-fields can be planted to provide supplemental forage adjacent to naturally occurring forbs and browse, which should represent highly attractive nesting, brooding, bedding, loafing, and/or escape cover (see *Chapter 11*).

Crop rotation

It is never a good idea to plant a specific plant in a particular plot year after year. Over time, insect pests, fungi, bacteria, and viruses can build-up and negatively impact productivity for select plantings. It is common to see clovers decline from fungal and viral diseases and grub damage. Brassicas are susceptible to fungal and viral diseases. Various plantings also use nutrients at different rates and may drain certain nutrients from the soil. It is common to see corn, grain sorghum, or sunflower production decline over time if planted year after year in the same area, even when a fertilization program is followed. Chufa tuber production often declines significantly one to two years after planting, usually because of plant density. Crop rotation not only can help reduce pathogens, but also help manage nutrient availability and site productivity.

The most popular rotation is following legume crops (such as soybeans) with grass crops (such as corn). Surplus nitrogen produced by bacteria within nodules on the roots of legumes remains available for nitrogen-hungry grasses (see *Inoculating legume seed* on page 54). This strategy obviously helps reduce N fertilization when a grass is planted. Grasses, such as corn, grain sorghum, or wheat, as well as brassicas, provide a good rotation when perennial clovers begin to decline. Not only can legumes be rotated with grasses, but other forbs, such as buckwheat, chicory, sesame, and sunflowers, can be included in the rotation as well.

Fig 2.8; Do not overlook the value of fallow fields! This was a corn food plot. There was a fair amount of corn left uneaten by spring. Instead of replanting, allowing such a field to sit fallow for a year or two provides outstanding forage and cover for many wildlife species. Even if the corn (or other crop) was all eaten, fallow rotation can be an excellent strategy.

When rotating crops, it might seem intuitive to rotate a cool-season plot with a warm-season plot, and vice versa, so the rotation allows a crop to immediately follow another at the end of its growth period, especially when using no-till technology. However, another strategy is to allow plots to remain fallow for one to two years, especially when managing annual plots (see *Chapters 9* and *10*). Residual grain can be used by wildlife as an early successional plant community develops. **Fallow plots should not be viewed as wasted space.** Early successional plants growing in fallow plots can provide nutritious forage, seed, soft mast, and valuable cover for songbirds, game bird broods, and fawns. Fallow fields also present an opportunity to remove specific undesirable weed species, with selective herbicide applications or disking.

Three

Chapter 3 — Soil fertility and amendment

Soil fertility is an important factor in crop production. Plants provide food for animals, and if soil nutrients are limiting, then plant growth and yield will be limited, which can limit the nutritional carrying capacity for wildlife. Soil fertility and amendment is particularly important for several plants that are grown for crops (such as soybeans or grain sorghum) or food plots (such as clovers and alfalfa) because they are not native to the region or soil conditions where they are grown and their production may be limited by soil pH or nutrient availability.

Soils within the United States have been classified into 10 taxonomic orders, based on soil origin and physical properties. Several soil orders are found throughout the South, but ultisols predominate. Ultisols are ultimately leached of nutrients with a high clay content and low pH. Ultisols can be very productive, however, if limed, fertilized, and managed properly. Soil pH is often 5.0-5.7 and phosphorus (P) and potassium (K) levels are typically low (less than 20 pounds P per acre; less than 90 pounds K per). Amending ultisols with proper amounts of lime and fertilizers is necessary to realize optimum plant growth and yield and provide needed nutrition to wildlife.

Other soil orders commonly found in the Northeast, Midwest, and further west include inceptisols, alfisols, and spodosols. Inceptisols are weakly developed soils found along the Mississippi River and near the Gulf and Atlantic coastlines. These soils often are excellent for agricultural production. Alfisols are moderately weathered soils with a high base mineral content. They are probably the most naturally productive soils without irrigation or fertilization and are prevalent in the corn belt of Indiana, Ohio, Michigan, and Wisconsin. In the South, alfisols occur along Mississippi River drainage areas, within the Black Belt of Alabama and Mississippi, and in northern Kentucky, northern Missouri, central Oklahoma, and east-central Texas. Spodosols are acidic sandy soils usually well-leached of nutrients. The majority of spodosols in the South are found along the Atlantic Coastal Plain. Mountainous areas (especially the Appalachians) may contain several soil orders as a result of variable drainage patterns and parent material. Therefore, soil fertility in mountainous areas is variable and may be very high or quite low within a relatively short distance, according to topography, geology and other factors.

There is a common misconception that mountain soils are poor and that they result in relatively low and less productive wildlife populations. Some areas, such as ridges and steep slopes, do have shallow, rocky, relatively unproductive soils, but other areas, such as lower slopes, coves, and bottomlands, have very deep, productive soils. **Wildlife populations respond to land-use practices** that influence habitat composition and structure (through various vegetation types and successional stages).

Indeed, land-use influences food availability as well as survival by providing suitable cover. **Therefore, the reason white-tailed deer, wild turkey, cottontail, bobwhite, or mourning dove populations may be lower or less productive in some mountainous areas (such as national forests in the southern Appalachians) is not because of soils, but because the land-use practices and habitat composition is not suitable** to support or produce populations found in the Piedmont or Inner Coastal Plain, for example. Vast closed-canopy forests do not support white-tailed deer, wild turkey, eastern cottontail, mourning dove, or northern bobwhite like broken woodlots with agriculture and high-quality early successional cover.

Soil pH

Soil pH has a great influence on nutrient availability and plant growth because it affects the solubility of minerals. When soil pH is low (less than 5.8), many nutrients (such as N, P, K, S, Ca, and Mg) needed by plants become less available. Phosphorus, for example, forms insoluble compounds with aluminum (Al) at soil pH less than 5.5 and with calcium (Ca) at soil pH greater than 7.5. Strongly acid soils (pH 4.0-5.0) can have toxic concentrations of soluble aluminum and manganese (Mn), which adversely affect growth and development of many food plot plantings, such as corn, grain sorghum, alfalfa, and clovers. This condition is corrected by liming. By simply adjusting soil pH to 6.1-6.5, without any fertilization, plant growth may improve substantially because nutrients are released through chemical reaction and organic decomposition. After pH has been corrected, applications of the appropriate fertilizers will boost plant growth and yield.

pH
The "power of hydrogen" (pH) is a measure of the hydrogen ion (H+) concentration in soil and is measured on a logarithmic scale from 0 (extremely acid, or sour) to 14 (extremely basic, or sweet). What this means is a soil of pH 5 has 10 times more H+ in solution than soil of pH 6 and 100 times more H+ in solution than a soil of pH 7. The extreme pH range found in soil is approximately 3.5-9.5.

Alkaline soils may be common in some areas, especially the western U.S. They usually are a result of alkaline parent material and can lead to reduced nutrient availability, especially micronutrients. Iron deficiency is a common problem in alkaline soils. If soil pH is above 8.0, it may be necessary to reduce pH by adding organic material (such as leaves, sawdust, sphagnum peat moss), elemental sulfur (about 250 to 500 pounds per acre), or acidifying fertilizers (such as ammonium sulfate).

Soil pH also influences microorganism activity. In acid soils, nitrogen-fixing bacteria and other bacteria that decompose organic material are not active. This lack of activity negatively affects legume production and reduces the amount of nutrient release through the decomposition process. Soils become acid as basic cations (such as Ca++, Mg++, K+ and Na+)

are replaced (exchanged) on soil colloids by hydrogen ions (H+). Hydrogen ions are formed from water and dissolved carbon dioxide and are increased when ammonium-containing or other acid-forming fertilizers are applied. The replaced cations then are moved deep into the soil profile or carried away by water (leached) from the site. Thus, plant-available nutrients can be reduced drastically.

Cations (positively charged ions) are attracted to (or adhere to) the surface of clays, humus (decomposing plants), and the surface of plant roots, which are negatively charged by electrostatic attraction. Adsorbed cations can be exchanged by other cations through mass action and move into the soil solution. The soil cation exchange capacity is influenced by the amount and type of clay present and the amount of organic material in the soil. Clay particles are much smaller than silt particles, which are much

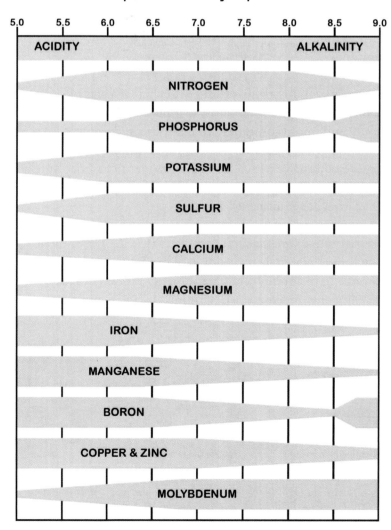

Effect of pH on availability of plant nutrients

Fig 3.1; Soil pH should be adjusted to 6.0-7.0 to ensure soil nutrients are available to plants. Adapted from International Plant Nutrition Institute.

smaller than sand particles. Furthermore, less-weathered clay particles are layered (like a deck of playing cards); thus, they have much more surface area than silts or sands, and provide soils with a high clay content a much higher cation exchange capacity and buffering value than other soils. What this means is that it requires substantially more lime to raise soil pH to 6.5 in a clayey soil that has a pH of 5.5 than a sandy soil that has a pH of 5.5.

Benefits of correcting soil pH with aglime

Low soil pH is increased by adding liming materials, most commonly carbonates, but sometimes oxides,

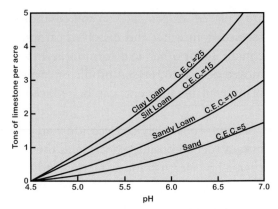

Fig 3.2; More lime is required to increase pH in clayey soils as opposed to sandy soils. This graph shows the approximate tons of lime needed to raise the pH of a 7-inch layer of soils with different cation exchange capacities. Adapted from R.L. Donahue, R.W. Miller and J.C. Shickluna, Soils: An introduction to soils and plant growth, Prentice Hall, Inc.

hydroxides and/or, where available, industrial byproducts, such as silicates of calcium and magnesium. The vast majority of agricultural lime (aglime) used is calcitic limestone (ground limestone) and, more commonly in many areas, dolomitic limestone (ground limestone high in magnesium). Agricultural limestone, or "aglime," refers to those materials registered usually with a state's Department of Agriculture as having a liming value meeting the minimum specifications for fineness of grind and neutralizing value (calcium carbonate equivalent).

Various liming materials have a different calcium carbonate ($CaCO_3$) equivalent (CCE). The CCE of several liming materials is shown in Table 3.1. Most states require liming materials offered for sale have a minimum CCE, usually 70-80. The relative neutralizing value and overall quality of aglime is determined by CCE and fineness

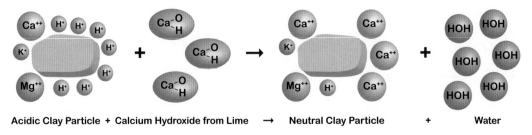

Fig 3.3; Calcium hydroxide (and some calcium carbonate) is formed when lime is added to acidic soils. The H+ from exchangeable sites combines with the OH– of the solubilized lime to form water, thus effectively neutralizing H+. Adapted from R.L. Donahue, R.W. Miller and J.C. Shickluna, Soils: An introduction to soils and plant growth, Prentice Hall, Inc.

of grind. Most states also require liming materials be ground to a particular fineness. Fineness is important because lime particle size influences how quickly it begins neutralizing soil pH. In Tennessee, for example, aglime must be ground so that at least 85 percent passes through a 10-mesh sieve (100 openings per square inch) and at least 50 percent passes through a 40-mesh sieve (1,600 openings per square inch). Aglime coarser than 60-mesh requires several months to produce significant changes in pH,

Table 3.1 Acid-neutralizing values for several liming materials.

Aglime material	Calcium carbonate (CaCO₃) equivalent	Pounds needed to equal 1 ton of pure CaCO₃
Calcium carbonate	100	2,000
Calcitic limestone	85-100	2,000-2,350
Dolomitic limestone	95-108	1,850-2,100
Calcium oxide (burnt or quick lime)	150-175	1,145-1,335
Calcium hydroxide (hydrated or slaked lime)	120-135	1,480-1,670
Calcium silicate	86	2,325
Basic slag	50-70	2,860-4,000
Ground oyster shells	90-100	2,000-2,220
Cement kiln dusts	40-100	2,000-5,000
Hardwood ashes[1]	40-50	4,000-5,000

[1]Hardwood ash contains roughly 5% potash with insignificant amounts of N and P.

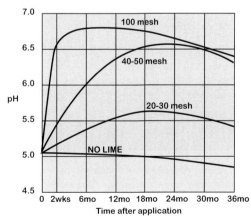

Fig 3.4; The finer the lime, the more quickly it reacts with the soil to raise pH and the quicker the calcium becomes available to the plant. Adapted from International Plant Nutrition Institute.

whereas 60- to 100-mesh aglime will produce changes within two to three weeks if weather conditions are favorable and recommended amounts are applied. Regardless, the full neutralizing effect is not usually realized for at least six to 18 months. Some states also use a "relative neutralizing value" (RNV) to describe and quantify the combined effects of fineness of grind and calcium carbonate equivalent on effective neutralizing value.

Aglime increases soil pH by replacing adsorbed hydrogen ions on soil colloids with calcium ions, but there are many other benefits of liming acid soils besides increasing pH.

Dolomitic lime adds calcium and magnesium to the soil. Liming increases nitrogen availability by increasing soil microbial activity, which speeds organic decomposition and improves nitrogen fixation on the roots of legumes. Liming makes phosphorus more available by reducing the solubility of aluminum and iron, which tie-up phosphorus. Liming reduces excessive uptake of potassium by plants. Liming increases plant-available molybdenum. Liming improves fertilizer efficiency by 50 percent or more. And finally, as nutrient availability and plant growth are improved, the efficacy of herbicides is increased.

Lime moves very slowly through the soil profile; therefore, liming efficacy can be improved by incorporating lime by disking or tilling into the soil column a few inches where plant roots will be growing. Because the full effect of liming is not realized until at least six months after application, it is important to plan ahead and lime plots in spring before fall planting or lime in fall before spring planting. Nonetheless, if you are

Fig 3.5; These iron-clay cowpeas were planted in soil with a pH of 5.1. Lime, phosphate, and potash were added as recommended by a soil test and incorporated into the soil in May, just prior to planting. Notice how the lower leaves are crinkled and curled with yellowish coloration. Some are scorched along the leaf edges and have become ragged. This condition is symptomatic of Mn toxicity (as opposed to Mn deficiency). Also associated with Mn toxicity is Al toxicity, which, along with K deficiency, results in poorly developed root systems and slow growth. Poorly developed root systems may lead to several deficiency problems. However, in this case, as soil pH began to neutralize and nutrients became available to the cowpeas, additional growth (on top of the plant) appeared normal. By late July, the cowpeas looked completely normal and were growing vigorously. This example illustrates the importance of liming several months in advance of planting if possible.

Fig 3.6; Site and nutrient availability are major considerations when planting food plots. Liming acid soils is absolutely critical to increase soil pH, improve availability of nutrients, improve nitrogen fixation among legumes, and increase herbicide effectiveness. Hiring a lime truck from the local fertilizer supplier is much more efficient and economical than buying and applying bagged lime.

ready to plant and haven't limed yet, go ahead and lime just before planting — that's much better than not liming at all. Growth and production will improve through the season as soil pH slowly increases. Over a few to several years (according to soil type and management), soil pH will gradually return to original levels. That is why it is important to soil test each plot every two to three years and top-dress as necessary.

If you are no-till planting, surface applications of lime will increase soil pH, but perhaps not as deep or as quickly as incorporated lime. That's okay. The benefits of a no-till planting (increased organic material and associated nutrients, top-soil creation, soil microbe activity, and moisture retention) far outweigh any slight decrease in pH adjustment.

Many soils will require 1-2 tons of lime per acre to increase soil pH to 6.5. Some soils require 3 tons per acre, and there are a few areas that will require 4 tons per acre or more. **The entire recommendation can be applied at once. There is no benefit to making split applications.** Aglime can be purchased in bulk or bagged. Agriculture supply dealers can be contracted to spread bulk aglime or it can be delivered and you can spread it yourself, usually with a rented lime buggy. Bagged lime is available pulverized (usually in 50-pound bags) or pelletized (usually in 40-pound bags).

Facts about pelletized lime and liquid lime

A common misconception persists that pelletized lime reacts more quickly in soil and that less pelletized lime is needed than aglime or other liming materials. Although this is possible, it is largely inaccurate. Some companies, such as those in the marble industry, began making pelletized lime decades ago as they tried to figure out what they could do with a byproduct, which was extremely fine-mesh lime (such as 200-mesh). Extremely fine-mesh lime is similar in texture to talcum powder and blows away with the wind if not formed into pellets. It reacts more quickly in soil and does not remain in soil as long as coarser lime. However, what began as utilization of a byproduct soon became an intended product as the lime industry realized there was a market for such material. Homeowners, in particular, were especially interested in lime that could be applied with a lawn spreader. Thus, lime companies began processing aglime into pellets to facilitate spreading. Today, the majority of bagged pelletized lime sold in stores is not extremely fine-mesh lime. The neutralizing value of bagged pelletized lime, bagged pulverized lime, and aglime spread with a lime truck is largely the same, depending on the stated mesh-size. The main difference is price. Pelletized lime is much more expensive because it is formed by compressing smaller particles into larger granules that are easy to spread onto yards with a push-type lawn spreader. Do not let a salesperson tell you less pelletized lime is needed than pulverized lime (because pelletized lime costs more than pulverized). That is not true unless the pelletized lime is formed from extremely fine-mesh lime, which will be stated on the label on the bag. If the pelletized lime has been formed from very finely ground limestone (such as 100% passing through a 100-mesh sieve), then it will react more quickly in the soil.

Similarly, there are lots of questions about liquid lime. Liquid lime (or fluid lime) is processed by suspending very finely ground aglime in water or liquid nitrogen fertilizer with a suspending agent (such as attapulgite clay). Typically, 100% of the liming material in liquid lime will pass through a 100-mesh sieve, and 80 – 90% will pass through a 200-mesh sieve. The aglime content of the material may be 50 – 70% aglime. The very finely ground limestone reacts quickly in the soil. However, the effect of change in pH is shorter lived than that from a traditional aglime application.

Improving soil fertility

Plants require non-mineral nutrients and mineral nutrients. Plants obtain non-mineral nutrients — hydrogen (H), oxygen (O), and carbon (C) — from air and water. Mineral nutrients are obtained from the soil. Macronutrients are classified as primary and secondary. The primary nutrients include nitrogen (N), phosphorus (P), and potassium (K), known as the big three fertilizer nutrients. The secondary nutrients include calcium (Ca), magnesium (Mg), and sulfur (S). The micronutrients include boron (B), chloride (Cl), copper (Cu), iron (Fe), manganese (Mn), molybdenum

(Mo), zinc (Zn), cobalt (Co), and nickel (Ni). Availability of micronutrients for maximum plant production is essential, but not as often limiting as macronutrients. Other micronutrients have been found essential for at least some plants, but information is limited and they are almost never deficient in soils to limit plant growth.

The main source of nutrient anions (nitrate nitrogen, P, S, B, Cl, and Mo) is soil organic matter. As organic material is broken down and decomposed, it becomes organic matter and ultimately humus, which slowly release these nutrients for plant uptake, thus reducing fertilizer costs, which is a primary benefit of no-till planting. Do not overlook this important consideration. The main source of nutrient cations (ammonium nitrogen, K, Ca, Cu, Fe, Mg, Mn, Zn, Ni, and Co) is cation exchange with soil particles and humus. In many soils, these processes are not adequate to supply enough nutrients for optimum plant growth; thus, fertilizers are needed. Soil organic matter plays a very important

Fig 3.7; The yellow discoloration on the older leaves of this corn show N is limiting. It is common for yellowing to begin at leaf tip in corn and extend along the midribs. Photo by International Plant Nutrition Institute.

role in plant nutrition, and increasing soil organic material by the addition of plant and animal material, such as wood chips, leaves, and manure, should be a major consideration for many nutrient deficient soils.

Nitrogen

Nitrogen (N) is the key nutrient in plant growth. It is a part of all proteins (plant and animal), chlorophyll, and nucleic acids. Nitrogen is supplied by atmospheric reactions (especially electrical storms), decomposition of organic material (about 5 percent by weight), and N fertilizers. Adequate nitrogen leads to increased plant growth and produces more palatable and higher quality forage with thinner cell walls, increased protein, and increased water-use efficiency. Increased plant growth and quality provide additional food and nutrition for wildlife. Crops usually contain more nitrogen than any other nutrient; however, various crops require more nitrogen than others. Corn, for example, is a heavy nitrogen user, whereas wheat requires considerably less. Inadequate nitrogen usually leads to pale yellowish-green plants and reduced production. Crude protein levels also are lower when nitrogen is limiting.

Nitrogen is unique in that plants can use it in both the cation form (ammonium, NH_4^+) and anion form (nitrate, NO_3^-). Both of these are very soluble in soil water and are readily taken in by plants if available during the growing season. With heavy rainfall, nitrogen may be lost through leaching or surface run-off. Nitrogen also may be lost or made unavailable through denitrification and volatilization. Soil bacteria can

Fig 3.8; *P-deficient plants, such as this dwarf essex rape, may appear purplish-red.*

Fig 3.9; *The yellowish discoloration and scorching along the edge of these soybean leaves suggests K is limiting. Photo by International Plant Nutrition Institute*

Fig 3.10; *Acquiring minerals, especially calcium and phosphorus, is essential for bone growth, including antlers. Proper habitat management, which may include food plots, ensures adequate nutrition is available for bucks to express their potential. Photo by Joe Gizdik.*

change nitrate nitrogen to gas when certain soil conditions prevail (primarily wet periods in clay soils). Volatilization is common when nitrogen fertilizers containing non-stabilized urea are broadcast and not incorporated into the soil by disking or if it doesn't rain soon after application. It can lead to as much as 30 percent loss of available nitrogen within one to two days after application. Incorporation after application, as well as rain or irrigation, prevents volatilization of urea fertilizers. Using urea fertilizer that has been coated is another option to help avoid volatilization. Volatilization of ammonium nitrate and ammonium sulfate is not a concern, unless applied on soils where heavy lime applications were recently top-dressed and not incorporated.

Availability of nitrogen in a usable form is critical for most plants. Only legumes can obtain nitrogen from a gaseous form (see *Inoculating legume seed* on page 54).

Phosphorus
Phosphorus (P) is the second key plant nutrient. Wildlife managers should be particularly concerned about phosphorus availability because of its role in bone and tissue growth, including antlers and teeth, which are mostly calcium phosphates, and its role in fruiting and seed production. Phosphorus also is involved in metabolizing fats, carbohydrates, and amino acids. Phosphorus influences plant use of other nutrients, such as nitrogen, and plays an essential role in cell division and growth, in muscle contractions, and in the function of DNA molecules.

Young plants absorb phosphorus quickly if it is available. In fact, some crops accumulate approximately 75 percent of their phosphorus requirement by the time they produce 25 percent of their dry weight biomass. A deficiency of available phosphorus interferes with the ability of plants to use water and regulate internal temperature. Soil moisture, temperature, and soil pH greatly influence phosphorus availability. The main source for plant-available phosphorus is decomposition of organic material. Cold, wet weather slows bacterial action, which reduces decomposition of organic material and reduces phosphate availability, causing plant leaves to turn purplish-red. In mineral soils, phosphorus is largely unavailable when soil pH is below 5.5 and above 7.0, as phosphate ions react with iron and aluminum ions in acid soils and adsorb to calcium carbonate surfaces in alkaline soils.

Phosphorus availability is improved by adjusting soil pH between 6.1 and 6.5, increasing soil organic matter, and applying phosphate fertilizers. Phosphorus reacts quickly with the soil and does not move appreciably from the point of application (if not washed away).

Potassium

Potassium (K) is the third primary plant nutrient. Potassium is critical in nearly every aspect of plant growth. Potassium regulates water and nitrogen uptake, aids in photosynthesis, allows adequate sugar and nutrient transport, helps build plant proteins and carbohydrates, increases root growth, and improves drought resistance and disease resistance. These factors play an important role in terms of stand longevity among perennial forages and where plant cold-hardiness is a factor in forage persistence. In animals, potassium is important for nerve and muscle function, and it regulates sodium and fluids. Potassium-deficient plants experience poor seedling vigor, have weak stems, poorly developed root systems, and often exhibit leaves that appear scorched along the margins.

Although potassium is relatively abundant in many soils, it is not readily available because it is tied up primarily in micas and feldspars that weather very slowly. The addition of potassium fertilizers (such as potash, K_2O) is necessary to increase potassium availability. When managing soil potassium, it is wise to use split applications (avoiding heavy single applications) and maintain soil pH near 6.1-6.5 to reduce losses from leaching. Returning crop residues (organic matter) also can help retain potassium present on the site and reduce leaching.

Secondary macronutrients

Calcium influences cell wall strength and cell division (especially rapidly growing root tips), protein synthesis, and carbohydrate movement. In animals, calcium is a major component of bones (including teeth and antlers) and is important for proper nerve and muscle functioning. Magnesium is essential for the production of chlorophyll. In fact, at the heart of each chlorophyll molecule is a magnesium atom. Therefore, without magnesium, there would be no green plants. In animals, magnesium allows the nervous and various enzyme systems to function properly. Sulfur is required for the production of three amino acids found in plants and animals. These amino acids are

Fig 3.11; Micronutrient availability is important for various crops, such as this alfalfa plot in northern Missouri, which has just begun to bloom in early July.

necessary for synthesizing proteins, which are vital for both plants and animals.

Calcium is easily supplied through liming (calcium carbonate). Magnesium levels also are increased when dolomitic lime is used. In most areas, sulfur comes naturally from decomposition of organic matter and rainfall. The addition of fertilizers containing sulfur or calcium sulfate (gypsum) easily can correct any deficiencies.

Micronutrients

Micronutrients are essential for plant growth, but they are needed and used in minute quantities relative to the macronutrients. With the exception of boron, micronutrients are not often limiting for forage growth. However, sandy soils are more likely than clay soils to be deficient in available micronutrients. Several micronutrients (as well as several macro- and secondary nutrients) also are less available in cold, wet soils.

Alfalfa, clovers, and brassicas may benefit from 1-2 pounds of boron per acre per year (about 10-20 pounds of Borax per acre will work). Other micronutrients, such as zinc, are particularly important for other plants, such as corn and grain sorghum, and may improve yields if amendments are made. Wheat, oats, grain sorghum, and clovers have responded positively to an addition of copper, especially in sandy soils where leaching is common. Of course, the only way you will know if a micronutrient application is needed is through soil testing. Proper micronutrient availability is ensured by carefully managing pH. A soil pH of 6.1-6.5 helps ensure micronutrients are available, but not at toxic levels.

Crop oils and solutions

Although not fertilizers, various crop oils and solutions are commonly promoted to spray either on the seedbed prior to planting or on plants for increased plant growth and development. Unfortunately, few published data exist to verify these claims. Before purchasing, ask the dealer to provide data (not just advertising literature) that support claims of increased growth and development.

Soil testing: A cheap source of knowledge

Correcting soil pH and providing adequate nutrition is important for food plot success. Do not skimp on soil amendment if you want lush, productive food plots. The only way to know how much lime, fertilizer, or organic material is needed is to

Key to nutrient deficiency symptoms in crops[1]

Nutrient	Nutrient color change in lower leaves (translocated nutrients)
N	Plants light green; older leaves yellow (chlorosis); yellowing begins at leaf tip and extends along midribs in corn and sorghum.
P	Plants dark green with purple cast; leaves and plants small.
K	Yellow/brown discoloration and scorching along outer margin of older leaves; begins at leaf tip in corn and grain sorghum.
Mg	Older leaves have yellow discoloration between veins; finally reddish-purple from edge inward.
Color change in upper leaves (nutrients not translocated); terminal bud dies	
Ca	Emergence of primary leaves delayed; terminal buds deteriorate; leaf tips may be stuck together in corn.
B	Leaves near growing point yellowed; growth buds appear as white or light brown dead tissue.
Terminal bud remains alive	
S	Leaves, including veins, turn pale green to yellow; young leaves first.
Zn	Pronounced interveinal chlorosis on citrus and bronzing of leaves; on corn, broad white to yellow bands appear on the leaves on each side of the midrib; plants stunted, shortened internodes; new growth may die in some bean species.
Fe	Leaves yellow to almost white; interveinal chlorosis to leaf tip.
Mn	Leaves yellowish-gray or reddish-gray with green veins.
Cu	Young leaves uniformly pale yellow; may wilt and wither without chlorosis; heads do not form or may be grainless on small grains.
Cl	Wilting of upper leaves, then chlorosis.
Mo	Young leaves wilt and die along margins; chlorosis of older leaves as a result of inability to properly use nitrogen.

[1]Information courtesy the International Plant Nutrition Institute and other plant scientists.

It is important to note deficiency symptoms are not often clearly defined. Masking effects from other nutrients, secondary diseases, or insect infestations can prevent accurate field diagnosis. Deficiency symptoms always indicate severe nutrient availability. Many crops start losing yields well before deficiency symptoms appear. Positive identification of nutrient deficiencies requires soil testing and analyzing plant tissue samples.

Is soil amendment really needed?

All too often, I hear "I'm not worried about fertilizing or spraying weeds — I'm just planting for wildlife, not for crop production." Sadly, what this really means is "I'm ignorant about fertilization and weed control, and I'm too lazy to learn anything about it and do what is necessary to have a productive food plot." This is unfortunate for a lot of people. These are the folks who are always frustrated with their food plots because their plots don't produce much and wildlife activity around them is low. You will never get much out of your plantings unless you are serious about the process and make the appropriate amendments and manage accordingly. Yes, you are planting for production! In fact, anyone "just planting for wildlife" should want to get the absolute highest yields they possibly can out of their plantings because they are usually planting limited acreage. Your efforts will be much more efficient, and less acreage will need to be planted, if you manage accordingly and crop yields are high.

collect soil samples and have them tested. Soil tests are cheap (usually around $8–$15; free in some states) and provide much-needed information. Unbelievably, relatively few people take advantage of soil testing. It is unfortunate because nutrient availability is often a limiting factor in forage production. It is amazing that food plot enthusiasts will pay $70 or more per acre on seed, but will not take the time to collect soil samples and have them tested to determine how much lime, fertilizer, or organic material are necessary to enable the plot to produce all it can.

Soil tests are no better than the sample collected. A soil sample should be representative of the field being planted. Each field should be sampled and tested separately. Where there is a considerable change in soil color or texture within a field, multiple soil samples might be needed for a single field. Regardless of color or texture, soil nutrient availability can vary considerably across a field. Therefore, a soil sample should actually represent a number of evenly distributed subsamples (10-20 per acre) collected across the field. For large fields, a general recommendation is to collect separate soil samples per 5 acres.

Soil samples are best collected with a soil probe, but a small garden shovel works fine. Samples should be taken to the depth used for fertilizer calibration trials. For most food plot plantings, that is approximately 6 inches. Mix these subsamples together in a bucket and pour a sample of this material in a soil test box (available from your county Extension office). Remove rocks and organic debris from the sample before filling the box. Soil samples should be dry. If your sample is moist when collected, allow it to dry before boxing and mailing. Label the box and accompanying soil test sheet as directed and return it to your county Extension office or send it directly to the soil test lab. Results may be sent to you via regular mail or email by most labs.

Table 3.2 Soil test ratings[1] for phosphorus and potassium in pounds per acre.

Rating[2]	Phosphorus (P)	Potassium (K)
Low	0–18	0–90
Medium	19–30	91–160
High	31–120	161–320
Very High	120+	320+

[1]These ratings are based on research data collected under various soil conditions and cropping systems from locations throughout Tennessee by University of Tennessee AgResearch and University of Tennessee Extension personnel.

[2]Low: Crops are probably yielding less than 75 percent of their potential and should respond to application of that nutrient.

Medium: Crops are probably yielding 75 percent or more of their potential and may or may not respond to application of that nutrient.

High: Crops are probably producing at or near 100 percent of the soil's potential without addition of that nutrient. Any amount recommended is to maintain present soil test levels.

Very High: Nutrient is well in excess of the amount needed to produce 100 percent of the soil's potential. Application of the nutrient is not recommended; further addition may create nutrient imbalances.

Soil tests vary among soil testing labs, but a basic test usually provides soil pH, current levels of phosphorus (P) and potassium (K), and recommendations for lime, nitrogen (N), phosphate (P_2O_5), and potash (K_2O) per acre according to the requirements of the stated crop. Available N generally is not included in a soil test because soil N is not stable and fluctuates over time. N recommendations are given for each stated crop. Soil samples also can be tested for micronutrients, organic matter, and soluble salts for an additional charge. Nutrients are normally reported in pounds per acre present in the soil. Ratings for P and K usually are reported as low, medium, high, or very high (Table 3.2). Ratings for micronutrients usually are reported as sufficient (S) or deficient (D). A basic soil test usually provides all the information you need. However, requesting information on micronutrients and organic matter

Fig 3.12; Fertilizer applications are most easily accomplished with a cyclone spreader. Calibrating the spreader prior to fertilizer applications is just as important as calibrating it prior to sowing seed! Here, fertilizer is being spread prior to planting a warm-season food plot.

is a good idea when planting a plot for the first time or if plot production is less than desirable, yet pH and levels of macronutrients are sufficient. If you have any confusion reading the soil test, call or visit your county Extension agent for free professional advice and consultation. For best results and continued successful plot production, it is best to soil test annually until pH and nutrient deficiencies have been corrected, then at least every two to three years for maintenance, while applying additional lime/fertilizer as recommended.

Fertilizer applications

The three numbers on a fertilizer bag refer to the grade, or the percentage of total nitrogen (N), phosphate (P_2O_5), and potash (K_2O) in the bag. For example, a 50-pound bag of 15-15-15 will contain 7.5 pounds of N, 7.5 pounds of P_2O_5, and 7.5 pounds of K_2O. It is important to note that P and K contained in fertilizers are in the oxide form (P_2O_5 and K_2O), not actual P and K. Phosphate (P_2O_5) contains 44% P and potash (K_2O) contains 83% K. Soil tests typically provide recommendations for the amount of phosphate or potash fertilizer needed, not actual pounds of P or K.

Although commonly used, complete fertilizers often are not necessary for many food plots. A more sensible approach may be to apply a specific-nutrient (or high-analysis) fertilizer. For example, a 50-pound bag of ammonium nitrate (34-0-0) contains 17 pounds of N. One bag of a phosphate fertilizer, such as triple super phosphate (0-46-0), contains 23 pounds of P_2O_5. A bag of muriate of potash (0-0-60) contains 30 pounds of K_2O. One bag of these fertilizers usually costs a little more than a bag of 15-15-15, but you get more than twice the amount of the nutrient needed per bag and you don't apply unneeded fertilizer. Using high-analysis fertilizers can be particularly important when planting legume plots (such as clovers, alfalfa, cowpeas, soybeans, lablab, and jointvetch). Properly inoculated legumes assimilate nitrogen from bacteria attached to their roots; thus, relatively little, if any, N fertilizer is necessary and N applications may lead to increased weed pressure. Also, if one nutrient is already rated high in availability, additional application of that nutrient may create a nutrient imbalance. Secondary nutrients and micronutrients typically are added separately if needed, but occur in several blended fertilizers. Fertilizer analysis is printed on the bag.

Fertilizers are most often applied with a three-point, hitch-mounted or pull-behind spreader. When planting with conventional tillage techniques, incorporating fertilizer into the top few inches of soil will reduce nutrient runoff and volatilization of N fertilizers containing urea (unless coated or applied just before rain). Incorporation of fertilizer is not necessary however, especially when using no-till planting techniques. It is important to realize good fertilizer management is possible only with adequate moisture. Split fertilizer applications, rather than one large application, may allow plants to use added nutrients more efficiently and lead to increased production. Slow-release fertilizers also help.

Balanced nutrition through fertilization refers to the continuous availability of all required nutrients in the proper amounts. Balanced nutrition is not a simple process. All plants do not require all nutrients in the same amount. For example, when grown together, some grasses may absorb various nutrients more readily than some legumes, leaving the legumes deficient. Some crops (such as oats) may not need as much fertilization as others (such as alfalfa) because of an ability to absorb particular nutrients (such as P) without additional fertilization. When planting or managing mixtures (for example, oats/clover/chicory), follow soil test recommendations for fertilization. As a general rule, you should fertilize according to the plant with the greatest nutrient requirements.

Soil nutrients are removed from a site through harvest and by grazing animals not fenced within an area. As nutrients are removed by wildlife foraging on plants, they must be replenished or soil fertility and productivity will decline. Balanced plant nutrition through fertilization is possible only by soil testing regularly (at least every 2-3 years).

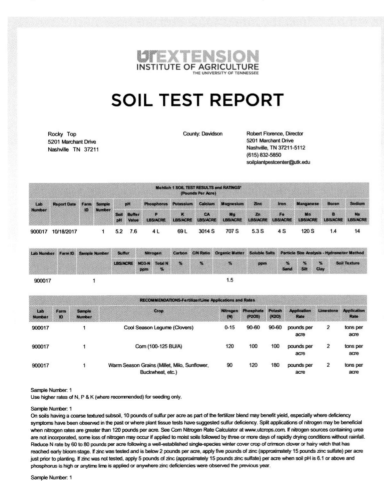

This is a soil test report as provided by UT Extension. Your name and address, county where soil sample was collected, and the date tested are identified at the top of the report. The lab number identifies this sample so it can be referenced later if needed.

The Soil Test Results and Ratings identify soil pH and the actual amount of nutrients in pounds per acre extracted in the soil sample. Mehlich 1 refers to the technique used to extract and determine nutrient availability. Different soil labs may use different techniques. Nutrient availability is rated as low, medium, high, or very high for P and K (see Table 3.2), and satisfactory or unsatisfactory for secondary nutrients and micronutrients. This soil test report provides recommendations for only one sample, identified on this report as Sample Number 1. You can identify the sample number with up to six digits or letters.

Water pH refers to the actual pH of the soil. It is determined by mixing the soil with deionized water, then measuring the pH of the deionized water solution. The buffer value is a measurement obtained in the lab after adding a buffer solution to the soil sample and allowing it to sit for approximately 10 minutes. The buffer solution slurry then is measured for pH. This measure represents how much cation exchange took place within the soil sample and is highly influenced by the percentage of clay in the sample. Thus, the buffer value explains how much lime is needed, as a clayey soil of a given pH will require more lime than a sandy soil with the same pH. Buffer pH has no practical value other than it represents the cation exchange capacity (CEC; see Soil pH on page 20) of the soil. As you can see, available P and K are low. However, Ca, Mg, Zn, Cu, Fe, and Mn are sufficient for the crops identified.

A basic soil test from most labs provides soil pH and levels of P and K. For an additional charge, soil samples can be tested for micronutrients, organic matter, soluble salts, and nitrate nitrogen. Levels of nitrogen are not provided in a basic test by most labs because N fluctuates so much over time, depending upon site conditions and weather. A recommendation for N fertilization is given for each crop even though current availability is not shown in the report. A rating is not provided for B or Na because data still are being collected to determine satisfactory levels for various crops. Organic matter is 1.5 percent. This information is important when using preplant or preemergence herbicides. Some of these herbicides have different recommended rates depending on percent soil organic matter. Usually, application rates will be higher if soil organic matter is greater than 3 percent.

Following the Soil Test Results and Ratings are the Recommendations (on a per acre basis). For this field (Sample Number 1), separate recommendations are provided for establishing clovers, corn, and other warm-season grains.

Prior to planting either a cool- or warm-season plot, soil pH should be amended by applying 2 tons of lime. To establish clovers, 0-15 pounds of N, 90 pounds of phosphate, and 90 pounds of potash are recommended. These amounts could be accomplished with approximately four bags of 0-46-0 and three bags of 0-0-60. N (no more than one bag of 34-0-0) may be incorporated prior to planting or top-dressed after germination and initial growth. If you are maintaining a perennial plot of clovers (as opposed to establishing an initial plot), the phosphate and potash recommendations can be lowered by 30 pounds each. That is because more phosphate and potash are needed during establishment for initial root development. N fertilization is not needed when establishing legumes if the seed have been properly inoculated. If the seed were not inoculated, N fertilizer may be applied.

If planting corn, 13 bags of 15-15-15 could be applied prior to planting. After the corn germinates and grows approximately 12-18 inches tall, two bags of 34-0-0 could be applied. If legumes are being rotated with corn, the N recommendation can be reduced as indicated.

Prior to planting various warm-season grains, 120 pounds of phosphate and 180 pounds of potash should be applied. P and K could be amended by applying six bags of 0-46-0 and six bags of 0-0-60. Half or more of the N recommendation (90 pounds) may be applied prior to planting and the other half could be top-dressed approximately three weeks later. Buckwheat does not require as much N as millets, grain sorghum, or sunflowers, so only 50 pounds are needed if planting buckwheat alone.

If you have questions interpreting your soil test, contact your county Extension office for free assistance and advice.

Acts 8:30-35

Fig 3.13; It is impossible to know how much lime or fertilizer is required without collecting a soil sample and having it tested. It is the best few dollars you will spend on your food plots.

Four

Chapter 4 — Plot preparation

Some basic plant definitions

An understanding of plant types and phenology is helpful when growing and managing food plots. Plants grown in food plots include forbs, grasses, and one sedge. Forbs are broadleaf herbaceous plants. Forbs commonly planted in food plots include the *Brassicas* (or "greens," such as the rapes, kale, and turnips), buckwheat, burnette, chicory, sesame, and sunflowers, as well as various legumes, including alfalfa, alyceclover, the true clovers, cowpeas, jointvetch, lablab, lespedezas, partridgepea, perennial peanut, and soybeans. Grasses commonly planted in food plots include corn, grain sorghum, various millets, oats, cereal rye, and wheat. The one sedge often planted in a food plot is chufa, which is a variety of yellow nutsedge.

All of these plants are further defined as annual, biennial, or perennial, and warm- or cool-season (see *Appendix 1*). Annual plants germinate, grow, flower, and produce seed in one growing season and, depending on the plant, variety, and management strategy, may or may not reseed. Biennial plants normally require two growing seasons to complete their life cycle. Perennials continue living after flowering and producing seed and, depending upon management, may persist for several years.

Warm-season plants typically germinate in the spring and grow through the summer. Except for perennial peanut, all of the warm-season plants planted in food plots are annuals. Most cool-season plants produce the majority of their growth during fall, spring, and early summer; however, some species and in various regions produce considerable forage throughout the summer and some continue to grow during the winter, especially mild winters. Both annual and perennial cool-season plants are commonly grown in food plots.

These same plant category definitions also apply to weeds, which can include forbs, grasses, and sedges. Weeds are defined as *unwanted plants for a specific purpose*. For example, ragweed, pokeweed, beggar's-lice, and tropic croton produce seed relished by northern bobwhite and mourning dove, and deer graze the foliage. These plants are promoted when managing early successional communities for those species. However, they may be considered weeds in a clover plot grown for white-tailed deer because they exhibit a different growth pattern than clovers and, according to density, may reduce clover production. **Separate acreage should be devoted to early successional plant communities and food plots, just as separate acreage should be devoted to warm- and cool-season food plots, depending on your objectives. Furthermore, realizing the growth patterns (warm- or cool-season) and life cycles (annual or perennial) of weeds is absolutely critical when trying to manage them.**

Fig 4.1; Understanding plant phenology is critical to manage successful food plots and control various weeds, such as this musk thistle, which is a biennial and is about to bolt.

Preparing the site — Controlling existing vegetation before planting

Preparing the site before planting is extremely important, especially if perennial weeds are present and you are planting a perennial food plot such as ladino clover and chicory. Annual weeds are not a huge concern if they are disked or sprayed before they produce seed. However, perennial grasses (especially tall fescue, bermudagrass, johnsongrass, and bromegrasses) and some specific perennial forb weeds (such as curly dock and horsenettle) may present major competition problems in perennial food plots if not sprayed prior to planting. If you plan to use conventional tillage and plant an annual crop year after year, eradicating perennial weeds with herbicide applications is not as critical. Planning ahead is important when preparing a site for a successful food plot.

An application of glyphosate is commonly used to kill existing vegetation prior to conventional seedbed preparation and prior to no-till planting (see discussion on postemergence applications under *Herbicide applications* on page 64). One to 2 quarts (depending on weed species present and age of weeds) of a glyphosate herbicide will kill a majority of weed species. You should wait at least two weeks after spraying before preparing the seedbed. Waiting allows you to see any areas you missed and

40

*Fig 4.2; Before planting perennial food plots, eradicate perennial cool-season grasses by spraying the appropriate herbicide. **Preparing the plot by plowing and disking does not get rid of these grasses!** If you do not kill them by spraying before planting, many of the root systems will remain alive, and you will have to fight them later, as evident in this ladino clover food plot where residual tall fescue is coming back seven months after planting the clover.*

spray them before disking or planting. Before applying a glyphosate herbicide, or any other postemergence herbicide, be sure to identify if the product contains a surfactant (see *What are surfactants?* on page 71). Most glyphosate herbicides contain surfactant, but some do not. If the postemergence herbicide you are using does not contain a surfactant, be sure to add one according to herbicide label directions.

If perennial cool-season grasses, such as tall fescue or orchardgrass, are present where you intend to plant, spray 2 quarts of a glyphosate herbicide per acre in the fall before a spring planting or 2 quarts per acre in the spring before a fall planting. Fall spraying is most effective because nutrients are being transported from the leaves to the roots in preparation for winter senescence. Spring spraying is more likely to require repeat applications the following year. Regardless, it is best if you burn or mow the field before spraying if there is dead thatch and vegetation from past seasons still present. These methods provide a "clean" field for spraying, free of thatch and dead material that will block much of the herbicide from contacting the green, growing grass. The grass should be growing vigorously with leaves at least 6-10 inches long when sprayed for best results. Plan ahead and do this correctly. Perennial grasses should be killed at least a month before planting if you are using conventional tillage.

To control bermudagrass, burn the field in late winter. Spray bermudagrass the following summer about the time the bermudagrass begins to flower, often in June. Imazapyr (48 ounces of Arsenal per acre) or glyphosate (5 quarts of Roundup per acre) are options. Imazapyr is more effective killing bermudagrass than glyphosate; however, there is a 12-month minimum crop rotation restriction following application of imazapyr. Therefore, you would not plant the field anytime soon. If you want to plant the field in the fall after spraying, use a glyphosate herbicide. Do not incorporate lime or fertilizer prior to spraying grass — wait at least four weeks after spraying.

To control bahiagrass, prepare the field by mowing or burning in spring, then spray in May before bahiagrass flowers. Metsulfuron methyl (0.3 ounces per acre of Cimarron or Escort) works well if you intend to plant a perennial plot the following fall. However, if you intend to plant soon after spraying, you should use 4 quarts of a glyphosate herbicide per acre.

If johnsongrass is dominant over the field, wait until it gets 18 to 24 inches tall and spray with 2 quarts of a glyphosate herbicide per acre. Later, if appropriate, use a preemergence herbicide when you plant that will help control seedling johnsongrass (see *Herbicide applications* on page 64).

Warm-season forb weeds are best controlled with glyphosate and/or a forb-selective herbicide, such as 2,4-D, Banvel, or Clarity (see *Appendix 2*). Several forb-selective herbicides can be tank-mixed with a glyphosate herbicide for increased broad-spectrum control. Refer to herbicide labels for tank-mixing application rates. Planting Roundup-Ready crops for one or more seasons is another option for controlling stubborn perennial forb weeds. Using Roundup-Ready crops annually, however, should be approached with caution. Continued use of glyphosate alone can lead to

Fig 4.3; For optimum results, perennial cool-season grasses, such as tall fescue and orchardgrass, should be sprayed in the fall with glyphosate. Fields should be "clean" before spraying. This field was hayed in early October to prepare for spraying in early November. This timing allowed the herbicide to come in contact with growing grass, not dead thatch and stems from previous years' growth. This timing is a good opportunity for no-till top-sowing small-seeded species, such as clovers and alfalfa. Top-sow prior to spraying. As existing cover dies, the planted seed will germinate and begin to grow. New plant growth is supported with nutrients released as dead plant material decomposes and as soil moisture is retained without disking. Increase seeding rate by 25% when no-till top-sowing.

physiological resistance in certain weeds. Using herbicides other than glyphosate every two to three years will provide better long-term control of problem weeds.

Preparing the seedbed
Plowing/disking or drilling?
Once problem weeds have been controlled, it is time to prepare the seedbed. Lime may be incorporated into the root zone by disking at this time if you are planting via conventional tillage. If you intend to plant with a no-till drill, you may need to mow or burn the dead vegetation before drilling, according to how much thatch is present. Otherwise, no further

Fig 4.4; This field of bermudagrass was effectively killed with Arsenal AC. Refer to Appendix 2 *for crop rotation timing when using imazapyr.*

preparation is needed. If you are planting via conventional methods, and if fertilizers have been disked into the seedbed, the plot may be ready to plant (top-sow or drill with a conventional drill). If not, disking or tillage (with a rotovator) will incorporate lime/fertilizers and create a clean seedbed. Mowing or burning may be necessary to reduce vegetative cover on the field prior to plowing/disking/tilling. Moldboard plowing should be avoided. Using a chisel plow does not bury organic material and will not create a hardpan. After plowing, disking with a tandem disk or tilling prepares a fine seedbed. A clean, smooth, firm seedbed is desirable for sowing small seed, such as clovers, jointvetch, and alfalfa. These small-seeded forages also can be established successfully by light disking after the existing vegetation has been killed and top-sowing. No-till top-sowing is especially applicable when planting woods roads and mowed paths where plowing or heavy disking may not be desirable. No-till techniques are discussed in *Chapter 5*.

After many years, continued plowing and disking can create a hardpan (compacted soil 8-15 inches deep that inhibits deep root systems and reduces soil moisture). A subsoiler or chisel plow (with shanks penetrating 16-20 inches deep) can be used to break the hardpan and conserve soil moisture. Of course, no-till techniques avoid creation of a hardpan and associated problems and are recommended in most situations when possible over conventional tillage techniques.

Fig 4.5a and b (below); Burning thatch and other dead material is an excellent way to prepare a field for disking, drilling, or top-sowing. Burning dead material is not necessary when no-till drilling relatively large-seeded crops, such as corn, soybeans, cowpeas, or grain sorghum.

Soil moisture and level of preparation – what to look for
The correct amount of soil moisture is critical when preparing a seedbed for conventional tillage, especially in clay soils. If the soil is too dry, dirt clods will be large. If too wet, discs will clog with mud, and large clods will result once the soil dries out. When worked at the proper moisture level, dirt clods fall apart in relatively loose, small particles. Of course, this is not as much of a concern in loamy soils and not a concern in sandy soils. Soil moisture can be deceiving by looking at the surface. Only by digging down to the depth the soil will be worked can soil moisture be determined. Soil moisture should be checked in a few spots across the field, similar to collecting soil samples for testing. By digging down approximately 6 inches with a shovel, the moisture level can be checked by squeezing a handful of soil. If the soil sticks together in a ball, it is too wet. If it looks and feels moist, yet crumbles, it is perfect. If no moisture is felt and it is fairly difficult to crumble because it is hard, it may be too dry to prepare a fine seedbed for small-seeded species such as alfalfa and clovers. Soil moisture also is important when using a no-till drill, especially if drilling large seed, such as soybeans, cowpeas, and sunflowers. A moist seedbed is desirable that will allow proper seeding depth (1 – 2 inches). A hard, dry seedbed will not allow many drills to plant at the correct depth, and seed will be at or just below the soil surface.

Larger seed, such as soybeans, cowpeas, and lablab, do not require as fine of a seedbed when planted conventionally. They will germinate well in a relatively coarse seedbed provided they are covered an inch or so and receive adequate rainfall. If adequate soil moisture is present, a previously plowed seedbed should only need a couple passes with a disc harrow prior to planting large seed. Germination and initial growth of smaller seed is *considerably* better in a finer seedbed. Additional disking or tillage is usually needed for these species. A rotovator is an excellent implement for preparing a seedbed for small-seeded species, especially in clay soils.

Fig 4.6; A chisel plow with shanks 16-20 inches long does a great job incorporating lime and breaking hardpans created after many years of shallow plowing and disking.

Five

Chapter 5 — Planting

It is best to plant when adequate soil moisture is present to improve germination and establishment, or just prior to rain. Germination and growth are less than desirable when it is dry for an extended period after planting. If seed germinate and seedlings do not get adequate moisture soon, they desiccate and die. Therefore, planting by a certain date is of little concern unless there is projected rainfall.

Food plots are planted either by broadcast spreading seed or using a drill or other type of planter. Both techniques can be used with conventional tillage or no-till technology. Broadcast seeding is normally done over a well-prepared seedbed using tillage techniques. Drilling seed is normally accomplished with no-till drills. However, some drills and planters require a cultivated seedbed. Also, some small-seeded forages, such as clovers, can be top-sown without tillage (with proper weed control).

Broadcast seeding

Most broadcast spreaders are mounted to a three-point hitch behind a tractor. These spreaders are efficient when planting relatively large plots (at least 2 acres), as well as spreading fertilizer. Hand-held broadcast spreaders are effective and efficient when planting smaller plots. Calibrate the seeding rate by marking off a 1/10-acre area and weighing the appropriate amount of seed. **Seeding rate calibration is a critical step in successful planting.** Start at a low setting and adjust up as necessary. Practice with additional plantings to get the seeding rate just right. When broadcast seeding with a hand-held seeder, walk at a slow-to-moderate pace. When using a tractor-mounted broadcast seeder, be sure to record the gear, rpm, and seeder setting when the correct rate and setting is determined. You will need to do this separately for different species because the seeder setting will be different for different-size seeds.

Drilling seed and calibration

Calibrating drills and planters prior to planting is absolutely critical. A few methods exist for calibrating drills. One method involves counting the number of revolutions the drill wheel makes to cover 1/10 acre (accounting for the width of the drill). Once determined, elevate the drill with a jack and turn the wheel the predetermined number of times, which should be covered in detail in the owner's manual. Collect the seed in buckets or on a tarp and then weigh the seed. Then you can set the drill as needed. Another method involves removing the seed drop tubes and attaching plastic sandwich bags to the bottom of the tubes with rubber bands. Then operate the drill over 1/10 acre, or elevate the drill with a jack and turn the wheel the predetermined number of times that would represent 1/10 acre. Remove the seed bags, weigh the seed, and set the drill as necessary. Planters operate by allowing seed to pass from seed boxes through plates that have certain-sized openings. Plates with different-sized openings can be used to increase or decrease planting rate or to plant larger or smaller seed.

Fig 5.1; Drilling seed is a most reliable planting technique. Seed is planted at the desired depth, and germination rates and seedling survival are higher.

Advantages and disadvantages with drilling and tillage

Conventional tillage and no-till drilling both have advantages and disadvantages. Depending on the existing plant cover and the type of plot planted, herbicides may not be necessary with conventional tillage — but they usually are. When using a planter with conventional tillage techniques, considerable weed control is possible by cultivating between rows. Drilling or planting requires approximately 25 percent less seed sown per acre than with broadcast seeding because seed placement is precise and germination rate and seedling survival are greater. On the other hand, conventional tillage disturbs the soil and stimulates germination of weed seed in the seedbank. Tillage can release problematic weeds and decrease available soil moisture, increasing the chances of seed desiccation if a prolonged dry spell ensues. Soil moisture is why it is so important to watch the weather forecast and plant just ahead of rain. Exposed soil also is subject to erosion. Relatively steep slopes should not be planted with conventional tillage techniques. Disturbed soils often are more compacted and reduce water infiltration, which causes increased runoff of soil nutrients and reduces soil moisture for plants.

No-till drilling seed obviously requires a drill, which is expensive, and most drills are quite large, requiring large trucks to transport them from site to site. Also, because of their large size, it may not be possible to get some drills into remote locations. Prior to planting, herbicide applications are necessary to kill existing plant cover. If the existing cover is not sprayed, planted seed have little chance of germinating and growing amid existing growing vegetation. A huge advantage of no-till planting is soil conservation. Organic matter and the associated nutrients accumulate and create top soil. Soil microbial activity is increased when the soil is not disturbed, providing more nutrients to plants. Also, no-till planting requires much less time and work than conventional tillage. Soil moisture is conserved when using a no-till drill, and the possibility of soil erosion is virtually eliminated by no-till planting.

Don't give up hope if you would like to plant with a no-till drill, but cannot afford one. Check with your local Natural Resources Conservation Service and Farm Service Agency offices. In many states, they rent drills at a very reasonable rate. For most applications, a drill allows you to plant more precisely and with less time required than conventional tillage techniques and top-sowing.

No-till top-sowing

Another option in no-till planting is top-sowing with the aid of herbicides. You must kill existing vegetation with herbicide (such as Roundup). Sow small-seeded species just prior to spraying. (Note: increase your seeding rate approximately 25 percent above the normal broadcast rate). Doing so allows rain to carry the seed down to mineral soil while the existing vegetation dies and decays. The amount of existing plant cover and soil condition will influence success. If there is not much plant cover and the soil is crusted over, germination usually is less than desirable. In such situations, disking or dragging prior to seeding is recommended. If there is grass sod on the site, no-till top-sowing usually works well. If the vegetation is relatively tall, mowing once the vegetation dies will help germination and seedling survival. No-till top-sowing works well with clovers, alfalfa, brassicas, American jointvetch, and alyceclover. It does not work as well with larger seed that require coverage. An exception can be when larger seed are top-sown onto an area where there is considerable vegetation coverage that is relatively tall, and the taller vegetation is mowed after it dies, and then cultipacked or rolled. Frost-seeding is another example of top-sown no-till plantings. Sow small-seeded, cool-season species on top of snow and frost-heaved ground prior to thawing. A selective herbicide application may be necessary once spring green-up begins, depending on the seedbank and what you plant.

Seeding depth

Regardless of planting method, seeding depth is a major consideration. Planting or covering seed too deep is a common reason for plot failure. Grains and other relatively large seed (such as corn, grain sorghum, Austrian winter peas, cowpeas, soybeans, lablab, and sunflowers) should be drilled or covered by disking approximately 1 inch

Fig 5.2; *This corral was covered with bermudagrass. The horses were removed and the bermudagrass was sprayed in late summer with glyphosate at 5 quarts per acre. Ladino clover was no-till top-sowed prior to spraying. After the ladino clover established, incoming weeds were controlled with an application of Pursuit (4 ounces per acre) and Clethodim (12 ounces per acre) tank-mixed.*

Fig 5.3; *Large-seeded species, such as cowpeas, soybeans, grain sorghum and cool-season grains, can be covered by disking. However, don't disk-in clovers — you will cover them too deeply! Always cultipack the seedbed after top-sowing small seed.*

deep. The cool-season grains (oats, wheat, and cereal rye) germinate better when lightly covered — especially oats. Small-seeded species, such as clovers, should be covered no more than 1/4 inch. When mixtures of both large- and small-seeded species (wheat or oats with clovers) are planted using conventional tillage methods, the large seed should be planted first. After seeding, cover the large seed by disking, then plant the small seed, and then cultipack.

Seeding depth can be a real problem with some premixed commercial seed blends. With both large and small seed in the bag (such as lablab/soybeans/cowpeas with grain sorghum/jointvetch/alyceclover), the small seed tend to gravitate to the bottom of the seed box and are sown before all the large seed are sown. A much better approach is to sow the seed separately, or to drill large and small seed in separate boxes, sowing each at the appropriate rate simultaneously.

Cultipacking

Cultipacking improves the germination rate and seedling survival of top-sown seed, especially small seed that require firm seed-to-soil contact. A cultipacker is the implement of choice for firming a seedbed. The results are far superior to dragging a chain-link fence or some other crude object around the field behind a tractor or ATV, which is more appropriate for breaking up clods and smoothing the seedbed. Cultipacking is necessary prior to seeding small seed if the seedbed is fluffy and is always recommended after seeding small-seeded species (such as alfalfa, clovers, brassicas, jointvetch). Cultipacking prior to seeding usually is necessary when a rotovator is used to prepare the seedbed. As a general rule, if you leave a boot imprint an inch deep, the seedbed should be cultipacked before planting small-seeded species. Small seed usually can be sown on top of lightly disked soil without cultipacking beforehand. After sowing small seed, the seedbed should be firmed using a cultipacker. Cultipacking also will improve germination of large-seeded species. Cultipacking is an important step in successful food plot establishment when seed are top-sown.

Fig 5.4; A cultipacker is the best implement to firm a seedbed prior to sowing small seed, such as clovers, and to establish firm seed-to-soil contact after sowing. Some folks use a section of chain-link fence to smooth a plot and reduce clods prior to planting. This method can lead to a better seedbed, but it is not a replacement for a cultipacker, which is needed to provide firm seed-to-soil contact.

Planting large and small seeds together with conventional tillage

Broadcast planting mixtures of large and small seed following conventional tillage methods is best accomplished by the following procedure.

1. Prepare seedbed by disking or tilling. Lime and fertilizer may be incorporated at this time. Also, if you are applying a preplant-incorporated herbicide, it should be applied prior to disking or tilling.
2. Sow large seed — such as soybeans, cowpeas or oats — onto prepared seedbed.
3. Cover seed by disking approximately 1 inch deep (wheat and oats no more than 1/2 inch).
4. Firm the seedbed using a cultipacker. Cultipacking is especially important when planting small seeds, such as clovers and alfalfa.
5. Sow small seed, including clovers, alfalfa, chicory, brassicas, jointvetch, and millets.
6. Cultipack seedbed once again to ensure firm seed-to-soil contact and improve the germination rate and seedling survival.
7. Apply preemergence herbicide, if appropriate.

This procedure allows good germination and seedling establishment of both large- and small-seeded plants. Also, it leads to an even distribution of seed. When large- and small-seeded species are mixed together, the small seed tend to gravitate to the bottom of the seed hopper or box and are sown before all the large seed are sown. Thus, an uneven distribution results.

Seed selection and calculating pure live seed (PLS)

After you have decided what you are going to plant, you should purchase high-quality, certified seed with a high germination rate. Each bag of certified seed should have a seed tag attached.

If there is no seed identification tag attached to the bag, do not buy it. This tag should identify the variety of seed in the bag, seed origin, percentage of pure seed, percentage of inert material, germination rate, test date, and the presence/amount of weed seed. If the seed is a legume and it is preinoculated, the percentage weight of the seed coating also should be identified on the seed identification tag. If you are ordering seed, the salesman or seed representative should be able to provide this information over the phone or email. Seed germination rates of most food plot seed typically decrease over time. If it has been more than a year since the seed was tested, true germination probably will be lower than that reported on the seed identification tag. High-quality seed should have a germination rate of at least 80 percent.

Calculating Pure Live Seed (PLS) can be an important step in successful food plot establishment. Percentage of PLS can be calculated from information provided on the seed tag. The percentage of pure seed usually is the first thing listed on the tag and represents the percentage of the contents in the bag that is actual seed. Some of the actual seed is not going to germinate. The percentage of seed that can be expected

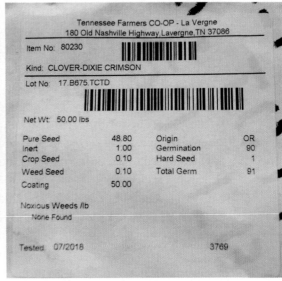

Fig 5.6a and b; Certified seed will have a seed tag attached that tells you exactly what you are buying, and gives you information regarding seed quality. Preinoculated seed is coated to protect the bacteria that will inoculate the seed once it is planted. The coating may be blue, pink, yellow, or white (such as this crimson clover seed). Regardless of color, the coating represents a certain amount of weight, and the weight of the coating should be accounted for when calculating PLS.

to germinate is provided as the germination rate. The germination rate usually is around 80%, which means 20% of the seed in the bag is not going to germinate. The percentage of "hard" seed also is provided. Hard seed includes dormant seed that will not germinate until after freezing and thawing or some other form of scarification. If the seed in the bag is a legume (such as clovers and alfalfa) and has been preinoculated, there will be a coating around the seed to protect the inoculant. The coating may represent 33 - 50% of the contents of the bag and will be listed on the seed tag as percent coating material. The amount of inert matter (which may include any organic material other than seed, such as seed hulls, pieces of leaves or stems, etc.), other crop seed, and weed seed also is given on the seed tag. The collective percentage of inert matter, other crop seed, and weed seed is almost always less than 1%. If the percentage of weed seed approaches 1.0%, that would indicate there is a fair amount of weed seed in the bag.

Lots of factors influence seed germination and seedling success. You probably will not see a difference in plant density or crop biomass if you plant raw (uncoated) seed with an 80% germination rate vs. a 95% germination rate. However, if 50% of the contents in the bag is seed coating and the germination rate is only 60%, then you are effectively planting much less seed than the recommended seeding rate per acre. Therefore, adjusting your seeding rate can be critical to establishment success.

Various perennial clovers may become more dense over 1 - 3 years as stolons and rhizomes (depending on the clover) spread. Alfalfa may become more dense over time as individual plants mature and produce additional stems. Therefore, perennial plants may fill-in over time if your effective seeding rate was relatively light. However, annual plants, such as crimson, berseem, or arrowleaf clovers, may not fill-in, and in many situations, it is important to have a full stand initially to sustain grazing pressure. Bare soil resulting from an insufficient seeding rate allows increased weed pressure and can further lead to poor stand establishment. Therefore, adjusting your seeding rate according to PLS is especially important for annual plantings, but also can be important for perennial plantings. The procedure for calculating PLS is shown below.

Variety: Crimson clover (see Fig. 5.6a and b)
Pure seed: 48.80%
Germination: 90.0%
Hard seed: 1.0%
Coating: 50%
Inert matter: 1.00%
Other crop: 0.10%
Weed seed: 0.10%

Pure seed (0.488) × Germination (0.90) = 0.4392 (43.92% PLS)
Desired planting rate (20 pounds per acre) ÷ 0.4392 = 45.54

Therefore, in order to plant 20 pounds of this crimson clover per acre, you should sow 45 pounds of material in this bag. Note: although some people may contend that hard seed should be included in the germination rate, I do not. Most food plot enthusiasts want their planting to respond and begin growing soon after sowing, not the following year. Hard seed are dormant and will not germinate until months after planting, if at all. Alfalfa and perennial clovers may fill-in the open spaces over time as the plants mature. However, these open spaces also allow more weed competition. Consider this when planting your forage food plots.

Fig 5.7 and 5.8; Inoculated seed, such as these soybeans (above), should be allowed to dry in the shade. The seed should be ready to plant in an hour or so. If allowed to dry in direct sunlight, the bacteria may be killed. Properly inoculated seed (below) is obvious. You can see the black peat from the inoculation mixture stuck all over these lablab seeds.

Inoculating legume seed

Legumes are plants that bear seed in a pod and have a symbiotic relationship with certain species of nitrogen-fixing bacteria (such as *Rhizobium* spp. and *Bradyrhizobium* spp.). These bacteria attach themselves to the roots of legumes and form nodules. From these nodules, the bacteria extract nitrogen from the atmosphere. *Rhizobia* and others obtain energy from the plant, and the plant receives nitrogen produced by the bacteria. Thus, both bacteria and plant benefit from the relationship. Nitrogen fixation is influenced by many factors, including soil pH, nutrient availability, soil moisture, temperature, and plant health.

Nitrogen fixation is important when planting wildlife food plots for three reasons: 1) minimal nitrogen fertilization is required, thus you save money, 2) nitrogen is not a limiting factor to properly inoculated plants, and 3) weed competition is reduced because little or no nitrogen fertilizer is applied. Depending on the legume planted, properly inoculated seed may produce up to 200 pounds or more of nitrogen per acre, which is significant in terms of reducing fertilization and herbicide costs, especially when rotating a legume crop with a grass, such as corn, grain sorghum, wheat, or oats.

Particular legumes require specific bacteria. That is, no one kind of bacteria will properly inoculate all legumes. Therefore, you must use species-specific inoculant. Although bacteria, such as *Rhizobia*, are found naturally in the soil, it is important to inoculate seed prior to planting to ensure the *proper* bacteria are *in contact* with the seed. Legume seed are inoculated with *live* bacteria and there is a shelf life associated with each bag of inoculant or preinoculated seed. Inoculant should be stored in the refrigerator and never placed in direct sunlight (such as the dashboard of a truck). Bacterial survival is greatest in soils with a relatively neutral pH (6.0-7.0). Acid soils (pH less than 5.8) do not support bacteria as well and inoculation efforts often fail in these conditions. Ideally, inoculated seed should be sown in a moist seedbed or just before rain. Dry conditions extending a few days after planting will reduce inoculation success significantly and nitrogen fertilization then may be necessary.

You can buy preinoculated seed of several legumes, meaning the seed have been inoculated with the proper bacteria prior to bagging. Preinoculated seed are coated to protect the bacteria. The coating usually is off-white, yellow, pink, gray, or blue (see Fig 5.6a). Preinoculated seed should be sown before the inoculant expiration date, as indicated on the seed identification tag. If the inoculant surrounding the preinoculated seed has expired, inoculate the seed before planting.

Inoculating legume seed can be somewhat messy. It is convenient to buy preinoculated seed. However, the cost per acre of preinoculated seed is greater than raw seed because the coating around the seed often represents 35 - 50% of the contents in the bag. If you purchase raw seed, inoculation is an important step in establishing successful legume food plots. Improper inoculation methods, acid soils, and planting in dry seedbeds are reasons why many attempts at establishing and maintaining healthy legume forage food plots fail. At the same time, it is important to realize inoculation is not necessary when planting a particular legume in a field where that legume was successfully established in the past few years. If successfully inoculated previously, those bacteria should remain in the soil on that site for several years. That is why most soybean producers, for example, do not inoculate soybean seed every year in those fields where soybeans have been grown in the last two or three years.

Steps for inoculation
1. Buy specific inoculant for each legume planted. Inoculant has a limited life span (it contains live bacteria), so check the expiration date and store inoculant in the refrigerator prior to planting. Do not expose inoculant to heat or direct sunlight. Always use fresh inoculant. Do not use leftover inoculant from last year.
2. Inoculants are packaged in a medium of peat, which is black. Pour the inoculant over the seed in a bucket. Be aware a bag of inoculant will inoculate a lot of seed (50 pounds or more of some plant species). Therefore, relatively little inoculant is needed for an acre's worth of clover, for example.
3. Add a commercial sticker or sugar-water solution (4 parts water to 1 part sugar) to the seed/inoculant. Commercial stickers may be available in powder or liquid form, but they can be difficult to find. Sugar water works just as well. **Soft**

drinks should not be used as a sticker because the pH of most soft drinks is very low and the acid solution may kill the bacteria.

4. Add just enough water to form a "slurry" (you don't want it too wet—just enough to stick the inoculant to the seed). Mix the inoculant/sticker/seed slurry well, making sure all seeds are coated with inoculant. This step is critical, and it can be done by hand. Although there are live bacteria in the inoculant, it is no different than picking up a handful of soil (which contains millions of bacteria). If the inoculant does not stick to the seed, the entire process is of no value.

5. Spread inoculated seed out on some newspaper, cloth bags, or a sheet to allow the inoculated seed to dry **in the shade** (it will take no more than an hour). Once dry, you can sow the seed. If you do not sow the seed right away, you can store inoculated seed in a cool, dry place for no more than a couple of days or re-inoculation will be needed. Likewise, planting on a moist seedbed or just prior to rain is important to ensure inoculation success. Inoculated seed should not be mixed with fertilizer, as the salts in fertilizer may kill the bacteria.

Seed coatings

Although not inoculants, seed coatings may be promoted for increased seedling survival, plant growth, and development. Increased root development has been documented following application of some products, as well as slight increases in crop production. However, substantial increases in plant growth or yield have not been published. Additional research is needed to clarify the benefit, if any, of these treatments.

Seeding rates

Seeding rates for individual plant species are normally given on a per acre basis, and thus represent the amount of seed necessary to plant and cover a 1-acre area. Seeding rates are based on a PLS rating of 100 percent (see *Seed selection and calculating pure live seed (PLS)* on page 52). That is why it is often important to calculate PLS and adjust the seeding rate as appropriate.

Often, food plots comprise multiple species to form a combination, blend, or mixture. When mixing species for a blend, the seeding rate for each species is adjusted according to the number of species or varieties in the mixture, the composition preferred, timing of maturation of each species, and the growth form and desired structure of the resulting stand. The individual seeding rates for each species are not combined! This would result in overseeding and money wasted. It also would result in some species being crowded out and underrepresented in the plot. For example, small grains such as wheat and oats often are planted with ladino clover and chicory, primarily as a fall attractant and a nurse crop while the clover and chicory develop. The individual seeding rate for wheat and oats is 120 pounds or more per acre. If combined, this would result in 240 pounds of small grain planted (a double seeding rate) and little or no clover and chicory would germinate and grow through the fall and spring because the plot would be completely filled with wheat and oats. By reducing the rate of wheat and/or oats to 40-50 pounds, the small grains will complement the developing clover/chicory very well and help them establish with less weed pressure and less grazing pressure. The seeding rate of ladino clover and chicory should be reduced as well (their individual seeding rates are 6 and 10 pounds, respectively). A good mixture would include 20 pounds wheat, 20 pounds oats, 4 pounds ladino clover, and 4 pounds chicory.

The recommended seeding rates for various plants in *Appendix 1* are accurate and should be followed fairly closely to expect planting success. Overseeding is a common occurrence. Most people think more is better. More is not better when seeding food plots, especially grain plots. When forage plots (such as clovers, cowpeas, and chicory) are overseeded, seedlings are crowded and the plants compete with each other. Some die, some live, and the plot produces forage — only money is wasted. However, when grain plots (such as corn or grain sorghum) are overseeded, seedlings are crowded and little space is available for seed production. Vegetative growth may be acceptable, but seed production is significantly less than that produced by properly spaced plants. If the seeding rate of grains is questionable, it is better to sow less than more. Doing so will allow more seed production per plant. However, it also means additional space between plants will allow more weed response. This response may be good or bad, depending on the "weed" and the type of plot you are growing (this subject is discussed thoroughly in *Chapter 6* and with the focal wildlife species in subsequent chapters).

Sowing the precise amount of seed is not difficult. First, you must accurately determine the area to be planted. Second, you must accurately calibrate your seeding equipment. And third, you must calculate PLS and weigh the seed! Do not guess the weight of the seed. Imprecise measurements, when coupled with disregard of the first two steps, lead to grossly inaccurate seeding rates. Hand-held scales for weighing seed are inexpensive and important equipment when establishing successful food plots.

Recommended seeding rates for various food plot plantings are normally given as broadcast rates. Because of seed desiccation and less precise placement of seed, broadcast rates are greater than those when using a drill or planter. When using a drill or planter, the seeding rate may be reduced approximately 25 percent, depending on what you are planting.

Mark 4:26-28

Six

Chapter 6 — Weed and pest control

Weed control is a huge factor in food plot success. In many cases, weeds overtake the crop before planted seed can begin to grow or they overwhelm the crop through the growing season, robbing the crop of nutrients, moisture, and sunlight and causing a significant reduction in crop production. Weeds arise from the seedbank — those seeds occurring naturally in the top few inches of soil — and from seed hitchhiking on equipment, such as the grill of a tractor or top of a rotary mower. Seedbank composition varies tremendously in different areas and from site to site, however, you should expect undesirable plants on most sites and some level of control will be necessary to maximize growth and yield of planted crops and the resulting food value for wildlife.

Insect damage and plant diseases are not nearly as common as weed problems, but they still can be a problem with any food plot planting. Aphids, armyworms, grubs, cutworms, stalk borers, rusts, molds, and other insects and diseases can reduce production of your food plots. If your planting does not look healthy, it is most likely related to nutrient availability, but you should look carefully and check for insects and signs of disease. Sometimes, they can be difficult to see. If you think you have an insect or disease problem, call your Extension agent and ask them to come and look at your planting, or you might send them some high-resolution photos. I know this

Fig 6.1; Armyworms can be a problem in wheat, especially if planted in late summer. Photo by Ryan Basinger.

can be frustrating. It has been for me. However, with a little patience and persistence, you can find out what the problem is. You can go to http://www.utcrops.com for detailed information on crop insect pest control. Also, check the websites of the crop and forage professionals at your land-grant university. Most provide a wealth of information. This chapter concentrates on weed control. However, recommendations for treating some of the more common insects and diseases are provided for several planting recommendations in the following chapters, as well as in *Appendix 4*.

Learn to identify your plants!

The ability to identify plants and understand their life cycle is requisite if you hope to be successful controlling weeds and managing early successional communities. Unless you know which weeds you have, it is difficult, and sometimes impossible, to determine whether a cultural, mechanical, or chemical weed control method is needed. If an herbicide application is needed, you must know which weeds are present to identify the correct herbicide. Different herbicides control different species of weeds. It also is very important to understand the seasonal life cycles of weeds. Just as there are warm- and cool-season food plot plantings, there also are warm- and cool-season weeds. Understanding plant phenology one of the most important steps in successful weed management.

The most common naturally occurring plants in food plots (warm- and cool-season, annual and perennial) include crabgrass, broadleaf signalgrass, foxtail grasses, goosegrass, fall panicum, barnyardgrass, johnsongrass, bermudagrass, yellow nutsedge, sicklepod, spurges, carpetweed, cocklebur, jimsonweed, common lambsquarters, morningglories, spiny amaranth, pigweeds, ragweeds, pokeweed, horseweed (marestail), prickly sida, catchweed bedstraw, curly and broadleaf dock, horsenettle, narrowleaf and broadleaf plantain, maypop passionflower, white-heath aster, Pennsylvania smartweed, yellow nutsedge, thistles, sowthistles, tall ironweed, sumpweed, tall fescue, orchardgrass, bluegrass, velvetgrass, bromegrasses, wild garlic, chickweeds, Carolina geranium, henbit, purple deadnettle, speedwells, ground ivy, and hairy bittercress. I do not call this collective list "weeds" because many of them are quite beneficial in various food plots for various wildlife species. **If you are not familiar with these plants, refer to the *Plant Identification Guide* beginning on page 267 and purchase a weed identification guide with good color pictures and learn to identify them and other plants common in your area** (see *References and recommended reading* on page 260).

Of course, many plants produce highly desirable forage and/or seed for wildlife. Whether some of these plants should be determined "weeds" or not is according to the type of food plot they occur in, the wildlife species you are managing for, and your objectives. For example, pokeweed, ragweeds, and tropic croton produce

Fig 6.2; A weed is an undesirable plant. Although deer may graze tropic croton, and quail, doves, and other birds relish the seed, you may consider it a weed in a clover food plot, depending on your objectives. Separate acreage should be devoted to food plots and early successional areas that contain tropic croton and other desirable plants.

seed readily eaten by doves and bobwhite, and deer eat the young leaves of these plants. As a result, they would complement a corn/sunflower plot planted for doves, but they may be discouraged in a clover plot for deer because they may shade-out clovers and lead to a diminished stand that would be more susceptible to other, less desirable plants. **Do not confuse managing food plots with old-fields!** Managing and maintaining naturally occurring early successional plant communities should be a top priority (above food plots) if you are interested in white-tailed deer, northern bobwhite, eastern cottontail, and many other species. Old-fields and other early successional plant communities provide much more than food plots, including cover for nesting, brooding, fawning, and escape. However, food availability in old-fields and food plots may differ greatly by season and amount, according to what is planted and management. Separate acreage should be devoted to naturally occurring early successional plant communities and food plots.

Cultural, mechanical and chemical weed control methods

Three primary methods are used to control weeds in food plots: cultural, mechanical, and chemical. All three are useful in various scenarios, but best results often are realized with an integrated approach. Be prepared to use all three as appropriate.

Figs 6.3 and 6.4; An integrated weed control program is usually needed when managing food plots. Mowing, for example, will help get rid of many annual weeds, such as horseweed (above), if mowed just before they produce seed; however, herbicides will be required to get rid of curly dock (below), which is perennial.

Cultural methods involve using sound agronomic practices and common sense, such as matching the seed you are planting with the correct soils, preparing the seedbed correctly, and planting seed at the correct time, rate, and depth. Amending soils with the proper amounts and kind of lime and fertilizer also help give your planting a better chance to compete with weeds.

Another common sense approach that most people don't consider is integrating cultural considerations with other control methods. Matching your crop with the weeds present in the seedbank is possibly the single best way to eventually eliminate even the toughest-to-control weeds from your food plots. For example, planting an annual cool-season plot in a field that has an overwhelming warm-season weed problem, or planting grasses only (such as wheat or oats) in a plot that has a tremendous broadleaf weed problem, which will enable you to use the most appropriate herbicide, provides you with more management flexibility and a much greater chance of success. Additional integrated cultural/chemical approaches are discussed below.

Mechanical methods usually involve mowing, disking, or cultivation. Mowing can help control weeds in perennial forage plots (perennial clovers, chicory, and alfalfa), but mowing does not get rid of weeds, especially perennial weeds. Continued mowing can effectively reduce annual weed competition if the weeds are mowed just before they produce seed and if they do not flower below the minimum desired mowing height.

Fig 6.5; Mowing annual weeds before they produce seed in perennial forage plots can reduce weed seeds in the seedbank and future weed pressure. However, selective herbicide applications are more effective and efficient in controlling weeds.

Many people mow perennial forage plots too often because they don't understand chemical weed control, resulting in an overall reduction in forage production.

Mowing annual food plots is not recommended because of the annual forage's life cycle and primary production period. However, it can be prudent to mow an annual plot if undesirable plants have completely overtaken the plot. When this happens, mow before the weeds produce seed and add to the seedbank.

Disking is most commonly used when preparing a seedbed. Repeated disking after allowing the seedbank to germinate (usually a few days after a rain event) helps reduce the weed seedbank considerably. However, this takes precious time if you intend to plant in that season. While waiting for additional weed flushes, a planted crop could be growing. On the other hand, if you are not going to plant until fall, repeated disking as soon as the seedbank germinates through spring and summer will help reduce the weed seedbank prior to planting a perennial food plot in the fall. Planting successive annual plots over two to three years with conventional tillage also can help reduce perennial weed problems, especially perennial grasses.

Cultivation is an excellent way to reduce weed pressure when using a planter to plant row crops. Crops such as corn, grain sorghum, sunflowers, sesame, soybeans, cowpeas, lablab, and chufa may be cultivated when planted in rows with a planter. You also can cultivate after planting with a drill, but you must close the appropriate seed ports to obtain sufficient row spacing to accommodate a cultivator.

Although it looks great to see those weeds disappear under the dirt behind a disk or cultivator, don't be fooled — other weeds will come back, even after repeated disking and cultivation. To enhance weed control with disking and cultivation, you must consider herbicide options.

Chemical methods involve herbicides and are by far the most effective way to control weeds. That is why they are used with every commercial field crop grown in the U.S. Herbicides are available for a myriad of applications (see *Appendix 2*), and four primary types of herbicides are used in food plot applications. *Broad-spectrum herbicides* kill all plants. The most common of these is glyphosate, which is found in a number of commonly used herbicides, such as Roundup. *Broad-spectrum selective herbicides* kill specific species of plants in various plant groups, including grasses, forbs (broadleaf herbaceous plants), sedges, and rushes. *Grass-selective herbicides* only kill grasses. *Forb-selective herbicides* (or *broadleaf-selective herbicides*) kill forbs, and some may kill various sedges and rushes, but not grasses.

Herbicide applications
There are three primary herbicide applications: *preemergence, preplant incorporated,*

and *postemergence*. Preemergence herbicides may be applied just prior to or just after planting and kill weeds as soon as they germinate. They also may be applied over an existing perennial crop, such as clovers and chicory, without a surfactant, prior to weeds germinating. Preplant incorporated herbicides are applied preemergence, but must be incorporated into the seedbed by disking, tilling, or cultivation. Therefore, preemergence and preplant incorporated herbicides are *soil active*. When planting via conventional tillage methods, preplant-incorporated herbicides often are incorporated when seed are covered by disking or tillage. Postemergence applications are sprayed on the leaves of growing plants. Most postemergence herbicides are not soil active, but many are. You must read the herbicide label to know for sure (also see Comments in *Appendix 2*).

Soil activity of preemergence and preplant incorporated herbicides is advantageous because weeds are killed as they germinate and they do not rob the crop of nutrients as it is trying to develop. However, soil-active herbicides used in food plot applications do not kill *all* weeds; you must read the label to know which weeds are controlled. Therefore, with several row crops, a postemergence herbicide also can be used to kill weeds resistant to preplant or preemergence herbicides.

Fig 6.6; Spraying warm-season annual weeds in late summer is a waste of time and money. This plot of iron-clay cowpeas was not sprayed, either pre- or postemergence. As a result, redroot pigweed has become established and is about to produce seed. Goosegrass also is about to produce seed in another area of the plot. Nonetheless, there is still plenty of high-quality forage available (iron-clay cowpeas) and deer are visiting the plot regularly. Rather than disturbing the deer's feeding pattern by mowing the plot, you would do best to either hand-pull the pigweed or leave the plot alone. Mowing would only destroy available forage and herbicide applications at this point would not be effective. If deemed necessary, the weed problem can be tackled next spring with a preemergence herbicide application prior to planting another warm-season plot.

When making postemergence applications, timing is critical. **Postemergence applications are most effective on most weeds when they are young, soon after germination.** For most selective postemergence herbicides, effectiveness is greatly reduced after weeds reach 4-5 inches tall. This statement cannot be overemphasized, and it is particularly important when managing perennial cool-season plots. With broad-spectrum applications (Roundup-Ready crops), timing is less important; however, even glyphosate applications are more effective when weeds are sprayed when young. This brings up a common problem with a lot of people — waiting until weeds are waist-high before realizing they have a weed problem, and then wondering what can be sprayed to kill the weeds. Don't worry — it's too late! The best thing that can be done at that time is mow the plot before the weeds flower and produce seed.

Integrating cultural and chemical control methods

Obviously, it is a huge advantage if you can identify weeds and know the composition of the seedbank. If you cannot identify weeds, you cannot determine which herbicide to use. If you know the weed composition, you then can decide which herbicides can be used to control them and match that with a planting labeled for that particular herbicide. This approach is opposite of how most people determine what they plant! You will be much more successful in your weed control efforts if you determine which

Fig 6.7; Weed control is often necessary for successful food plots. Preemergence and postemergence applications may be necessary for maximum production on some sites.

herbicide is needed (based on your knowledge of the seedbank), then decide what to plant, as opposed to planting something and finding out later there is no labeled herbicide that will control the problem weed(s) and not harm what you planted.

If you know perennial broadleaf weeds are going to be especially problematic, the easiest solution is to plant a grass crop, such as wheat, millet, corn, or grain sorghum. Then you can use the appropriate broadleaf-selective herbicide to control the problem weeds. If you know grasses are going to be a real problem, plant a forb (such as cowpeas, sunflowers, clovers, or chicory). Then you can use a grass-selective herbicide. If you know you will have a problem with various warm-season weeds, plant an annual cool-season plot. Then, you can use whatever herbicide needed to kill the warm-season weeds once the annual cool-season plot dies. The opposite is true if you have major cool-season weeds, such as purple deadnettle or ryegrass. Plant a warm-season plot, then use the appropriate herbicide to control the cool-season

Figs 6.8, 6.9, and 6.10 and the graph left; These three photos show the effectiveness of preemergence weed control in warm-season forage plots. The top picture shows a plot of iron-clay cowpeas that was not sprayed. The middle photo shows a plot of iron-clay cowpeas (grown adjacent to the plot in the top photo) that was sprayed preemergence with Pursuit. The bottom photo shows the two plots side by side a few weeks later. Johnsongrass and yellow nutsedge have overtaken the unsprayed plot.

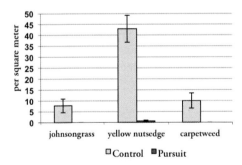

Fig 6.11; This graph shows the weed control possible with a preemergence application of Pursuit (4 ounces per acre) when applied to iron-clay cowpeas (as well as other labeled applications). Control plots received no herbicide application and can be seen in the pictures above.

weeds once the annual warm-season crop is dead. Make no mistake, annual plots can help you tackle tough-to-control weed problems. This is a major consideration and will save you time, money, and frustration.

Various herbicide considerations

Mode of action and chemical resistance

Herbicides kill plants in various ways or through different modes of action (see Comments in *Appendix 2*). Some reduce or block the

Fig 6.12; Planting a grass crop, such as this wheat, allows you to get rid of tough-to-control broadleaf weeds, such as curly dock and horsenettle, which were problem weeds in this plot.

production of amino acids, which are the building blocks of proteins. Others disrupt cell membranes, regulate growth, inhibit photosynthesis, inhibit lipid formation, inhibit lateral root development, and prevent shoot growth immediately following germination. Sometimes it is necessary to use an herbicide with a different mode of action, especially when plants become resistant to certain herbicides. A common example is where glyphosate-resistant pigweeds and horseweed have developed over time. Using a broad-spectrum selective or forb-selective herbicide with a different mode of action is required to control these problematic weeds.

Weed control varies greatly among different herbicides (see *Appendix 3*). The most effective long-term chemical weed control programs often involve multiple herbicides with different modes of action. This approach reduces the potential for weed resistance to a particular herbicide and increases control on many hard-to-kill weeds. Refer to the UT Extension publication *PB 1580 Weed Control Manual for Tennessee* (http://weeds. utk.edu) for more complete information on various herbicide-use strategies.

Crop-rotation restrictions

Many herbicides have crop-rotation restrictions. It is critical to plan ahead and keep notes on what you have sprayed because some of these herbicides may remain active in the soil for a few to many months. For example, imazethapyr, the active ingredient in Pursuit, can remain active for many months after planting. According to the Pursuit label, you should not plant alfalfa, clovers, rye, or wheat for four months after applying Pursuit. You should not plant corn for at least eight months after spraying Pursuit (unless you plant Clearfield varieties of corn), and you should not plant oats, grain sorghum or sunflowers for 18 months after spraying Pursuit. Seedlings of all these crops are susceptible to imazethapyr. This example is just one of many. Refer to the herbicide labels (also see Comments in *Appendix 2*) for crop-rotation restrictions with various soil-active herbicides.

Surfactants
Postemergence herbicides require a surfactant in order to be effective (see *What are surfactants?* on page 71). A surfactant is a "surface-active agent" that helps the herbicide stick, spread, wet, penetrate, and disperse on the surface of plants. Therefore, surfactants are not used with preemergence or preplant-incorporated herbicides. Nonionic surfactants and various crop-oil concentrates that contain nonionic surfactants are most commonly used. Some postemergence herbicides, especially most formulations of glyphosate, already contain a surfactant. Therefore, you do not have to add surfactant to the herbicide-water mixture. To know if a surfactant (and what kind) should be added to the postemergence herbicide you are using, read the herbicide label.

"Rain-fast" and "required moisture" time
Postemergence herbicides have a "rain-fast" time. That means if it rains within one to four hours after application, effectiveness will be reduced. To know the rain-fast time of the postemergence herbicide you are spraying, you must read the label (also see Comments in *Appendix 2*).

   Several preemergence and preplant incorporated herbicides have a "required moisture" time. That means in order for full effectiveness of the herbicide to be realized, it needs to rain or irrigation within two to three days after application (depending on which herbicide you use). Therefore, it is important to plan ahead and apply preemergence or preplant incorporate herbicides when rain is in the forecast.

Cost
Many herbicides seem ridiculously expensive. Although a jug of herbicide might cost $200-$500, the application cost per acre usually is within reason. If the initial cost is too high for you to consider, contact some friends and neighbors and split the cost with them. Paying $15-$20 per acre for effective weed control is cheap when you consider the expense of fertilizer, lime, seed, fuel, and your time.

Concerning herbicide labels …
Read them! They contain critical information that will help you tremendously. Always follow herbicide label instructions. Before using any herbicide, you should read the

herbicide label, be familiar with all restrictions and applications, and follow the directions specific for its application. Not using an herbicide in accordance with its label is a violation of federal law, not to mention the risk of crop damage, money and time wasted, and other potential negatives. To access herbicide labels prior to purchase, visit the **CDMS agrochemical database at http://www.cdms.net/manuf/manuf.asp**. And remember, professional advice concerning herbicide weed control is available *free of charge* at your county Extension office.

Last ditch effort?

OK, you haven't exactly done a stellar job at controlling weeds in your food plot. You've mowed and they keep coming back, you don't have or know the correct herbicide to use, and even if you did, they are too tall now to be susceptible to the herbicide application. It might sound crazy, but your last-ditch effort for controlling some weeds in a food plot might be pulling them up by hand. Actually, this technique is quite effective and efficient for some specific weed problems in relatively small plots. It is particularly true with stubborn annual weeds that have grown too tall or mature for herbicide applications, and when they are growing in various annual plots where mowing is not desirable. Some examples include horseweed, jimsonweed, pigweeds, sicklepod, passionflower, and morningglories. Even when these weeds are mature and about to produce seed, they can be pulled out of the ground easily within a couple of days after a rain. This technique also can reduce some hard-to-kill perennial weeds, such as horsenettle (wear your leather gloves!). Pulling weeds may take a few hours, depending on plot size and number of weeds, but it can be very effective, and it's a good way to get your kids outside and let them make some money at the same time!

Fig 6.13; It might seem ridiculous, but hand-pulling weeds that have escaped herbicide application can be very effective. Here, a manager is hand-pulling Palmer amaranth before the seed mature from a soybean plot to prevent it from spreading. Each plant may produce >500,000 seeds. Hand-pulling weeds such as this is easy a day or two after rain.

What are surfactants?

Surfactants are water- or oil-soluble surface-active agents added to postemergence herbicides to modify or enhance the effectiveness of the active ingredient. Surfactants help herbicides stick, spread, wet, penetrate, and disperse on the surface of plants. Hence, surfactants are not added to preplant incorporated or preemergence applications, only postemergence. In short, surfactants make postemergence herbicides more effective by helping them penetrate the plant.

Nonionic surfactants (NIS) commonly are used with both broad-spectrum and selective herbicides. NIS are soluble in cold water, are outstanding dispersing agents, do not foam much, and have low plant and animal toxicity. Therefore, NIS are almost always used with selective herbicides where desirable plants also occur. Further, NIS do not ionize in water, so they do not form insoluble salts and can be used with hard water. NIS usually are added to herbicide solutions at 0.25 percent by volume of spray solution (0.32 ounces per gallon of solution or 32 ounces per 100 gallons).

Crop oil concentrates (COC) are petroleum- or vegetable-based oils that contain surfactants and increase the absorption of herbicides into plant leaves. Methylated Soybean Oil (MSO), for example, is a commonly used vegetable-based COC. COC usually contain 80 percent oil and 20 percent NIS. Depending upon the application, some herbicide labels may recommend COC rather than NIS because COC alone can alter the structure of cell membranes and cause damage to plants. That is why NIS are often used with selective herbicide applications, whereas COC are often used with "burn-down" applications where the intention is to kill all vegetation present. Nonetheless, it is important to use a high-quality surfactant and follow the herbicide label instructions as some herbicides perform better with COC than NIS.

Liquid nitrogen fertilizers, such as urea-ammonium nitrate or ammonium sulfate, may increase the uptake of postemergence herbicides. However, they are not surfactants, even though they may be recommended on some herbicide labels as an additive to the spray mixture.

Seven

Chapter 7 — Final thoughts before planting

Food availability and wildlife use

The number of planting combinations that can be used in food plot mixtures seems infinite. Various wildlife species will eat a wide variety of forages and grains. And though some species, such as white-tailed deer, may differ somewhat in plant selectivity from one area to another, hungry animals with little to choose from will eat many plants and seed they wouldn't otherwise if something else was available. For example, if you see white-tailed deer eating tall fescue, dead leaves, rhododendron, or bark, rest assured high-quality forage is limited! Just because an animal eats something doesn't necessarily mean the animal is gaining adequate nutrition from it. Some of the food items consumed might even be detrimental to the animal. Don't let this happen. The word "preference" is commonly used when referring to what deer select to eat on a particular property. Technically, "preference" relates to a ranking if all options are available. Of course, all forage options are never available on any property. Therefore, deer "select" various plants over others, based on what is available. Their selection is based on palatability (plants that taste good), digestibility (plants and portions of plants that are most tender), and nutrient requirements (plants that have nutrients that deer require at that time). If you want to benefit wildlife populations with maximum nutrition, make sure highly palatable and nutritious foods that wildlife readily eat are available year-round, but especially during nutritional stress periods, which can vary on different properties even within a particular region.

There are many plantings available that are not recommended in this book. Some individuals and companies may recommend some of the plants in the list at right, but I have not found them worthy of recommendation. In fact, several of these should be considered invasive weeds that should be eradicated whenever encountered. None should be considered "selected" by wildlife and several are incompatible for use with other plantings. There are better plants to use. **Regardless, you should realize what I recommend in this book is based on research.** None of the recommendations in this book are influenced by marketing or advertising.

Remember from the introductory chapter, much more important than food plots is your holistic habitat management approach. Do not try to manage wildlife with food plots alone! **In most**

Plantings not worthy of consideration

Birdsfoot trefoil
Blue lupine
Crown vetch
Hairy vetch
Sweetclover
Velvetbean
Sainfoin
Small burnette
Ryegrass
Matuagrass
Orchardgrass
Timothy
Bluegrass
Sorghum-sudan

cases, food plots should be the last step in a habitat management plan. Much more important than food plots is providing and maintaining the appropriate vegetation types and successional stages in a suitable arrangement for the desired wildlife species across the property and landscape. Until you address the overall habitat limitations, the impact of a food plot program will be minimal and, in fact, could be detrimental.

Planting recommendations listed for various wildlife species in the following chapters incorporate high-quality forages and grains and have produced highly successful food plots when planted and managed appropriately. However, keep this in mind: planting and managing food plots correctly (using sound agronomic practices) is more important than the exact planting you use. I remind people of a little motto I have: **Keep it simple and do it right**.

A word about commercial seed blends

Many commercial food plot mixtures are available. You can't pick up a hunting magazine, or catalog, or internet site without seeing a barrage of advertisements. Most commercial mixtures contain high-quality seed and can produce excellent food plots if planted correctly on the appropriate site at the appropriate time. Many commercial mixtures contain seed that are readily available, but some use improved varieties of seed that have been developed over years of testing for drought tolerance, resistance to insects/disease, etc. As a result, many commercial mixtures may be advantageous

Fig 7.1; A problem with some commercial food plot mixtures is they contain various combinations of seed for which no selective herbicide can be used. When you plant a legume (such as cowpeas) with a non-legume (such as sunflowers or buckwheat) and a grass (such as grain sorghum), you are at the mercy of the seedbank. And you cannot count on a broad-spectrum herbicide and mechanical weed control! This plot was sprayed twice with a glyphosate herbicide at six weeks and three weeks prior to planting and disked prior to planting, killing additional plants that had germinated after the second herbicide application. Still, additional plants from the seedbank outcompeted this popular commercial mixture.

and worth a higher price. In fact, many are not that much more expensive than buying and mixing seed yourself. However, be aware that marketing is a powerful tool. Commercial blends cannot guarantee trophy bucks (even if some of the advertisements say they do!), and it takes a lot more than a food plot or mineral supplement to produce a mature buck or a healthy deer herd.

A problem with a few commercial seed blends is they contain odd mixtures. Some commercial blends contain both large seed and small seed — seed that require different seeding depths. Some contain seed suited for moist areas as well as seed suited for dry areas. Some blends contain seed for both warm- and cool-season plants. Many seed companies create such blends intentionally because they want to make sure something comes up and grows regardless of when or where the buyer plants it. Obviously, such a mixture is a waste of money for the educated food plot manager because much of the seed is destined to fail, wherever and whenever the mixture is planted. Before purchasing a commercial seed blend, always read the seed tag attached to the bag, identify the plant species included, then decide if that is what you want to plant. Always follow the steps outlined in the preceding chapters for plot preparation and planting; they are critical to success, whether you decide to form your own mixture or buy a commercial blend. Do not be misled by advertisements stating all you have to do is spread the seed on the ground with no preparation necessary; the results are usually disappointing.

Measuring food plot success

Unfortunately, few people actually evaluate the success of their food plot program. Time spent keeping records on your food plots and the wildlife that use them is very rewarding and will provide important information that will make future efforts more successful. For obvious reasons, it is important to record what and when you planted and what and how much lime and the types and amounts of fertilizers you applied. Doing so simply helps you remember what you did. The next step is to record your management efforts: when you mowed, disked, or applied herbicides (and which herbicides and rates you used). This information, along with precipitation data, should be coupled with observations on plant growth and wildlife use.

Plant growth and wildlife use are easily measured using exclusion cages (see *Monitoring food plot success with exclusion cages* on page 139). These cages protect plants from wildlife and allow you to see what the plot would look like if there was no use (this is most applicable with forage food plots as opposed to grain food plots). Not only does this help you identify which forages wildlife eat, but it also helps you identify the ones they don't eat! I like to use at least one exclusion cage per acre to provide an estimate of growth and use across the field. Cages 4 feet square and 4 feet tall work really well. These cages also let you see the effects of your liming, fertilization, and weed control efforts. Overgrazed plots usually are weedy and may

Fig 7.2 and 7.3; Infrared cameras can give you a good idea of the species and number of animals using your food plots. Photos by Andy Wulf and Jim Phillips.

not be able to respond to soil amendments. Other tidbits of information you should record include days until germination after planting, days until maturity (especially for seed- or grain-producing plots), rainfall through the season, effects of weed control (a general rating on how weeds responded after being sprayed), and forage growth/availability through the season.

Wildlife use also can be estimated using infrared-triggered cameras. Placing these cameras around and within your food plots can provide informative pictures, showing you things you might not have imagined if there wasn't a picture to prove it. Although placing a few cameras around a food plot does not constitute a wildlife survey, the cameras will show you many of the species visiting the plot and can give you a reasonable idea of how many animals are using the plot.

The final consideration when evaluating the success of your food plot program is deciding if your objectives have been met. If added nutrition/wildlife health were your objectives, do animals appear healthier? Have average weights and antler size increased? Has grazing/browsing pressure on native food resources been reduced? If your objective was hunting/observation, did you kill or observe more wildlife as a result of your food plot(s)? Are you happy with your efforts? Do you feel the benefits of your food plots were worth the cost/effort you put into them? Only you can answer these questions and make the final determination if your food plot program is successful or not.

My thoughts on the ethics of food plots

As mentioned in the preface, there are people who disdain food plots and anyone who plants them. They will try to claim that you are only baiting wildlife and that you only want to plant a food plot to help you kill something. They will say that there is enough agriculture, and that those species for which you are planting do not need additional management. They will say that natural communities should be managed, that lime, fertilizers, and herbicides should not be used. They will go on and on. That's OK. They are entitled to their opinion and they may bring up some points worth considering. Now, give them your opinion, and do so with thought, honesty, and conviction. If you are planting a food plot just to help you kill something, explain that, and be proud of it. Tell them that you enjoy the hunt and you enjoy eating the meat and sharing it with others. If you are planting for increased nutrition, explain how you are managing all of your property and that food plots are only one component in your

Fig 7.4 and 7.5; Many species benefit from food plots, not just those that are hunted. Nongame birds and pollinators are huge benefactors of various types of food plots. Conservation of pollinators is a major focus in natural resources management to help crop production concerns. Few plants provide pollinators more food than some of the clovers commonly planted in food plots, such as red clover (with a black swallowtail, left) and arrowleaf clover (with honeybee, right).

habitat management plan. Tell them that you enjoy growing things, similar to growing a garden, but your food plots are for wildlife, and how you enjoy seeing the many different wildlife species that use the crops you grow. Tell them how many species you regularly see using your food plots. Then, ask them what they do for wildlife. Ask them if they have a bird feeder in their backyard. If so, explain to them how it could be viewed hypocritical for someone with a little box of supplemental seed for birds in their backyard to complain about a land manager who has provided a field full of food for an array of species that you will never contemplate shooting, but only enjoy watching. Also, ask them how much money they spend annually on land conservation (which means managing land wisely). Ask them how much they pay on property taxes each year. Don't apologize for your efforts; the world needs more people like you.

Although there are chapters for only six species or groups of wildlife, many other species will benefit from your plantings. Some are game species (such as gray and fox squirrel, black bear, elk, groundhog, raccoon, and ruffed grouse); some are nongame species (such as indigo bunting, northern cardinal, blue jay, northern flicker, several species of sparrows, junco, goldfinch, and bluebirds). There will be others. Allow your food plot efforts to help you understand nature better. Don't be hesitant to put forth some effort and learn your plants and learn to identify all those little birds you see in your food plots. Trust me, you'll be glad you did!

Eight

Chapter 8 — Recommendations and strategies for White-tailed Deer

More food plots are planted for white-tailed deer than all other species combined. The primary usefulness of food plots for deer is providing additional food during nutritional stress periods (lactation, antler development, late summer, and mid- to late winter, depending on location) and helping deer remain at a high nutritional level. High-quality food plots can provide increased nutrition that can lead to increased body weights, increased reproduction and fawn development, and larger antlers. Of course, food plots also can be used to influence deer movements and facilitate hunting and observation.

Food plots should not be used to sustain an overabundant deer herd that is negatively impacting food and cover for other wildlife species. If deer density is so great that it is having a deleterious effect on habitat for deer or other species, you are mismanaging the property and you should re-evaluate your objectives and management strategies. Food plots can, however, be used to help improve this situation where it already exists. For example, a forest understory cannot recover from chronic overbrowsing until deer density is lowered and additional sunlight is allowed to enter the forest canopy. Deer density is not lowered overnight. Nor does

Fig 8.1; Unnaturally high deer densities decimate the forest understory and degrade habitat for other species. Food plots alone do not remedy this situation. The deer population should be lowered, the forest should be thinned, and additional habitat management should be implemented.

a decimated understory recover in one or two years. As serious efforts are undertaken to lower deer density, and as the plant community is recovering, well-managed food plots can help buffer browsing pressure on the recovering forest understory. Of course, during this process, other habitat management practices should be implemented as well. This example highlights how food plots can help alleviate browsing pressure when habitat restoration efforts are underway concurrent with serious efforts to reduce deer density.

Fig 8.2; Increased responsibility comes along with providing increased nutrition. When deer have increased nutrition, increased reproduction may follow. Be prepared to kill the appropriate number of does to keep the deer population in balance with their habitat.

Food plots can influence body weights, reproduction, and antlers. Everyone is happy with larger deer and larger antlers, but not everyone is ready for increased reproduction. **It is important to realize there is a responsibility that comes with increasing nutritional carrying capacity.** More deer must be removed from the system or overabundance will occur. You should not increase available nutrition if you are not prepared and able to harvest the appropriate number of deer.

This chapter highlights planting and management recommendations for food plots designed for white-tailed deer. Most of these recommendations are from 20 years of research and demonstration conducted with my graduate students, including 10 years of cafeteria-style plot work completed across several fields where we compared many forages with respect to germination, growth, deer selectivity, resistance to grazing, and nutritional quality (see *Appendix 5 and 6*). Additional research efforts focused on developing the best mixtures possible with respect to complementary forage

Fig 8.3a,b; Here are two of several demonstration/research fields established and maintained across Tennessee, 1999–2008. Data were collected to determine germination and growth rates, deer selectivity, resistance to grazing, nutritional quality, and herbicide recommendations for a wide variety of forages. Yield and consumption were monitored through stationary and mobile exclusion cages, placed at random within each 0.10-acre plot (20 plots per field) at the end of each month. Photo 8.3b by Chris Shaw.

growth and maturation and herbicide compatibility. Further testing has focused on management strategies for various plantings. This information should help you in your management efforts.

Forage quality considerations for white-tailed deer

Deer select the food they eat based on availability, palatability, and nutritional content. Deer cannot eat a particular food if it is not available. If it is available, deer selectivity is based on some combination of palatability, digestibility, and nutritional content. Nutrition alone does not dictate selectivity. For example, total digestible nutrients in sicklepod may exceed 70 percent with 32 percent crude protein. The problem is, deer won't eat sicklepod unless they are about to starve. The presence of tannins, alkaloids, toxins, and other compounds influence forage selectivity.

Forage quality is based on digestibility and nutritional content. Foods that are highly digestible provide lots of energy, which is obtained primarily through nonstructural carbohydrates (or plant cell contents, such as sugars, starch, and protein). Structural carbohydrates compose cell walls and include lignin, cellulose, and hemicellulose. Deer cannot digest lignin; however, bacteria in a deer's stomach enable it to obtain some energy from cellulose and hemicellulose. Therefore, digestibility is highly correlated not only to the type of plant, but also to the part of the plant eaten and plant maturity.

Young, leafy material is highly digestible, whereas stems are less digestible because they contain more structural components. Older plants have more stems and more structural components than young plants. Thus, a given plant part is more digestible in young plants than older plants. Food plot programs designed to provide maximum nutrition for deer should contain forage that is highly digestible with food available year-round, but especially during nutritionally stressful periods. Meeting these nutritional requirements may require different types of plantings as described in the pages that follow.

Fig 8.4; Forage quality decreases over time as the plant matures and lignin content increases. As this process occurs, deer will eat the more palatable portions of a plant, such as the leaves of these soybeans, as opposed to relatively large stems. Deer are considered "concentrate selectors," meaning they select certain plants and select portions of a plant that are most palatable and digestible.

Cellulose and hemicellulose take much longer to digest than sugars and starch; therefore, they stay in the stomach longer. When deer are forced to eat poor-quality foods (when high-quality foods are not available), the total amount of food the deer can consume is reduced because it takes longer for poor-quality food to pass through the digestive system. Obviously, nutritional quality can have a direct impact on weight, milk production, antler growth, and survival.

Protein is found in plant cell contents and cell structures. Proteins found in cell contents are readily digestible, whereas those in cell structures are not. Digestible protein is strongly influenced by plant maturity and available nitrogen. As plants mature, cell walls become more lignified and constitute a larger percentage of the plant, which reduces the amount of digestible protein and energy within the plant. Available nitrogen may influence plant protein content because nitrogen is a component of all proteins.

Protein requirements for white-tailed deer vary by sex, age, and season. Adult deer require 9-10 percent protein to maintain body condition, whereas fawns may require 20 percent or more for optimum growth and development. Maximum growth and development of yearling and adult bucks may be realized when available forage provides approximately 16 percent protein. Lactating does may need 20-22 percent protein for optimal milk production.

Forage quality is estimated with a variety of laboratory techniques. Most common are neutral detergent fiber (NDF) and acid detergent fiber (ADF) analyses. NDF represents all cell wall material, including cellulose, hemicellulose, and lignin. Thus, NDF distinguishes the relatively indigestible cell wall material from the soluble cell content. As mentioned earlier, some of the hemicellulose and cellulose may be digested with the help of bacteria in a deer's stomach. ADF represents the cellulose and the indigestible lignified portions of the forage sample (NDF minus hemicellulose). Thus, the NDF value is negatively correlated with the total fiber content of the forage, and the ADF value is negatively correlated with the digestible portion of the forage. Forages with ADF values below 35 percent generally are highly digestible for white-tailed deer. Digestible energy often is estimated as total digestible nutrients, or TDN, using equations that incorporate NDF or ADF values.

If you have questions about the quality of your food plots, visit your county Extension agent. Forage samples can be analyzed for just a few dollars. When forage quality information is coupled with a soil test, you can get an excellent idea of the amount of nutrition you are providing through your food plots.

Food plot size, shape, location, and other considerations

Food plots for deer are usually between ¼ and 3 acres. Food plots planted specifically for hunting are usually ¼-1 acre, whereas those planted with the intention of providing maximum food availability and increased nutrition are often 2 - 5 acres or

larger. These larger plots are often called feeding plots. Relatively large food plots often are necessary in areas with high deer densities. Where food plots this large cannot grow because of overgrazing, deer density should be lowered and additional habitat management implemented.

The primary determinants of food plot size are your objectives, habitat quality, deer density, shade effect, and distance to cover. Food plots should be large enough and wide enough to allow at least four hours of direct sunlight, and the distance to cover from the middle of the field should not be more than about 80–100 yards. Therefore, plots larger than 10 acres receive more use by deer when they are rectangular rather than square (this would be appropriate for properties with relatively high deer densities). Fields larger than this can be made more attractive to deer and huntable for you by breaking the field into sections using hedgerows of trees and shrubs that provide hard and soft mast and evergreen cover that provide visual breaks across the field. Regardless of plot size, high-quality food resources should be available throughout the property (except within sight of roads and in some cases property boundaries) to ensure adequate nutrition is available to all the deer.

Food plot shape is another consideration, especially those created specifically for hunting. Various-shaped food plots can be created to influence how deer travel through the food plot. Most-noted are hourglass-shaped plots and boomerang-shaped plots, made popular several years ago by Neil Dougherty. When positioned properly with respect to prevailing winds, pinch-points are created at the narrow portion of the hourglass and at the bend of the boomerang, providing relatively high-percentage areas for close shots. Debris from newly created food plots can be positioned along food plot edges to influence where deer enter and exit food plots. Don't worry about creating food plots with wavy edges. A wavy edge is not going to make the food plot

Figs 8.6a and 8.6b; Here are two examples of hunting plots. Fig 8.6a shows an hourglass-shaped plot designed specifically for bowhunting. Deer feeding on one side of the hourglass are instinctively drawn to walk into the other side. As they do, the restricted area is relatively narrow and affords a bowhunter a high-percentage shot. Fig 8.6b shows a narrow plot positioned in dense cover for a rifle hunter. Deer in dense cover feel relatively safe to walk into a narrow lane and feed before dark, offering a rifle hunter a shot at deer that otherwise might not wander into a larger field during daylight.

more attractive to deer. Food plots are made attractive by planting nutritious foods deer prefer, in or near areas they already travel or where they are funneled. Find the right spot, plant the right forage, and deer will use the food plot.

For optimum use, plots should be adjacent to good cover and where deer travel regularly (see Fig 2.4 on page 11). Certainly, you can influence where deer travel by creating attractive cover, primarily through forest management and old-field management. Nonetheless, increased use can be expected when food plots are planted where deer want to go as opposed to trying to entice them to an area where you want them to go. Topography, as well as natural and man-made funnels, can influence how deer enter and exit the food plot. Savvy hunters plan ahead and use this to their advantage.

Hunting food plots without educating deer can be difficult. The more you hunt near a food plot, the more deer realize what you are doing and adjust their timing of use (meaning: they stop using the food plot during daylight hours). Planning an access and exit strategy is critical. Of course, food plots should be hunted only with certain winds, according to where the deer are bedding, with appropriate deer stand locations. Do not hunt food plots (or any area) unless the wind is right, and don't walk through bedding areas in the afternoon to get to a food plot. If the plot is in the woods, the access road or trail should not lead "straight" into the plot. Make sure the road curves before getting to the plot so you are not visible from a distance. Paths you create from the road or trail should lead to specific locations you intend to hunt with limited visibility. Walking across an opening to get to a food plot is not advisable. You will be detected more times than not if you do this after dark, and all you are doing is educating the deer to your movements and helping them pattern you.

Even with the correct wind, getting into and out of a deer stand or hunting area undetected can be challenging and should be given considerable thought. If the stand is within sight of the plot, consider moving it back into adjacent woods and make sure there is a visual break that will not allow deer in the plot to see you enter or exit. Common scenarios include getting into the stand too late in the afternoon and getting out of the stand after dark when deer are in the plot. Planting a tall visual break at the edge of the plot can help. Placing a stand several yards back in the woods, not at the very edge of the field, helps to avoid spooking deer. If you want to plant perennial screening cover, I recommend a tall variety of switchgrass, such as *Alamo, Kanlow,* or *Cave-in-Rock*. I like switchgrass better than other native grasses for screening cover because switchgrass remains standing through winter much better than big bluestem or indiangrass, which commonly fall over with wind and rain by November. **Do not plant *Miscanthus*** for screening cover, or any other reason! It is terribly invasive and not as good for screening cover as the other options listed here.

If you prefer to plant an annual for screening cover, Egyptian wheat (15 pounds PLS per acre) or sunn hemp (*Tropic Sun* variety, 40 pounds PLS per acre) can be used. Both grow more than10 feet tall and if the strip is about 15 feet wide, both provide an excellent visual buffer. The only drawback from using sunn hemp as screening cover is that deer eat it. If deer density is relatively high, the sunn hemp may not grow tall enough to provide screening cover. After a couple years, I recommend rotating the visual break by planting a new strip beside the old one and allowing the previous one go fallow for a couple years. And you will probably find that you didn't really need to plant that screening cover after all. On most sites, fallow vegetation grows tall and dense and ultimately provides more permanent screening cover than planted grasses as brambles and woody sprouts begin to take over.

Figs 8.7a and 8.7b; Above, this strip of Egyptian wheat was planted specifically to allow a bowhunter to get in and out of a stand near the iron-clay cowpea food plot undetected. The picture of sunn hemp below was taken in early November. It provided excellent screening cover from the food plot on the other side.

The screening cover I like best is a strip 15-20 feet wide of various shrubs and small trees. I like wild plum, hawthorn, eastern redcedar, and white pine. White pine and eastern redcedar can be kept bushy by cutting the top out of the trees every few years. Plant the shrubs/trees as recommended from your supplier, but let the space in between grow-up in "weeds," brambles, naturally occurring shrubs/trees, etc. You want a visual buffer, not a manicured shrub planting.

Planting food plots too close to the woods edge is a common mistake. Unless the plot is designed specifically to hunt over, planting adjacent to the woods' edge is ill-advised because of shade effect and nutrient competition from the trees. It is much more sensible to allow a soft edge of forbs, grasses, brambles, and shrubs to develop 30-50 feet between the woods or tree line and the plot, even when you are planting plots with respect to orientation. If you are planting plots with a north-south orientation or along the southern side of an opening to conserve soil moisture (see *Where to plant food plots?* on page 10), you still should not plant within the drip-line of the trees' crown. Move away from the tree line 30-50 feet for better forage production.

Fig 8.8; When planting larger plots, allow a soft edge to develop between woods and the planted portion of the field. Trees rob the crop of water, sunlight, and nutrients. As a result, you waste money on seed, lime, fertilizer, and herbicide when planting too close to trees.

Thinning undesirable trees 100–300 feet into the woods around openings (especially relatively large fields) is an excellent way to create an ecotone from forest to field and provide additional browse (leaves and tender twigs of woody species) and better cover. The enhanced structure created with soft edges and thinned woods around a field may influence deer to use the plot earlier (before dark) than later. Regardless, if you continually hunt over the food plot, it won't be long before deer are using it only at night. Instead of trying to shoot deer in the food plot, it is usually a better tactic to hunt along trails leading to the field, or downwind of the food plot where bucks often scent-check for does in estrus during the rut.

Finally, as you determine what is best for your property, consider the following questions. Is the forage you are planting available during times of the year when naturally occurring forage and nutrition are limiting? Are you growing a food plot just to attract deer to facilitate hunting? Is the quantity of forage available enough to justify establishment and management costs? Are the deer actually eating the forage? Is forage quality sufficient for the deer to gain nutritional benefit? Answers to these questions will help you determine and evaluate your management strategy.

How much acreage should you plant?

Many landowners try to establish 1-5 percent of their property in food plots for deer. This amount may be a good ballpark figure but, in reality, there is no cookbook percentage of acreage that should be planted. Every property has unique characteristics. The amount of land you plant in food plots should be determined

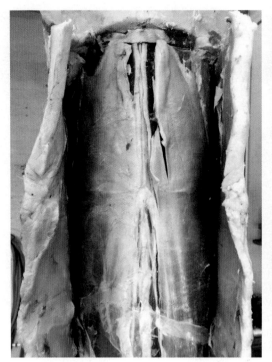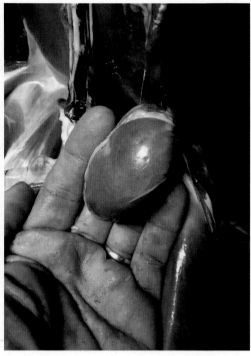

Fig 8.9a and b; Healthy deer have plenty of fat. The deer on the left came from a property in North Carolina that contained abundant early successional vegetation in old-fields across the property with scattered food plots. The kidney from the deer on the right came from a property where natural vegetation was limited because of high deer density, and food plots were about the only source of nutrition. As a result, growth of forages in the food plots was not able to get above a few inches, and the deer on that property had virtually no fat in December.

by your objectives, the surrounding landscape and habitat quality, and deer density. If food is a limiting factor for deer in your area, you should implement additional habitat management (including food plots) and/or lower deer density. Continue to do this until you provide ample forage for deer as well as other species. A simple rule of thumb is that your food plots should never be eaten down to the ground. If so, that is an immediate indication that food availability may be limiting potential weight and overall health. You should evaluate the amount of fat around the kidneys and around the back on all deer killed, as well as compare weights by sex and age class with other managers in your area. Remember, you should not try to "feed" deer solely with food plots. Instead, use food plots to supplement naturally occurring foods. Regardless of the quality or abundance of food plots, deer always continue to eat naturally occurring foods.

Dedicating separate acreage to warm- and cool-season plots

In many situations, planting both warm- and cool-season forage and grain plots should be considered to meet your management objectives. However, they should be planted in different fields or different sections of a field. That is, don't take away

Fig 8.10a,b; Separate acreage should be devoted to warm- and cool-season food plots. The picture above shows a jointvetch plot on October 1. The middle cage shows cumulative growth through the summer. The other two cages show growth and use just during September. Cool-season plots are establishing in the background. If the jointvetch was disked under in September to plant cool-season forages, food would have been taken away when it was needed. An exception is when you can no-till drill a new crop into an annual crop that has matured. For example, drilling soybeans into crimson clover after the clover dies saves soil moisture and increases organic material in the topsoil. Later, once the soybeans mature, the crimson clover reseeds naturally and provides excellent forage through winter (below).

available food in preparation to plant something else. This consideration is very important and should be followed in your field management plans. For example, iron-clay cowpeas provide nutritious forage until the first frost, and soybeans can provide a tremendous source of energy (beans) through fall and into winter. If a warm-season plot is mowed, disked, and planted to cool-season forages in September, high-quality forage is taken away at a time when natural availability often is low (late summer). Likewise, arrowleaf clover provides forage through July. If a plot containing arrowleaf clover is disked in May to plant a warm-season plot, a prime food source is removed when large amounts of high-quality forage is critical (just before fawning and during early antler development). An exception to this strategy is when you can no-till drill a new crop into an annual crop that has died. Examples include drilling soybeans into annual clovers and wheat when the plot is nearing maturity.

Deciding how much acreage in warm-season and cool-season plots to plant depends on your objectives, habitat quality, and deer density. If you desire to maximize body weights, antler size, and productivity, then careful consideration should be given to forage quality and quantity during spring and summer when does are preparing to give birth, and later for milk production to feed fawns, and for antler development. You must blend availability of naturally occurring forage with the timing of productivity of various food plot plantings to make sure adequate high-quality forage is available. Various food plot plantings peak in productivity and quality at different times of the year, and just because a food plot forage is classified as "cool-season," does not necessarily mean it is productive during winter (see graphs on page 91). Also, be aware that **everything green in woods and fields during spring and summer is not good deer food!** Some knowledge of naturally occurring plants is necessary to make the best decisions as to what is needed on your property. Plenty of high-quality forage in late summer and early fall helps deer accumulate fat going into winter (this can be a critically important period—see Fig 8.9a and b). Food plot forages that provide lots

Fig 8.11a,b; Perennial forages, such as perennial clovers, alfalfa, and chicory, are most productive during late spring and early summer. They are not productive during the winter (except in the Deep South). Here, this perennial clover plot in Zone 6 was lush and productive in early July (left), but dormant in late winter (right).

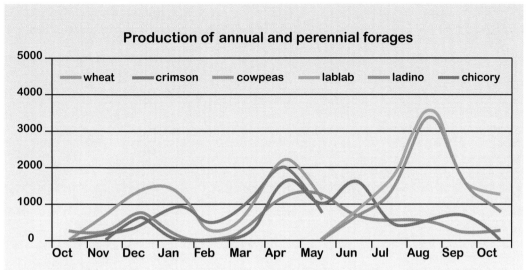

Fig 8.12a; This graph shows monthly production for two perennial and four annual forages averaged across three fields in Tennessee (all forages present in separate plots in each field) with varying deer densities over several years. Production of annual forages is shown from the time they were planted until maturity. Production of perennial forages is shown through the year. Production of annual cool-season forages continues through the winter, whereas production of perennial cool-season forages wanes. Nothing out-produces annual warm-season forages during summer. Refer to Appendices 5 - 8 *for information concerning deer selectivity and use of various forages.*

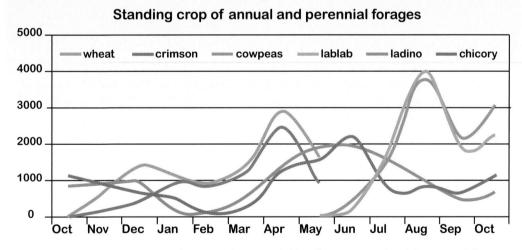

Fig 8.12b; This graph shows standing crop (forage available) of two perennial and four annual forages averaged across three fields in Tennessee (all forages present in separate plots in each field) over several years. Standing crop will vary depending upon deer density and habitat quality. Here, standing crop has been averaged across three areas where deer density ranged from 30-90 deer per square mile. Of particular interest is forage availability during the primary stress periods for white-tailed deer (late summer and winter). It is clear annual forages provide by far the most forage during these periods. Established perennial forages complement annual forages by providing nutrition in fall while annual cool-season forages are developing and in spring while annual warm-season forages are developing.

Fig 8.13; Perennial forages are most productive from late spring through midsummer. However, by summer's end, perennial forages lack the productive potential of warm-season annuals. The drought during the summer of 2007 was the worst on record in many areas of the U.S. By late August 2007, this alfalfa and red clover in North Carolina (on the left in the photo) were largely senescent, but soybeans (right) continued to grow and produce forage. This scenario further highlights the benefit of multiple forage types to provide high-quality nutrition throughout the year.

of food in spring and early summer do not provide much in late summer and early fall. Forage availability during winter is almost never a limiting factor down South, but can be up North where persistence of deep snow and cold temperatures can lead to starvation. Food intake by deer is considerably less during winter as opposed to summer because deer are not growing as much and they may utilize fat reserves to meet nutritional demands.

Geographic location also influences how much warm- vs. cool-season acreage to plant. In the Deep South, various warm-season plots can be planted as early as March and receive use through November or December. Oppressive summer heat limits production of perennial cool-season plantings. Therefore, it is sensible to plant more acreage to warm-season plots than cool-season plots, and most cool-season plots could be annuals. Up North, a majority of the acreage planted should be cool-season plots (both annual and perennial) because of the relatively short summer growing season and because many cool-season forages can be planted in spring and perennial cool-season forages are very productive during summer up North. However, where possible, corn and/or soybeans can be an important planting because of the energy value during winter. While winter greens are covered in snow, corn ears hanging on the stalk are still available.

Annual or perennial plots — which is better?

A common dogma persists that perennial food plots provide more forage, cost less, and require less work than annual plots. If both are managed for maximum production, a detailed analysis shows this may not be accurate. If portrayed by season, it is clear annual forages may produce more during late summer and winter than perennial forages (see sidebar on page 94). However, perennial forages are highly productive April through June, which is important for antler production and lactation. In most areas of the country, cool-season annual forages produce more than perennial forages during the winter and warm-season annual forages produce more than perennial forages during the summer. Obviously, production for different plantings varies by region. Perennial cool-season forages may produce through the winter in the Deep South, but often are severely stressed and sometimes killed during summer. Perennial forages produce well through the summer in the higher elevations of the southern Appalachians and up North, but there is no production during winter. Cool-season annuals — especially wheat, cereal rye, and brassicas (including rapes, kales, turnips) — and warm-season annuals, especially corn and soybeans, provide food up North during winter. Although perennial clovers, chicory, and alfalfa are classified as cool-season plants, production of these forages is inadequate to feed deer during winter from the Mid-South northward.

Although seed for annual plots can be more expensive over time than seed for perennial plots, if managed correctly, you may spend more money managing perennial plots (mowing and spraying) than planting annual plots. Considering workload, perennial plots may require more visits to the field than annual plots, depending on how many times you need to spray or mow. So, which is better? Only you can make that decision for your property. Both annual and perennial forages have advantages when developing a sound food plot management plan to meet forage production needs for white-tailed deer throughout the year. To realize the full potential for white-tailed deer on most properties, both annual and perennial plots may be needed.

Warm-season plots for white-tailed deer

Warm-season plots are unique in that, according to what you plant, they can provide an important food source for deer during summer and/or winter. Warm-season forages, such as soybeans and cowpeas, provide nutrients necessary for milk production, fawn growth, and antler development from late spring through early fall. Later, corn grain and soybean seed can provide a high-energy food source that can be important during winter, especially in areas where available forage is limited or during years with little acorn production.

Various forbs (broadleaf herbaceous plants), if available, comprise the majority a white-tailed deer's diet during spring and summer. Warm-season forage plots should

Annual vs. Perennial plots

An economic assessment of annual and perennial forage plots for white-tailed deer is shown below. This assessment considers 2 acres managed in annual food plots and 2 acres managed in a perennial plot over a two-year period. **Actual costs and production data from demonstration/research plots in Tennessee are shown.** This assessment implies you are doing the work yourself and does not include costs for labor, equipment, or fuel, which can be considerable. It also does not include costs for lime and fertilizer or initial broad-spectrum herbicide, as both types of plots would require roughly the same amendments and preparatory treatment.

Annual plots

Cool-season (1 acre over 2 years)
$25 seed (oats — first year) in September
$5 postemergence herbicide
  (Harmony Extra) in November to kill
  cool-season weeds
$25 seed (oats — second year) in September
$5 postemergence herbicide
  (Harmony Extra) in November to kill
  cool-season weeds

Total forage produced for both years
(October — early April):
5,918 pounds dry weight
(Note: this weight includes palatable biomass only; it does not include biomass data from the bolt/flowering/seed-forming stages.)

Number of visits to the plot: 4 (2 to plant; 2 to spray)

Warm-season (1 acre over 2 years)
$64 seed (iron-clay cowpeas — first year)
  in May
$20 preemergence herbicide (Pursuit)
  at planting
$64 seed (iron-clay cowpeas — second year)
  in May
$20 preemergence herbicide (Pursuit)
  at planting

Total forage produced for both years:
  18,775 pounds dry weight (Note: this
  weight includes palatable forage only; not
  large stems)
Number of visits to the plot: 2 (to plant
  and spray preemergence)

Total forage produced from 2 acres of annual plots over two years: **24,693 pounds dry weight**
Total cost for both annual plots (2 acres; 2 years): $228
Cost per pound forage produced: <$0.01
Total visits to both annual plots: 6

Perennial plot

Cool-season (2 acres over 2 years)
$102 seed (ladino clover and oats) in
  September (because oats were added,
  weed pressure was reduced and herbicides
  were not necessary in the fall
  after planting)
Mowed in June after oats and clover had
  produced seed
Mowed in August (to knock back warm-
  season weeds before they
  produced seed)
$58 postemergence herbicide
  (Butyrac 200 and Clethodim) application
  in October
$58 postemergence herbicide
  (Pursuit and Clethodim) application in May
Mowed in August

Total forage produced from a 2-acre perennial plot over two years: **13,846 pounds dry weight**
  (includes approximately 2,000 pounds
  produced by the oats from October through
  early April following planting)
Total cost: $218
Cost per pound forage produced: $0.015
Total visits to the perennial plot: 6 (1 to
  plant, 2 to spray, 3 to mow)

Fig 8.14; Although warm-season plants, standing soybeans and corn provide a tremendous food resource for white-tailed deer and many other species through fall and winter.

Fig 8.15; High-quality warm-season forages, such as these soybeans, provide nutrients needed by does that are drawn down by nursing fawns. Photo by Michael McCord.

include high-quality legumes and other forbs that supplement naturally occurring forages and browse, especially during late summer when naturally occurring foods are considerably less digestible and palatable. Legume-dominated warm-season forage plots (such soybeans, cowpeas, and lablab) provide high levels of protein and total digestible nutrients — exactly what growing deer need during summer.

Warm-season food plots are commonly planted as single species (especially soybeans or corn) and in mixtures, which often contain several species. Soybeans and corn are most often planted separately to provide large amounts of food in relatively large plots. Warm-season attraction plots for hunting or viewing are more often planted in mixtures, though large plots also can be planted with mixtures.

Soybeans and corn are the primary warm-season food plots on most properties when managing for white-tailed deer. Planting and managing soybeans and corn are discussed separately from planting and managing warm-season mixtures below.

Soybeans and corn

Soybeans and corn are by far the most popular and important warm-season plantings for white-tailed deer — and for good reason. White-tailed deer prefer soybeans over all other warm-season forages, and acorns are probably the only other food deer prefer over corn during fall and winter. Not only are these crops palatable, but very nutritious as well. Soybean forage is extremely high in digestible protein and energy, and the beans are often overlooked as a source of energy during fall and winter. In fact, on a per pound basis, soybeans provide more energy from fat and sugars than corn (see Table 8.1 on page 108). However, on a per acre basis, corn provides more energy than soybeans because more bushels of corn grain are produced per acre than soybeans. Regardless, both corn and soybeans provide high-quality food that deer readily eat. Each year, most of the largest deer killed in the country have corn and soybeans within their home range, and usually in their rumen.

Growing productive plots of soybeans and corn can be complex. There are many decisions to make when choosing which variety of soybeans or corn you need to plant. Read the discussion below and contact your Extension agent for free professional advice if you need help deciding what to plant. Keep in mind the planting rates for corn and soybeans are provided as number of seed per acre rather than pounds per acre because seed size among varieties is variable.

Planting and managing soybean plots for white-tailed deer

It is easily argued that soybeans are the "best" warm-season food plot planting for white-tailed deer, and perhaps the most important overall, for a number of reasons. Soybeans are highly preferred by deer, provide outstanding nutrition through late summer and, depending on management, can provide an outstanding food source during winter. Soybeans are readily available and there are hundreds of varieties, which provide tremendous management flexibility with regard to maturity, weed control, planting dates, and methods. All of these factors make soybeans an excellent planting for deer.

Soybeans are categorized into 13 maturity groups (MG). Ten are commonly grown in the U.S. (MG 00-8; see Fig 8.16). Soybeans flower and mature in response to daylength, regardless of the age of the plant. Early-maturing groups are typically grown in the northern U.S. (most commonly MG 0-4), whereas those that take longer to mature are best adapted to the South (most commonly MG 4-7). If the later-maturing varieties are planted in latitudes above the Mid-South, they typically remain vegetative until frost and may not produce beans.

Fig 8.16; This map shows the approximate zones of adaptation for soybean maturity groups. Soybeans are sensitive to photoperiod, which influences the timing of flowering. During summer, photoperiod is longer in northern latitudes than southern latitudes. Maturity groups are adapted to different latitudes. If a soybean variety within a maturity group is planted farther south than it is adapted, it will grow shorter with less yield. If planted further north than it is adapted, it will grow taller and may flower too late to produce beans.

Soybeans are further classified as determinate or indeterminate. Determinate varieties produce most of their vegetative growth before flowering and pod set. Indeterminate varieties continue to develop leaves for a period of time even while flowering and developing pods. In general, MG 0-4 are indeterminate, and MG 5-8 are determinate. MG 4 soybeans may be planted early (mid-April) in the South so flowering can occur before late summer when dry conditions often prevail. Determinate varieties can be planted in late spring in the South because the growing season is so long. In recent years, several varieties of "forage" soybeans with an indeterminate growth habit have been developed. These varieties tend to produce more branches per plant (up to approximately 36 vs. 14) than standard agricultural varieties. Most of these are MG 7 or 8 beans, which perform best in zones 3-8 where the growing season is longer and they will produce beans.

Soybeans varieties exhibit one of three canopy types: bushy, intermediate, or narrow. Varieties with a bushy canopy should be planted on a wider spacing (30-inch rows or broadcast) with lower plant populations. If not, they will try to grow taller, but the crowding causes the plants to produce small stems that cannot support the increased height and they fall over (lodge). Intermediate varieties should be planted in 15-inch rows. Narrow canopy soybeans should be planted in narrow rows (7 inches).

Fig 8.17a (June), b (August), and c (October); Drilling soybeans into wheat and crimson clover after the wheat and clover die is a better strategy than top-sowing clover into soybeans when the soybeans begin to die. This is the perfect example of blending science with common sense!

In general, determinate varieties tend to have intermediate-to-bushy canopies and indeterminate varieties tend to have narrow canopies. However, if you don't know the canopy structure of the variety you are planting, it is best to use a 15-inch row spacing.

Soybeans are a legume crop that forms a symbiotic relationship with *Bradyrhizobium* bacteria to supply the nitrogen needs of the plant. Both native populations of bacteria and inoculant seed treatments (see *Inoculating legume seed* on page 54) will help supply adequate N for the crop. Seed inoculation is necessary if soybeans have not been grown in the field the past few years. Pre-inoculated soybean seed are more convenient to use than planter box treatments and often come treated with an insecticide and fungicide as well. A relatively low rate of N (30 pounds per acre) at planting may help soybeans establish more quickly in areas where weather is cool and wet. Check inoculation success by pulling up some plants and see if nodules are present on roots. If your soybean plants are not dark green, N may be limiting (see *Key to nutrient deficiency symptoms in crops* on page 31) and fertilizer and/or lime amendments may be needed. Always follow soil test recommendations to help ensure establishment success and optimum production.

Soybeans can be planted when soil temperatures reach 55°F at 2 inches deep, which may range from mid-March in the Deep South to early June in areas such as Iowa, Michigan, and Pennsylvania. However, the optimal soil temperature for soybean germination is in the mid- to upper 70s. Planting rate is determined by seed count and is usually 130,000-200,000 seeds per acre. Lower populations may be more productive than higher populations where soil fertility is relatively low or where rainfall may be limited or where beans are planted in wide rows (with more seed/row). In relatively small plots where soybeans are likely to be overgrazed, the seeding rate can be increased 2 – 4X the recommended rate. Increasing the plant population by such an amount can help the crop at least reach 12 inches or so in height and provide forage through the season. I have seen this work, but not in all places. Deer densities in some areas simply will not allow small plots to establish, regardless of seeding rate. Seeding with a drill or broadcast planting is generally less efficient than a planter and higher seeding rates may be used to offset seed loss and erratic depth control. At the higher population, plants are taller and pods form higher on the stem. Weed pressure typically is less with higher populations. Varieties with a bushy canopy also typically are more productive at lower populations than higher populations because the canopy is able to respond to the increased available space and produce more forage and beans. If you are broadcasting seed, sow 80-100 pounds PLS per acre (approximately 1 ½ to 2 bags per acre). Soybean seed should be planted or covered by disking 1-1.5 inches deep, not deeper than 2 inches.

There are numerous herbicides with various applications labeled for soybeans, including preplant incorporated, preemergence, and postemergence herbicides (see *Appendix 2*). Many soybean varieties are Roundup-Ready, which are tolerant to glyphosate, and LibertyLink varieties are tolerant to glufosinate. There also are varieties available tolerant of dicamba. Refer to herbicide labels for specific use of these and other herbicides. Continued and repeated use of glyphosate in the same field year after year has led to several weeds developing tolerance to glyphosate (as well as some other herbicides). It has become necessary to rotate use of herbicides that kill weeds with a different mode of action. Use of preemergence herbicides prior to planting in addition to postemergence herbicides after weed emergence is often necessary. Refer to *Appendix 2* for information on various herbicides and applications. Also refer to page 239 for summary of herbicide mechanism of action.

Soybeans are susceptible to various insect pests and fungal diseases (especially during rainy years). However, most seed are treated with a systemic insecticide that provides protection for about 25 – 30 days to help ensure a vigorous healthy stand. There is concern that such insecticide seed treatments could have negative effects on pollinators, especially honey bees, but this issue is still being researched. Various insects and diseases are usually most problematic during a particular time of the growing season, but one or more may be problematic at any time during the growing season. Holes and major defoliation are the result of insect damage. There are caterpillars (moth/butterfly larvae), beetles, and grasshoppers that eat the leaves and, in some cases, the seed pods as well. Fall armyworms, corn earworms, and loopers (all of these are moth larvae) can be problematic, especially in late-planted soybeans. Planting relatively early in the season and controlling weeds that may serve as host plants for the insects are important cultural considerations to limit insect problems. You should check your soybeans early and often through the growing season to notice insect problems before it is too late. If leaf defoliation exceeds approximately 20 percent, you might consider spraying your beans with an insecticide. A variety of insecticides are available for soybeans (see *Appendix 4*).

Fungal diseases may cause surface bronzing or leaf spots. Consider planting soybean varieties that are resistant to several foliar diseases, which can be more efficient and effective than spraying fungicides. Nonetheless, if you have a foliar disease problem, a fungicide application (such as Quadris or Headline at 6 ounces per acre) will help with frogeye leaf spot, brown spot, and Cercospora leaf blight. Fungicide seed treatments will help safeguard against disease problems as the stand is establishing.

Visit utcrops.com (as well as other university agricultural sciences websites) for detailed information on treatment thresholds and chemical recommendations on crop pest management. Finally, carefully consider your objectives and the extent of an insect or disease problem before applying insecticides and fungicides. The thresholds for insect/disease control in food plots usually are well beyond those for production agriculture.

Soybeans grown for deer should be allowed to stand through winter. That is, if the soybeans were not overgrazed and beans are present, do not mow or disk the soybeans

Fig 8.19; Whether early-maturing or late-maturing, bean production and availability through fall and winter may be just as important on your property as available forage through summer. This situation is true even if your property is relatively small and surrounded by row-crop agriculture. A corn or soybean food plot can provide a high-quality food source after surrounding crops are harvested and very little food is available in harvested fields. Here, a soybean food plot in Iowa provided deer a high-energy food source during winter after surrounding soybean fields had been harvested and other foods were scarce.

Can't get soybeans to grow?
You could if you managed your woods!

Soybeans provide a tremendous food source for deer. The foliage is high in protein and digestible energy and, after maturity, the beans provide a tremendous source of energy through fall and into winter. Many people are frustrated when they plant soybeans because they are not resistant to heavy grazing pressure and plots are often overgrazed soon after the plants germinate. The problem often is too many deer and, usually, poor-quality habitat. This problem is alleviated by reducing deer density and additional habitat management. When high-quality forage is available in woods and fields, grazing pressure on food plots is reduced considerably. Then, use of soybean plots is concentrated more during late summer when the quality of natural forages declines. This is one example of how food plots can supplement naturally occurring forages.

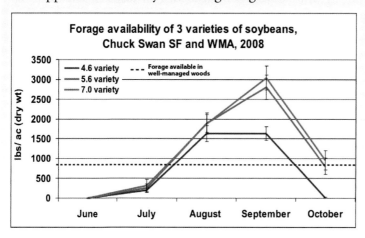

Fig 8.20a, b, c (graph above), and d; This soybean plot (MG 7.0) provided 3,000 pounds (leaves only) of highly digestible forage during a time (September) when naturally occurring forage quality was relatively low. Soybean plots such as this one are possible when deer density is managed accordingly and when adjacent woods and fields are managed appropriately. Grazing pressure on food plots is reduced when high-quality forage is available in the woods. Following an improvement cut and prescribed fire, the adjacent woods provided approximately 850 pounds (shown by dashed line on graph) of forage per acre (including only species selected by deer, not total biomass). The natural forage alleviated grazing pressure through the summer and allowed the soybeans to grow and later produce beans. During winter, this soybean plot provided more than 1,500 pounds of high-energy seed per acre.

Fig 8.18; Buckwheat can be added to soybeans to help alleviate grazing pressure on the soybeans while the soybeans are developing.

after they have matured. Allow deer to eat the soybean seed through fall and winter. If soybeans are overgrazed and no beans are present, cool-season forages can be planted in late summer or early fall. No-till drilling wheat and annual clovers over the overgrazed soybean plants is effective.

The most effective strategy to incorporate cool-season legumes into a soybean field is to plant a good reseeding clover, such as crimson, the fall before planting soybeans. This method is much more effective than top-sowing clover seed on top of dead soybean plants that still hold bean pods.

A drawback to planting soybeans is their relatively low tolerance to grazing while they are developing. This problem is alleviated by implementing additional habitat management strategies, planting more acreage (larger fields), and killing more does where needed. Planting buckwheat with the soybeans and periodic fertilization also can help soybeans grow tall enough to better resist grazing. Buckwheat germinates and grows faster initially than soybeans. Buckwheat is not nearly as preferred by deer as soybeans, but the addition of buckwheat will help buffer grazing pressure on the soybeans. The soybeans will later grow through the buckwheat. Do not add more than 6 pounds per acre or the buckwheat may outcompete the soybeans (unless you spray the buckwheat with an appropriate herbicide). Top-dressing soybean plots with 15-30 pounds N while the soybeans are establishing provides a boost of growth and helps them get through the critical period when they are most susceptible to grazing pressure. If buckwheat is added to soybeans, do not use preemergence or preplant incorporated herbicides. If Roundup-Ready soybeans are planted and sprayed with Roundup, the buckwheat will be killed, but that is not a problem as long as the soybeans have become established. Grass-selective herbicides also can be used in plots where buckwheat was added.

Temporary electric fences, both with and without area (smell) repellents, sometimes are used to keep deer out of warm-season forage food plots, especially soybeans, until they have established and can better withstand grazing pressure. The reason most soybean plots are overgrazed is related to deer density, availability and quality of naturally occurring forage, and plot size. In some situations, an exclusion fence is not needed as much as a reduction in deer density. This situation stimulates thoughtful consideration. Before installing some type of fence around a soybean plot, think about why you are planting soybeans. If the intention is to provide increased nutrition for antler growth and lactation, then keeping deer out of soybean plots until mid-July or

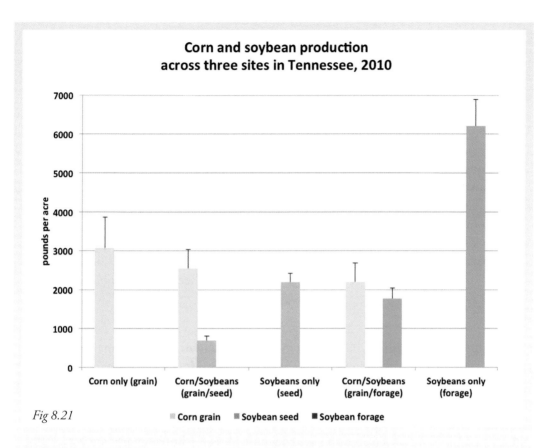

Fig 8.21

Is it best to plant corn and soybeans separately, or in a mixture?

People planting Roundup-Ready corn and soybeans often consider if they should plant them singularly or as a mixture. Normally, 40-50 pounds of soybeans are planted with 7-10 pounds of corn per acre, which equates to an average seed count of 100,000-125,000 soybean seeds and 14,000-20,000 corn seeds per acre. This planting rate has produced many successful food plots and fed many deer. However, is this the best planting strategy? For example, does 2 acres of a corn and bean mixture produce as much forage and/or grain as 1 acre of corn and 1 acre of beans planted separately?

We measured soybean forage and bean yield and corn grain yield following several planting rate combinations at three sites over two growing seasons (2010 and 2011). We planted 'Allen' soybeans, which is a Roundup-Ready MG 5 soybean with an intermediate canopy, on 30-inch rows at 120,000-185,000 seeds per acre (depending on treatment). We planted a medium-season (114-day) corn hybrid, which was Roundup-Ready, on 30-inch rows at 20,000-29,000 seeds per acre (depending on treatment).

Regardless of whether we used partial or full planting rates of corn or soybeans, we found soybean forage and bean production was more than three times greater when planted alone than when planted with corn. This result means more forage and beans can be produced in 1 acre of soybeans planted separately than in 2 acres of a corn and soybean mixture, regardless of the soybean planting rate in the mixture. As corn grows, sunlight reaching the soybeans is reduced. Anyone who has ever planted a corn and bean mixture has noticed "weak" beans where the corn is robust and relatively dense. And where the corn is relatively sparse (with less yield), the beans are taller and more robust.

From our data, we conclude it is not efficient from a yield perspective to plant soybeans and corn in a mixture. We recommend planting them separately. Where mixtures are desired, we recommend vining legumes, such as cowpeas or lablab with a "substrate" plant, such as peredovik sunflowers (see *Vining legume mixture* on page 112).

later is not helping anything in most regions because fawns are functional ruminants and 90% of antler growth is complete by early August. How large is the field? Small (less than an acre) soybean plots have little chance of establishing where deer density is moderate to high. Also, small food plots do not provide increased nutrition for many deer unless there are many small plots scattered throughout a relatively large area. Thus, small plots are primarily used as attractant plots for hunting and viewing. Other forages, such as oats, wheat, clovers, and even iron-clay cowpeas are more resilient to grazing and are better choices for small plots where deer density is relatively high. If soybeans are planted in small plots as an attractant during bow season, an electric fence may be needed to get the plot established. In larger fields, a repellent fence protecting the inner majority of the field, but allowing deer access to the outer edges of the field, aids in establishing the majority of the plot, but still allows access to a portion of the planting. A repellent or temporary electric fence also may be useful in "stockpiling" soybeans through summer to provide maximum bean production for a supplemental winter food source, which can be particularly beneficial up North.

If you cannot grow a food plot because there are too many deer, it is likely the population needs to be lowered and/or you need to provide more food, whether through food plots or other habitat management practices. Erecting a fence to keep deer out of the plot may only address the symptom, not the problem, of overgrazing. You must consider what the deer are eating while they are fenced out of the food plot. High-quality foods should be available throughout the year if you expect deer to express their potential, especially as related to fawn development (lactating does) and antler growth. If deer density is in check with the habitat, soybean plots should be able to support grazing and provide added nutrition through the summer.

The relative inability of soybeans to tolerate heavy grazing pressure can actually be useful. If soybeans cannot establish in relatively large plots because of overgrazing, it should be a sign to land managers that deer density may be approaching or has already exceeded nutritional carrying capacity of the surrounding habitat. Difficulty establishing soybeans can be a "wake-up call" that additional management is needed. Furthermore, overpopulated deer destroy habitat for many other species by overbrowsing. Without question, habitat management has to be coupled with population management to have healthy wildlife populations, and healthy wildlife populations are a product of healthy habitat.

Planting and managing corn and grain sorghum for white-tailed deer
Corn can be an important component in a food plot program, especially in relatively cold climates, or during years with relatively little acorn production. Growing corn can be expensive and difficult, but the benefit can be great and there are several options when managing a corn food plot. Plots of corn adjacent to forage plots make excellent sources of food and/or cover, not only for deer, but for many other species as well. Corn is highly attractive to many wildlife species, and that certainly includes white-tailed deer. However, attraction doesn't necessarily equate to need. Corn is a high-energy food. And availability of such food can be limiting in northern latitudes, especially where snow cover persists for relatively long periods and available forage to

Fig 8.22; Corn left standing can be an important food source during winter for deer and many other species, especially in relatively cold climates.

support deer is lacking. However, in more southerly latitudes, winter rarely, if ever, limits deer. Although deer may flock to corn when it is available, the prudent manager should carefully consider whether planting corn is worth the cost, especially when fall plantings of wheat, oats, or annual clovers, for example, provide adequate digestible energy to supplement naturally occurring foods.

As with soybeans, there are hundreds of varieties of corn available for various soils, climates, and applications. Selecting the best variety can be complex. Time to maturity is an important consideration. If you are planting up North, you will want to plant a variety that matures earlier (80-100 days) than later (120 days). Corn can be planted once soil temperatures at planting depth reach about 55° F. Germination rate increases with warmer temperatures, but many producers plan to plant once soils temperatures reach 55-56° F. Check with your Extension agent or local seed dealer and find which varieties perform best in your area. A Bt corn hybrid can provide protection from various stalk and leaf-feeding caterpillars. Bt corn produces a protein derived from bacteria (*Bacillus thuringiensis*) that kills certain species of insects. And yes, Bt corn is safe for human and wildlife consumption. Any corn hybrid purchased from a seed supply store has been treated with at least one fungicide and at least one insecticide.

Corn is normally planted with a corn planter in 20- to 30-inch rows with seed 6-12 inches apart to provide 16,000-30,000 seeds per acre for optimum grain production. The lower rates (16,000-22,000) are recommended for sandy loams and relatively dry areas, whereas higher rates (24,000-30,000) are recommended for heavy clay soils that hold moisture much better. Seed should be planted 1.5–2 inches deep. Slightly deeper is OK if soil moisture is limiting on top, but you don't want to plant deeper than 2 inches if moisture is plentiful. If you don't have access to a corn planter, 8-13 pounds PLS of corn seed per acre (depending on variety and soil type) can be broadcast and covered by disking 1- to 2-inches deep.

A significant drawback to planting corn is the expense. Corn is a heavy nitrogen user and it is especially important to manage soil fertility as recommended after soil testing when trying to maximize grain production. The amount of N is highly correlated to the number of bushels of corn grain produced. In most areas, a good yield of corn is about 160 bushels per acre (or about 9,000 pounds). Such yields will require about 160 pounds of actual N, or about 10 bags of ammonium nitrate (34-0-0) per acre, as well as considerable amounts of phosphate and potash, as recommended from a soil test. Agricultural producers commonly rotate corn and soybeans. Nitrogen manufactured in nodules on the roots of soybeans becomes available to other plants the following growing season (see *Inoculating legume seed* on page 54). Because corn requires so much nitrogen and nitrogen is so expensive, rotating these crops increases efficiency. Seed corn prices are variable and fluctuate year to year. However, seed corn is often available free through local chapters of conservation organizations or state wildlife agencies. A bag of seed corn (80,000 seed) will plant 2-4 acres, depending on planting rate and location.

Considering the cost of planting corn, if row cropping occurs on or around your hunting property, it may be cost effective to pay the producer to leave a section of the field unharvested for deer (as opposed to you spending time and money planting a food plot). Find out what the producer is getting per acre for the grain, then offer that amount per acre you would like left unharvested. Be sure and explain how profits actually will be increased because time and fuel will not be spent harvesting those acres.

Some people plant grain sorghum instead of corn because of the associated expense. Grain sorghum is not nearly as attractive to deer as corn, but deer often will eat the seedheads, sometimes during the milk stage, but most often after maturity. The variety of grain sorghum should be considered before planting. Tall varieties may compete better with weeds, and bird-resistant varieties help prevent seed depredation through the summer by house sparrows, starlings, and grackles. There are many varieties of grain sorghum (see sidebar *Grain sorghum or milo — which is it?* on page 191). Contact your county Extension agent and ask for recommendations in your area. Wild Game Food (WGF) sorghum is a popular, relatively short (3 feet) grain sorghum developed to resist bird depredation. Bird resistance is related to tannin content.

Typically, white grain sorghums (such as Hegari, see Fig 12.9 on page 189) contain fewer tannins than red or brown grain sorghums; therefore, white grain sorghums may be preferred by deer and other wildlife. Preference is a moot point for deer, however, if birds consume all the seed before they mature. Tannin content drops considerably after a couple of frosts, increasing palatability of red and brown sorghums during late fall and winter. Grain sorghum is normally planted with a grain drill or may be broadcast at about 10 pounds per acre once soil temperatures reach 60° F.

A variety of herbicide applications are possible when planting corn or grain sorghum. Glyphosate can be sprayed postemergence over varieties of Roundup-Ready corn. Postemergence applications may be applied anytime from emergence until the V8 stage (8 leaves with collars) for corn or until the corn reaches 30 inches in height, whichever comes first. An initial application of 1-4 pints (depending on weeds present) of a glyphosate herbicide should be applied when weeds are 2-8 inches tall. For best results on several perennial weeds, allow them to grow to 6 inches before spraying. A sequential application may be necessary if a new flush of weeds appear. Roundup-Ready technology is a great strategy to use when tough-to-handle weeds are present. However, after growing Roundup-Ready crops, there can be problems with volunteer sprouting if another crop is planted. If desired, residual Roundup-Ready corn can be killed with a grass-selective herbicide, such as Clethodim. Roundup-Ready technology is especially effective when one of the preplant/preemergence herbicides

Table 8.1. A comparison of nutritional values of the grain/seed produced among various crops[1].

Component -per 100g portion (Unit)	Corn	Wheat	Soybeans	Grain Sorghum
Energy (kJ)	1528	1369	1866	1419
Protein (g)	9.4	12.61	36.5	11.3
Fat (g)	4.74	1.54	19.94	3.3
Carbohydrates (g)	74	71	30.16	75
- Fiber (g)	7.3	12.2	9.3	6.3
- Sugar (g)	0.64	0.41	7.33	0
Calcium (mg)	7	29	277	28
Iron (mg)	2.71	3.19	15.7	4.4
Magnesium (mg)	127	126	280	0
Phosphorus (mg)	210	288	704	287
Potassium (mg)	287	363	1714	350
Sodium (mg)	35	2	2	6
Zinc (mg)	2.21	2.65	4.89	0
Copper (mg)	0.31	0.43	1.7	-
Manganese (mg)	0.49	3.99	2.5	-
Selenium (μg)	15.5	70.7	17.8	0
Vitamin C (mg)	0	0	6.0	0
Thiamin (B1)(mg)	0.39	0.30	0.87	0.24
Riboflavin (B2)(mg)	0.20	0.12	0.87	0.14
Niacin (B3) (mg)	3.63	5.46	1.62	2.93

[1] Table adapted from USDA Food Composition databases.

labeled for corn also is used.

Prior to planting non-Roundup-Ready corn and grain sorghum, atrazine, Bicep II Magnum, or Dual Magnum may be applied preplant incorporated to control a wide variety of grass and forb weeds (Note: grain sorghum seed must be treated with Concep seed safener before applying Bicep II Magnum or Dual Magnum). Basagran or Permit can be applied postemergence to control several forb weeds and yellow nutsedge. 2,4-D, Aim, atrazine, Banvel, and Clarity are other herbicides that can be applied postemergence to control forb weeds only. **Before spraying, it is important to realize many "weeds" can complement corn/grain sorghum plots.** Common ragweed, pokeweed, smartweeds, giant foxtail, crotons, and others can provide additional seed relished by many birds, including northern bobwhite, mourning dove, and several species of native sparrows. A weedy grain plot also attracts wild turkey and bobwhite broods (see *Chapters 9* and *10*). During helicopter surveys, I have noticed white-tailed deer select weedy sections of corn fields (compared to "clean" sections of fields without weeds) to bed during the day. Herbicides or cultivation may be used, however, if undesirable plants, such as cocklebur, curly dock, broadleaf signalgrass, goosegrass, horsenettle, jimsonweed, sicklepod, spiny amaranth, and yellow nutsedge, are expected.

Prowl can be applied preemergence for corn, but not preplant incorporated (see *Appendix 2*). Prowl also may be applied postemergence incorporated for corn and grain sorghum after crop plants are least 4 inches tall (that is, crop must be cultivated with at least 1 inch of soil thrown over the base of the crop prior to herbicide application). Pursuit can be applied preplant incorporated, preemergence, or postemergence in Clearfield varieties of corn only (do not apply Pursuit to non-imidazolinone varieties of corn).

Various insect pests have the potential to reduce corn yields considerably. An efficient and effective approach in managing insect pests and diseases is to plant corn varieties resistant to various pests and diseases and to use a seed treatment. Consult your Extension agent for information on which varieties are best suited for your area. Beyond planting treated seed, cultural methods should be considered as well. Planting corn early reduces chances of crop damage from some insect problems, such as armyworms. Corn fields should be scouted through the season to check for insect problems. An insecticide treatment may be warranted if armyworms are found on 25 percent of corn seedlings, when 5 percent of corn seedlings are damaged by cutworms, or when 10 percent of corn seedlings are infested by stinkbugs (see *Appendix 4* for insecticide options). However, as stated with soybeans, you should carefully consider your objectives and the extent of an insect or disease problem before applying insecticides and fungicides. The thresholds for insect and disease control in food plots are usually well beyond those for production agriculture.

A common "problem" when growing corn is the number of wildlife species that eat corn. Your attitude is important! **You should expect and welcome native wildlife species on your property to benefit from your food plots.** Even if you are focusing on one or two species, your habitat management efforts should benefit several others as well. That being said, there certainly are times when strategies for dealing with crop depredation by wild hogs, raccoons, squirrels, black bears, and blackbirds are needed. Grain availability into winter may depend upon field size and location, management, deer density, use by other species, and the acorn crop. Larger plots (2-3 acres, or more) may be warranted, but not necessary in all situations. Crop depredation is another reason planting corn can be expensive — you often have to plant extra acreage because of use by so many species. Small plots may be decimated in a night or two by wild hogs or raccoons. Sometimes populations of problematic species need to be lowered to match habitat limitations. Control can be accomplished through trapping or shooting, according to species and local regulations. Ask a biologist with your state wildlife agency to help you identify the best solution to reduce numbers of problematic species.

Another strategy is to allow corn plots to lay fallow. Corn plots may get relatively little use during years with a heavy acorn crop. All the time and expense growing corn can seem wasted when it is not eaten. Don't despair. If allowed to lay fallow, the plot will provide outstanding fawning cover and brooding cover for wild turkeys and quail. The corn will remain protected in the shucks. Deer and other species will eat it through winter and spring and you will provide outstanding early successional cover with the corn still standing (see Figures 9.4 and 10.9). Deer will be eating many of the forbs responding from the seedbank through spring and summer. This strategy is excellent when managing many fields and when trying to be as cost-effective as possible in your food plot program.

Other warm-season single-species plantings for deer

Although soybeans are the most highly selected warm-season forage for deer, other single-species plantings also can make excellent warm-season forage food plots. In particular, American jointvetch, Florida beggarweed, perennial peanut, and alyceclover can provide high-quality warm-season single-species plantings in the South. Buckwheat can be used as a single-species planting, especially where quick green-up is needed, and where you want to increase soil organic material. When planting vining legumes, such as cowpeas and lablab, it generally is best to use them in a mixture with a "substrate" plant, such as peredovik sunflowers, to enable the vining legumes to provide more forage per acre. Some good mixtures that we have used regularly with success are provided below.

If you are in the Deep South, from the Lowcountry of South Carolina around to coastal Texas, and are looking for a relatively low-maintenance, high-quality perennial forage, you might consider perennial peanut. It is a subtropical legume that grows best in sandy loams of the lower Coastal Plain where winter-kill is not a problem. Perennial peanut is resistant to disease and drought, and tolerates relatively low pH, low fertility, and heavy grazing pressure. Perennial peanut spreads vegetatively by underground rhizomes. It is best established by sprigging rhizomes (similar to bermudagrass) 1 ½ - 2 inches deep in a fine, weed-free seedbed during winter before spring green-up. Contact your Extension agent to locate a source for sprigs. Imazapic and imazethapyr are excellent preemergence herbicides when establishing perennial peanut. Clethodim can be used postemergence to control grass weeds. Cool-season weeds can be sprayed with glyphosate following a couple frosts when the perennial peanut is dormant. If cool-season weeds are not expected to be problematic, annual cool-season forages, such as crimson and berseem clovers and oats, can be sown in fall to provide winter forage while perennial peanuts are dormant. Perennial peanut is expensive

Fig 8.23a,b; Florida beggarweed does well as a single-species warm-season planting from the Mid-South southward. Not only do deer select Florida beggarweed, but it also performs relatively well under heavy grazing pressure. Deer density was approximately 70 per square mile at this site.

Fig 8.24a,b; Perennial peanut is a long-lived, low-maintenance, perennial warm-season legume that performs well in sandy loams of the Deep South. Photos by John Drummond.

111

to establish (approximately $300 per acre), but it is a good investment because perennial peanut is long-lived (10 – 20 years is common) and requires relatively little maintenance.

Warm-season forage mixtures for white-tailed deer

Beyond soybeans, there are several other warm-season forages that should be considered, especially when planting mixtures. Although not as preferable to deer as soybeans, lablab and cowpeas offer outstanding forage and both are more resistant to grazing than soybeans.

It is important to consider weed control when considering a warm-season mixture. Grasses often are a real problem in warm-season plots, especially johnsongrass, crabgrasses, goosegrass, and broadleaf signalgrass. Considering how grass weeds can be so problematic and how grasses are not eaten to any real extent by deer during summer, I do not include grasses in warm-season forage mixtures for deer. Some people like to include grain sorghum in warm-season forage mixtures, but I generally recommend against it, for a couple reasons. First, including sorghum in a forage mixture prevents you from controlling grass weeds with a grass-selective herbicide. Second, deer may eat grain sorghum seedheads, but not the growing forage (leaves) unless natural forage availability is extremely limited. Why take-up space in a warm-season forage plot with less-desirable forage at the expense of weed control? If you want to provide grain sorghum seedheads, grow grain sorghum as a single-species planting and get maximum seedhead production. Vining legumes may climb grain sorghum, but a better option where grass weeds are problematic is sunflowers, which provides you with more herbicide options. Some commercial mixtures combine a leguminous forb (such as cowpeas or lablab) with a non-leguminous forb (such as sunflowers or buckwheat) and a grass (such as grain sorghum). Such a mixture is not sensible because there is no herbicide that can be used with such a mixture (unless the grain sorghum seed is treated with a seed safener). There are better mixtures to use and some simple solutions.

Vining legume mixture (PLS/ac)
    50 pounds iron-clay cowpeas
    10 pounds lablab
    5 pounds peredovik sunflowers

Benefit and considerations: Iron-clay cowpeas and lablab provide excellent forage for deer, especially during late summer or early fall when the palatability of natural forage and browse has decreased. These legumes also withstand grazing pressure relatively well, even while they are developing, and grow on a wide variety of sites. Although deer may browse newly appearing heads of sunflowers, they are not added to this mixture for forage, but as substrate for the cowpeas and lablab to climb and grow upon later in the season.

You don't have to include both cowpeas and lablab in this mixture. There is nothing wrong with only using one. However, if you want to add variety in your warm-season forage plots, cowpeas and lablab complement each other well. If you only use one, plant 75 pounds of cowpeas or 25 pounds of lablab with the sunflowers. I do not recommend adding smaller-seeded species, such as American jointvetch and alyceclover, to this mixture. They are relatively slow to establish and are usually overwhelmed by the climbing legumes. If you want to plant jointvetch or alyceclover, they will do best if planted separately (see *Lowcountry mixture*).

Management: Prowl, Dual Magnum, or Treflan can be applied preplant incorporated to control various forb and grass weeds. A grass-selective herbicide can be sprayed postemergence for additional grass weed control if necessary. If grass weeds are not problematic and/or if deer density is so high sunflowers are overgrazed as they are establishing, grain sorghum (3 pounds) can be used in place of sunflowers. If grain sorghum is added, however, do not apply Prowl or Treflan. Dual Magnum may be applied if grain sorghum seed is treated with Concep seed safener.

Reseeding soybean mixture (PLS/ac)
40 pounds Quail Haven soybeans
5 pounds peredovik sunflowers

Fig 8.25; *This mixture produces a lot of tonnage and deer are readily attracted, especially in late summer when natural forages begin to decline in quality.*

Fig 8.26; *Iron-clay cowpeas show great resiliency to grazing. As opposed to soybeans, cowpeas continue to produce side shoots and additional leaves in the presence of heavy grazing.*

Fig 8.27; Quail Haven reseeding soybeans are selected by whitetails and they resist grazing pressure quite well. If allowed to produce seed, QH beans can be retained for at least two years with proper management. A variety of weed control options are available (see text). Photo by Chris Shaw.

Benefit and considerations: Quail Haven reseeding soybeans produce a tremendous amount of forage and resist grazing pressure quite well. QH beans are a non-GMO Group 7 soybean. QH beans do best in Zone 8 and below (see Fig. 2.3 on page 9), though frosts are delayed long enough in Zone 7 in most years to allow seed production.

Management: Prowl, Dual Magnum, and Treflan can be applied preplant incorporated for weed control. A grass-selective herbicide can be applied postemergence for additional grass weed control. Quail Haven soybeans have excellent re-seeding capabilities. If allowed to mature and produce seed, another stand can be stimulated by light disking. Often, after the plot dies in the fall, there is a lot of dead plant material in the plot. Just allow it to decompose through winter. Spray cool-season annual weeds with a glyphosate herbicide before they flower and produce seed. In early April, spray the plot with Prowl or Treflan, apply fertilizer and/or lime as recommended by a soil test, and then disk the plot. You have effectively reseeded your QH beans. You can use a grass-selective herbicide later (postemergence) if a grass weed problem develops. If you want to add sunflowers or grain sorghum to the plot, just sow them before disking or drill them into the plot after you have disked the QH beans. If you add grain sorghum, do not apply herbicides.

Lowcountry mixture (PLS/ac)
   10 pounds American jointvetch (*Aeschynomene*)
   10 pounds alyceclover
    5 pounds buckwheat

Benefit and considerations: American jointvetch (also known as deervetch) and alyceclover arc high-quality forages commonly planted in the Lowcountry of South Carolina and in other areas of the Deep South where the growing season is long. I call this the "Lowcountry mixture" because Joe Hamilton (founder of the QDMA) pioneered deer management in the Lowcountry and he has long promoted Aeschynomene, and for good reason. Deer love it! (Joe writes about Aeschynomene in his book, "Firepot Stories." You should get a copy. Excellent reading!) Jointvetch was selected over alyceclover by deer in every field where we grew the two forages together over several years. Alyceclover was eaten, but jointvetch was always preferred. Thus, you might ask, "Why include alyceclover in the mixture?" The reason is alyceclover is a good forage and it can help buffer grazing pressure on the jointvetch, where overgrazing is a problem. Both jointvetch and alyceclover are quite slow to establish. We sometimes add about 5 pounds of buckwheat to the mixture to provide quick green-up and buffer grazing pressure. Buckwheat grows fast, but goes to seed quickly, and is not preferred by deer over American jointvetch or alyceclover. There is nothing wrong with planting American jointvetch or alyceclover as a single-species planting, but expect them to establish relatively slowly.

Jointvetch will continue to grow and produce forage through a long growing season, making jointvetch an efficient plot in the Deep South as it produces over so many months (can be planted late March and produce forage until first frost in November). If growth is not resisted by grazing, the stem of jointvetch may grow to 5 or 6 feet tall and become quite woody. However, it is an indeterminate grower and the plant keeps producing tender, nutritious forage that deer continue to eat until first frost.

Fig 8.28a, b, and c; American jointvetch and alyceclover (top) complement each other very well with similar structure and growth requirements. Jointvetch (middle) and alyceclover (bottom) grow well even in sandy-loam soils. About 40 deer were feeding daily (late August) in this 5-acre alyceclover plot in the Lowcountry of SC.

Seed production and availability influences cost of American jointvetch greatly. Jointvetch seed may cost $10 or more per pound some years. As a result, some people plant this mixture without the jointvetch or plant alyceclover by itself, as shown in Fig 8.28c.

Management: A grass-selective herbicide can be used to control grass weeds postemergence if sprayed when weeds are young. If buckwheat is not added to the mixture, jointvetch and alyceclover can be established relatively well via no-till top-sowing. Kill existing vegetation with a glyphosate herbicide in the spring prior to seeding. If top-sown over existing vegetation that has been sprayed, sow 10 pounds jointvetch and 10 pounds alyceclover per acre.

Cool-season plots for white-tailed deer

Cool-season plots are the most popular plantings for white-tailed deer. Cool-season forages may be annual or perennial and peak in production at different times of the year, depending on latitude and the forage grown. Therefore, it is necessary to identify the specific periods when you want additional forage available and choose a planting that is productive during that period. If you intend to ensure forage availability from early fall (when warm-season forages are waning) through midwinter (when little or nothing else green is available) until midsummer (when warm-season plots begin producing tremendous amounts of forage), then a variety of forages will be necessary. There is no one species that produces forage year-round (see Fig.

Fig 8.29; Clovers and cool-season grains (especially wheat and oats) are the main cool-season forages planted for white-tailed deer — and for good reason. They are productive, nutritious, deer readily eat them, and there are multiple varieties available, providing several management options.

8.12 on page 91). Depending on where you are located, you may be able to plant a mixture that will provide forage year-round (see Year-round forage plot on page 129).

Some cool-season forages germinate and grow quickly soon after planting (most annuals), whereas others are slower to establish (perennials and a few annuals, such as arrowleaf clover) and may require months before considerable forage is available. Most cool-season plots are planted in late summer or early fall, but mid-February through early April also is a good time to plant several cool-season forages in the South and through May farther north. If planted by mid-September with adequate rainfall, several annual cool-season forages will provide a grazable stand by mid-October. Timing of planting cool-season annual grains (such as wheat and oats) and brassicas (such as forage rape) is important. If planted too early (such as mid-August in the South), some of the brassicas may bolt and produce seed prior to cold weather during a relatively warm fall with plenty of rain. Likewise, cool-season grains may grow quickly with such conditions, and forage quality will be reduced if stems and leaves grow relatively large prior to cool weather. On the other hand, if planted late (such as mid-October), growth and forage production will be limited until spring.

Use of cool-season plots during the fall is influenced greatly by acorn availability. However, cool-season forages can play an important role in providing plenty of digestible energy through winter, helping deer enter spring in good shape. All perennial cool-season forages are slow to establish. Thus, it usually is prudent to add one or more cool-season annuals to perennial forage mixtures. When planted in the fall, perennial forages begin substantial production the following late winter or early spring, which coincides with a period (March through May) when forage high in protein is needed for beginning maximum antler growth and meeting reproductive demands.

Cool-season food plots are vulnerable to a host of weeds, especially perennial plots, which will contain both cool- and warm-season weeds. There are several selective herbicide options, depending on what you plant and weed(s) present. Weeds should be sprayed when young for effective postemergence control (see *Chapter 6*). The planting recommendations in the following sections and chapters, as well as the *Plant Identification Guide* in the back of this book, provide specific information on weed control.

Wait! What are you trying to do?

Before you start considering all the possible mixtures and management strategies for various cool-season food plots, you need to carefully consider what you are trying to accomplish with cool-season forages and honestly evaluate your abilities and time constraints. Before deciding what you want to plant, answer the following questions.

Do you just want to attract deer for hunting or viewing?
Are you seriously trying to provide additional nutrition during winter (after hunting season)?
How much time do you have?
Can you attend to weed problems (spray) as they arise?

As you consider your answers, don't make things too complicated. Remember, keep it simple, and do it right. If you just want to attract deer for hunting or viewing, you don't necessarily need a complicated perennial mixture. Don't let your ego get in the way here. There is nothing wrong with just planting wheat or oats, for example. If you want to provide forage through the winter, you probably should concentrate on annuals. If you are trying to provide forage April-June, then you should consider various annual clovers as well as perennial forages. However, if you are not prepared to use selective herbicides at the correct timing, you will be frustrated trying to manage perennial plots. As stated previously, both annuals and perennials have a place, but do not overlook the productivity of annuals and the relative ease of management.

Single-species cool-season plantings

Many people have been led to believe they have to plant a mixture of forages if they are going to plant a food plot for white-tailed deer. This belief is not true and counterproductive in some situations. Of course, there are times when mixtures are warranted, and there are many forages that perform best when included in mixtures. However, there are some species that perform best when planted alone, and there are some situations when a single-species planting is best for a particular objective. Soybeans and corn are two warm-season examples. In some situations, cool-season grains, alfalfa, and brassicas are cool-season examples.

Fig 8.30; Oats were consistently ranked as a high-preference forage by white-tailed deer in every trial in which they were included.

118

Cool-season grains — which is best?

Wheat, oats, and rye often are planted in forage plots for white-tailed deer. All three provide high-quality forage for deer while green and growing; however, there are considerations for each. All three are high-preference forages by white-tailed deer. That is, deer "like" all of them and all of them will be eaten if planted, at least to some extent, regardless of what else is available. There are, however, slight preferences by deer among the cool-season grains. When grown in separate side-by-side plots, we found **oats** were selected consistently over the others. That is, deer would go to the oats first and graze more than the other cool-season grains. Second in preference was **rye**. Note that I am referring to cereal rye, not ryegrass. **Ryegrass is not** a selected forage by deer and is extremely aggressive and difficult to control. Therefore, I do not recommend ryegrass as a planting for white-tailed deer (or anything else). If you are relying on ryegrass as a "green plot" for white-tailed deer, you need to consider the alternatives provided here and use a different approach. **Triticale** is a hybrid of rye and wheat. Deer selected triticale similarly to rye. **Wheat** was not selected as strongly as oats or rye, but don't let that influence your opinion of wheat. Wheat is a selected forage by white-tailed deer. If oats or rye are not in the field, you won't know the difference! Also, in our trials, deer never grazed **barley** to a measureable amount. It was not a selected forage and I do not recommend it for deer (geese ate it, but not deer).

Although oats are selected slightly by deer over rye and wheat, a drawback for oats is they are not cold tolerant. If you plant oats from the Mid-South northward, choose a cold-tolerant variety of oats or your planting will experience considerable die-back or be killed. Ask your Extension agent for the variety of oats that performs best in your area. Both cereal rye and winter wheat are very cold tolerant and there are varieties suitable for the northern U.S.

Don't think you can just disk a field, sow a cool-season grain, and have a productive food plot. It still is important to get the pH between 6.0 and 6.5, adjust P and K levels at least to medium, and add 30-60 pounds of N per acre as the plot is establishing. When soils are amended correctly, oats, wheat, and rye all provide outstanding forage with high levels of protein and total digestible nutrients until the grasses begin to mature the following spring. Once the cool-season grains begin to bolt (produce shoots that will later produce seedheads), palatability and digestibility decrease significantly, and so does use by deer.

Other considerations for cool-season grains include attractiveness of the grain, growth height, and weed control. Wheat produces a seedhead that is readily eaten by deer (and other species), if an awnless variety is planted (discussed below). Seedheads of the other cool-season grains, especially rye and triticale, are not eaten to a large extent by deer (if at all). This makes wheat attractive to deer not only during fall and winter, but also in late spring and summer (May – June) when mature seedheads are available. Wheat and oats grow to 2-3 feet in height, whereas rye and triticale grow to 5-7 feet tall. The tall growth of rye and triticale is problematic because of the large amount of debris on the field when the crop matures. If grown with other forages in a mixture, the debris often smothers the other forages. It also influences future management in the field. If you are planting something else and using conventional tillage, you will have to wait on the debris to die-down or mow, burn, or disk it in. However, if you are planting with a no-till drill, the debris can be advantageous as it will increase organic material and add to the top soil.

To summarize, if you are in the South and want to plant a high-quality forage that will attract deer from fall through early spring, plant oats. If you are in the Mid-South and northward, choose a cold-tolerant variety of oats, or plant wheat. If you want forage as well as grain for an additional food source into summer, plant wheat. The only time I recommend rye is if you are up North and want a winter green patch in a plot that you intend to plant in a warm-season food plot, or if you are trying to increase soil organic material. In my opinion, **an awnless variety of winter wheat is the best choice overall among the cool-season grains**. Considering the attractiveness of the grain and management flexibility, I include it in most cool-season mixtures.

Fig 8.31; Winter wheat provides excellent forage for whitetails through the fall and winter. We consistently recorded crude protein levels above 25 percent and acid detergent fiber levels below 25 percent from germination through April when the wheat began to bolt.

Using wheat to your advantage

If you want to plant a cool-season forage that will benefit more wildlife species than any other, plant winter wheat. As green forage, it is highly desirable by all grazers. After it matures, the grain is attractive to even more species, including a multitude of bird species. Before you disregard this as too simple, and you "want to do something more than wheat," think about it. Winter wheat remains green and growing through winter, providing deer with digestible energy. We consistently recorded crude protein levels above 25 percent and acid detergent fiber levels below 25 percent for green, growing wheat from fall through spring until it began to mature. Thus, wheat provides outstanding forage throughout winter. In our forage tests, deer consistently grazed wheat in every trial it was included, regardless of the other forages available (see *Appendix 5 and 6*). When grown in the same trials, various clovers, such as crimson and berseem, and oats were preferred slightly over wheat, but wheat was always selected by white-tailed deer.

One of the biggest advantages to planting winter wheat is management flexibility. There are three common scenarios. After providing forage through winter, 1) wheat can be sprayed or disked in the spring to plant a warm-season crop; 2) it can stand through the summer, allowing deer and other species to eat the seedheads, and then you can spray incoming weeds in July or August prior to planting a cool-season plot in late summer or early fall; or 3) it can lay fallow through the year, accumulate organic material, and allow you to rotate to a warm-season plot the following spring. **Do not overlook the value of wheat seedheads for deer.** Seedheads can be considered "free" food through summer, especially when grown on sites or in areas that may be too dry to reliably grow warm-season plots. Additional management options are available when you add other forages with wheat (see *Cool-season forage mixtures for white-tailed deer* on page 127).

Fig 8.32a, b, and c; Not only is green, growing wheat attractive to deer, the seedheads also are eaten by whitetails. Therefore, one planting of wheat can provide food during fall, winter, spring, and summer. Do not overlook the attractiveness of wheat seedheads for deer. However, deer do not eat awned (or "bearded") wheat seedheads (a, top) as readily as they do awnless varieties (b, middle). Deer had eaten all of the seedheads in this field (c, bottom) by early July (deer density was about 30 per square mile).

Fig 8.33; Although there are herbicides that can control ryegrass without killing wheat, selective control can be difficult. A better option to eradicate ryegrass is to plant forbs only (see Southern bottomland perennial mixture on page 136), which allows a grass-selective herbicide to control ryegrass.

Fig 8.34; Insecticide applications may be needed to treat problems, such as this lesser corn stalk borer larvae in this oat stem.

Another advantage to wheat is it is available anywhere and there are varieties suitable for any region. However, **it is important to select an awnless (or beardless) variety**. Deer eat wheat seedheads without awns much more readily than those with awns.

Broadleaf weeds are controlled easily in wheat plots with numerous broadleaf-selective herbicides, such as Harmony Extra, 2,4-D, Aim, Banvel, and Clarity. For best results, spray before weeds are 4 inches tall. This strategy will help "clean-up" fields with tough-to-control broadleaf weeds and still provide high-quality forage. If you have a problem with cool-season grass weeds, such as ryegrass, it is best to consider a forb-only planting (such as crimson or berseem clover) instead of wheat, or any other grass, so you can use a grass-selective herbicide. However, there are herbicides, such as Osprey and Achieve, labeled for wheat to control several annual grass weed problems, including ryegrass.

Timing of planting should be considered carefully. Winter wheat planted in late summer is more susceptible to armyworms. If you plant late in the fall and cold weather arrives not long after the wheat has germinated, forage for winter grazing will be limited. In Zones 6 and 7, planting wheat in early to mid-September is a safe bet. Further north, planting winter wheat in August will work. In the Deep South, planting in late September/early October will provide good winter grazing. Also, be sure you are planting *winter* wheat in late summer/fall and **not** *spring* wheat because spring wheat will grow quickly if conditions are favorable and mature before winter, providing low-quality forage through winter.

Wheat, as well as other cool-season grains, is susceptible to several plant pests. Aphids can be sprayed when infestations are beginning to cause leaves to dry up and die, often in fall (about 30 days after planting) or late winter (before March). Fall armyworms should be sprayed in the fall when four or more larvae per square foot are present. Baythroid XL 1 (2.0-2.4 ounces per acre), Karate 2.08/Warrior II (1.3 – 1.9 ounces per acre), Mustang Max (4 ounces per acre), and Sevin XLR Plus 4 (32-48 ounces per acre) can be used to control these pests. Diseases, such as leaf spot, stem rust, and powdery mildew are not common problems, but fungicides, such as Headline SC (6-9 ounces per acre) and Proline 480 SC (5 ounces per acre), can be applied if diseases occur. However, as stated with corn and soybeans, you should carefully consider your objectives and the extent of an insect or disease problem before applying insecticides and fungicides. The thresholds for insect and disease control in food plots are usually well beyond those for production agriculture.

Wheat can be drilled or broadcast. Approximately 120 pounds PLS (or 2 bushels) is the suggested single-planting rate per acre but 200 pounds PLS per acre may be warranted if intensive grazing pressure is expected. If other forages, such as clovers, are planted with wheat, the rate of wheat should be reduced to 30-50 pounds PLS per acre.

Planting greens — rape, kale, turnips, and radishes

The "greens" belong to the genus *Brassica*. They have been used as livestock forage for many years, but were popularized as a deer forage when Dr. Grant Woods visited New Zealand and brought back the idea of planting these forages in food plots. Greens often are included in mixtures with clovers, but to realize the tonnage possible by greens, it is best to plant them singularly, or with other greens. When included in mixtures with clovers (see *perennial mixture* listed on page 133), seeding rates for brassicas should be no more than 1 or 2 pounds per acre to avoid overcrowding the clovers and other forages.

Fig 8.35; Plots of rape and kale, such as this one in New York, can provide substantial tonnage of highly digestible forage, and can be an important food source for whitetails, especially through winter.

When you mention brassicas to people who have planted them for deer, you usually get a very good or very poor response. Mixed opinions on brassicas abound, primarily because of the variability in deer selectivity across regions and even local areas. It is not uncommon to hear of deer eating brassicas to the ground in one area and, just a few miles away, they didn't touch them on someone else's property. This variation in selectivity may be attributable to deer density and forage availability, soils and nutrient availability, or nothing more than deer being finicky! This issue is not limited to brassicas. Sometimes deer have to get accustomed to a new food source and "learn" that they like it before they eat much of it. Timing also is important when evaluating deer use of brassica plots. Typically, palatability and use increase after a few frosts.

Rape (*Brassica napus* and also *B. campestris*) is the most common brassica used in deer forage plots. There are many varieties of rape; some developed for oil production and some for forage production. Those developed primarily for oil also may be used as forage. A common variety of rape is Dwarf Essex; however, improved forage varieties are more leafy, grow more upright, and typically produce more tonnage. "Canola" represents rape varieties developed and registered by the Western Canadian Oilseed Crushers Association to produce a food-grade oil with low (less than 2 percent) erucic acid (CANadaOilLowAcid).

Fig 8.36a and b; Turnips (left) and daikon radishes (right) can be attractive to deer. The green leaves are not typically highly selected, but the taproot can provide a good source of energy through winter.

Kale (*Brassica oleracea*) is primarily grown for its leaves (greens). Collards is a cultivar commonly eaten by humans. As a forage, it typically produces more tonnage and is more cold tolerant (down to 10° F) than the other brassicas. Both rape and kale respond well to grazing as long as they are not grazed continuously below about 6 inches.

Turnips (*Brassica rapa*) produce a leafy top as well as a taproot and both may be eaten by deer. Like the other brassicas, there are many varieties of turnips. An important consideration among varieties is the leaf-to-taproot ratio. Some, such as Seven Top, All Top, and Pasja, produce little or no bulbous taproot; whereas others, such as Purple Top, Green Globe, and Purple Top/White Globe, have a greater proportion of taproot than leaves. Taproot production is obviously important if you want to provide more food through winter and if you want to increase soil organic matter. Turnips tolerate close grazing as long as the growing point at the top of the taproot is not eaten.

Radishes (*Raphanus sativus*) often are planted as an agricultural cover crop, but some varieties, such as daikon, are planted for deer. Radishes are not as cold-tolerant as rape, kale, and turnips, and may be killed if temperatures drop into the low-teens. The taproot of some radishes may grow to 30 inches deep, or more, effectively breaking hardpans, aerating the soil, improving drainage, and providing organic material. Deer may eat the forage as well as the taproot, especially of white radishes, such as daikon, and radishes may be more palatable than other brassicas prior to frost.

All of the brassicas are cold-hardy and relatively drought tolerant. They do best on well-drained medium-textured soils (loams) with a pH of 6.0-6.8. None of the brassicas performs well on poorly drained sites. It is best to avoid heavy clays. Turnip production is best in loose, well-aerated soils. Good production is realized with medium levels of P and K (follow soil test recommendations); add 60 pounds of N per acre at planting and again approximately 60-90 days after planting to increase production. Soil tests also may recommend 1-2 pounds of boron per acre.

Brassicas usually are planted in late summer, but also are planted in spring up North in areas such as North Dakota, Minnesota, upper Wisconsin, and northern Michigan, when the soil temperature reaches approximately 50° F. In most areas, brassicas are planted from late July through mid-September, according to latitude (at least 70 days before a killing frost). Growth is most vigorous in the fall when daytime temperatures are 40-60° F. Most plots are planted via broadcast seeding on a well-prepared, fine seedbed. If broadcast, rape and kale should be planted at 6-8 pounds PLS per acre and turnips at 4-5 pounds PLS per acre. If planted with a no-till drill or planter, plant 4 pounds PLS of rape or kale per acre in 3- to 7-inch rows (higher yields may be realized with 3-inch row spacing); and plant 3 pounds PLS of turnips per acre in 6- to 8-inch rows. When broadcast seeding, it is important to cultipack after sowing. Do not plant or cover seed more than ¼-½ inch.

Crop rotation is important when planting brassicas to avoid problems with several fungal diseases. Sclerotinia stem rot can be controlled with a fungicide application, such as Proline 480 SC (5.0-5.7 ounces per acre). To help prevent fungal disease problems, brassicas should not be grown on the same site more than three years in a row. Insect pests can be controlled with various insecticides (see *Appendix 4*), such as Intrepid 2F (8 ounces per acre). Trifluralin herbicides, such as 1-1.5 pints per acre of Treflan, can be applied preplant incorporated to control several annual grass weeds

as well as some broadleaf weeds, such as henbit and chickweed. Grass-selective herbicides also can be used to control grass weeds in brassica plots.

Sugarbeets

Sugarbeets (*Beta vulgaris*) is an old favorite for deer up North. Deer may eat the leaves, but the primary attraction is the large conical, white taproot that contains up to 22 percent sugar during the winter after planting. Sugarbeets are biennial plants. Sugar is stored in the roots during the first growing season and, when grown for production, the roots are harvested at the end of the first growing season. If not harvested, the plant will bolt and flower the following growing season. Although typically grown up North, sugarbeets may not survive winters in the far North (such as North Dakota, Minnesota, and Wisconsin). Sugarbeets usually are planted in April-May up North, but can be planted August through early September in the Midwest and Mid-South. The best growing conditions to maximize yield and quality are daytime temperatures of 65-80° F. Growth ends with a hard freeze.

Fig 8.37a and b; Although the leaves may be eaten, sugarbeets are planted because of the high sugar content in the roots. Kermit Pittman plants them on his property in the mountains of North Carolina and says the deer can't stay out of them once the weather gets cold.

Sugarbeets can be difficult to establish, but with attention to detail, you can grow a productive crop. If deer in your area are not accustomed to sugar beets, it may be necessary to dig some up to help deer "discover" them. Sugarbeets should be planted in well-drained loamy soils. Heavy clays and sands should be avoided. Sugarbeets have high soil fertility requirements and do best in soils with pH between 6.5 and 7.5. Be sure to follow soil test recommendations carefully. Once the plot establishes, 60-90 pounds of N per acre will encourage production, but do not apply more N or beet quality may suffer and yield decrease.

Seeding usually is done by broadcasting on a well-prepared fine seedbed, but can be done with planters. Seed should be covered lightly, approximately 1/4- to 1/2-inch deep, which can be accomplished by cultipacking a relatively loose seedbed after sowing seed. If broadcast, sow 8 pounds PLS per acre, or about 50,000-60,000 seeds per acre, which should provide a spacing of about 1 plant per square foot.

Sugarbeets are susceptible to many fungal diseases and pests. Planting seed treated with a fungicide is a good idea and a four-year crop rotation is recommended. That is, do not plant sugarbeets in the same plot for at least three years. A fungicide application, such as Proline 480 SC (5.0 – 5.7 ounces per acre), can help with problems such as sclerotinia stem rot. An insecticide application, such as Intrepid 2F (8 ounces per acre), can be used for armyworms and webworms (see *Appendix 4*). Grass weeds can be controlled with a grass-selective herbicide and certain broadleaf weeds can be controlled with a postemergence application of clopyralid (such as Stinger at 8 ounces per acre).

Cool-season forage mixtures for white-tailed deer

The key to forming good mixtures is to match forages that complement each other, both in terms of maturation to extend forage availability and with consideration for weed control. Various cool-season forages germinate at different times and display different initial growth rates. Of course, various herbicides can be used with some forages and not with others. When you understand this, you will begin to put together mixtures that make sense. We have worked diligently over the years to come up with mixtures that extend forage availability and enable weed control. Consider these carefully when determining what would best help you meet your objectives.

The "best" annual forage mixture (PLS/ac)
    15 pounds crimson clover
    10 pounds arrowleaf clover
    40 pounds wheat

Benefit and considerations: There is no mixture that will attract deer better than this one. The mixture contains three forages that complement each other perfectly in structure and phenology. Wheat and crimson clover germinate soon after planting and continue producing through winter (see *Tables 8.2 and 8.3* on page 141). In Zones 6-8, crimson clover flowers, produces seed, and dies in late April or early May. When planted in late summer or fall, wheat bolts and produces seedheads in May. Arrowleaf clover comes on strong and vines up through the mature wheat about the time the crimson clover dies. Deer eat the arrowleaf clover through July when it dies. Deer eat the wheat seedheads May through July, which is why I typically use wheat instead of oats. This mixture should be used as far north as the bottom of Zone 5 (see

Fig 8.38a-d; High-preference, high-quality, cheap, easy, grows just about anywhere — that pretty much sums up this mixture. Crimson and arrowleaf clovers are excellent reseeders and good companion plants. This plot was planted in September 2009 (pictured in November 2009, a). It has been maintained for four years by spraying glyphosate (1 quart per acre) in late July (pictured in 2011, b) and mowing two weeks after weeds die. That's it. Nothing else. This plot has not been reseeded since it was initially sown (pictured November 2011, c). By mid-May each year, the crimson has died (notice dead crimson flowers), the arrowleaf is growing strong, and the deer are eating the wheat seedheads (pictured May 2012, d).

Fig. 2.3) and where annual precipitation is at least 25 inches. Crimson and especially arrowleaf clover may winterkill north of that latitude.

Management: This mixture is as close to foolproof as it gets. It establishes very quickly and both crimson and arrowleaf clover are excellent reseeders. You can retain these clovers for years without replanting if they are not overgrazed and are able to flower and produce seed. After the arrowleaf clover dies (mid- to late July), wait a week or two, then spray the entire plot with a glyphosate herbicide (1-2 quarts per acre) to kill all incoming weeds. After weeds die (a couple weeks or so), mow the plot. Before mowing, additional wheat can be top-sown or drilled into the plot. Top-dress with lime and fertilizer (if needed) as recommended from a soil test. If the clovers were allowed to flower and produce seed, you should not need to re-sow. If the plot was overgrazed and seed production was less than desirable, top-sow or drill additional seed as needed prior to mowing.

Year-round forage plot (PLS/ac)
    10 pounds crimson clover
    10 pounds arrowleaf clover
    10 pounds red clover
    40 pounds wheat

Benefit and considerations: Everyone wants a food plot that produces year-round, right? Well, if you are in the bottom of Zone 5 through Zone 8, here it is. This is the only food plot mixture that truly produces food for a full year, in all seasons, after planting. And there is no other cool-season mixture that will produce more forage on an annual basis as this one. Plus, all four components in the mixture are high-quality, preferred forages for whitetails. My graduate students and I evaluated many seeding rates for this mixture over two years and this is the one that works best. The components complement each other perfectly. As the crimson dies in late April or early May, the arrowleaf and red clover begin vigorous growth. When the arrowleaf dies in July, the red clover is well established. Expect the red clover to continue to provide highly nutritious forage through the end of August.

Management: As the red clover declines in productivity (late August), spray the plot with glyphosate. After all weeds have died, top-sow additional red clover seed and mow the plot. Top-dress with lime and fertilizer if needed as recommended from a soil test. If the crimson and arrowleaf clovers flowered and produced seed, no reseeding

Fig 8.39a and b; For Zones 5-8 (see Fig 2.3), this is truly a year-round food plot. By adding red clover in the mixture, high-quality forage is available through the end of August. This picture was taken in west TN at the end of August. The plot was planted the previous September. Wheat and crimson clover were available from late September through April. Wheat seedheads were available in May and eaten through the summer. Arrowleaf clover was available April through July, and the red clover from May through the end of August, providing high-quality forage at a time when naturally occurring forage is relatively low in quality.

should be necessary. If not, top-sow or drill additional seed prior to mowing. That's it. No continual mowing; no continual spraying. You work on the plot during one or two visits in August or September and it's good to go for the rest of the year.

Fig 8.40; *This annual clover mixture provides outstanding forage through winter all the way into July. In early May, the crimson clover has about finished blooming, balansa clover is in full bloom, and the berseem and arrowleaf clover have not begun to bloom.*

Full-on annual clover mixture (PLS/ac)
10 pounds crimson clover
8 pounds berseem clover
5 pounds arrowleaf clover
4 pounds balansa clover
40 pounds winter wheat or oats

Benefits and considerations: This mixture contains four annual clovers that are strongly selected by deer and complement each other with respect to growth and maturity. Crimson and berseem clover germinate and grow fast, providing grazable forage soon after planting through winter. Berseem is not cold tolerant and will winter-kill from the Mid-South northward. Avoid winter-kill by planting a cold-hardy variety, such as Frosty berseem. Balansa clover follows in production and begins producing considerable tonnage late winter through spring.

Arrowleaf peaks in production May – July. According to your location, crimson will flower first in mid- to late April, followed by balansa in late April/mid-May, then berseem in mid- to late May, and finally arrowleaf in July. The addition of wheat and/or oats provides an attractive quick green-up soon after planting, and if you use awnless wheat, seedheads are available for deer and other wildlife May-June. This mixture performs well from the bottom of Zone 5 through Zone 8.

Management: This mixture is very easy to manage. Crimson and arrowleaf are strong reseeders. Balansa also will reseed naturally. After the plot has died, incoming weeds can be sprayed with glyphosate in late July/early August. Lime and fertilizer can be added as recommended from soil test after spraying incoming weeds. Dead material can be mowed or lightly disked 2 – 3 weeks after spraying. Additional berseem clover seed and wheat/oats can be top-sown or drilled in late August – late September, depending on latitude and location.

Red clover and alfalfa — not just for cattle!

Red clover and alfalfa should be considered seriously when planting a perennial cool-season food plot for white-tailed deer. Red clover is classified as a biennial clover, but with proper management, it can be retained for three years in the Mid-South and northward. In the Deep South, red clover is an annual. Alfalfa is a perennial legume that can be retained for many years with proper management, including strict attention to pH and insect pest control. Red clover and especially alfalfa are tolerant of dry conditions and very productive where soils have been amended properly. There are many varieties of red clover and alfalfa. Check with your Extension office to learn which variety is best in your area.

Red clover and alfalfa do not get the credit they deserve as deer forages. They have been called "cattle forages," with claims that they are too stemmy to be considered deer forages. These claims are misleading. Red clover and alfalfa do produce relatively large stems as they mature, but this is not what deer eat. Deer eat the foliage produced on the ends the stems! This relationship is true of other forages also, such as American jointvetch, iron-clay cowpeas, and even soybeans.

The ability of red clover and alfalfa to produce high-quality forage through the summer months is noteworthy and an important consideration.

Fig 8.43a,b; The red clover (top) and alfalfa (bottom) shown here are both in their third year. Their ability to withstand relatively dry conditions and produce high-quality forage through summer is impressive. They are both preferred forages by white-tailed deer.

Winter greens patch (PLS/ac)
5 pounds daikon radish
3 pounds forage rape or kale
1 pound purple top turnip

Benefits and considerations: Brassicas can provide a great source of energy during winter. Even when covered in snow, deer will dig down to get at the nutritious forage. Typically, the foliage of rapes and kales is selected over the foliage of turnips and radishes. However, turnips and radishes provide a taproot that may be eaten through winter. All of these are cold-hardy plants, especially the rapes and kales. This mixture performs well as a fall-planted crop in Zones 5 – 8. In Zones 3 – 4, planting in spring is recommended.

Fig 42a and b; A winter greens patch can be highly attractive to white-tailed deer during fall/winter. The addition of radishes or turnips in the mixture provides an additional source of energy through winter. Right photo by Kip Adams.

Management: Brassicas should be planted relatively early compared to clovers and cool-season grains to allow substantial biomass production before hard frosts. Planting late July through August is appropriate in Zone 5, late August/early September in Zones 6 and 7, and early to mid-September in Zone 8. A trifluralin herbicide, such as Treflan, may be applied preplant incorporated to control various grass and forbs weeds, and a grass-selective herbicide can be applied postemergence to control cool-season grasses. Brassica plots should be rotated to help preclude development of fungal diseases. If desired for quick green-up, 50 pounds of winter wheat or cereal rye could be added to this mixture. Even though you amended the

132

soil according to soil test prior to planting, applying an additional 30-60 pounds of N per acre in late fall will give the plot an added boost for increased production through winter.

Standard cool-season perennial mixture (PLS/ac)
   4 pounds ladino white clover
   5 pounds red clover
   3 pounds chicory
   40 pounds oats or wheat

Benefit and considerations: This mixture performs very well, but do not expect to retain ladino clover on exposed sites that become excessively dry during the summer. A cool-season annual mixture is much better suited to those sites. There are several varieties of ladino and intermediate clovers to choose from. Some are better adapted to wet conditions; some are adapted to drier conditions; some are more resistant to diseases; some withstand competition from grasses better than others. However, all produce excellent forage for deer. Expect high use of this mixture for several years, provided the plot is managed correctly by top-dressing as needed with lime and fertilizer and with weed control. Red clover and chicory tolerate dry conditions fairly well (especially chicory). Wheat and oats are included because they germinate quickly

Fig 8.44; This is as good as a perennial plot gets! Preferred perennial forages are mixed with high-quality, fast-growing annuals. With proper management, the clovers and chicory can be maintained for several years. This plot was planted in early September; picture taken Nov. 1.

and provide forage as the clovers are establishing. Berseem clover is an annual clover that germinates and grows fast initially and is highly preferable to deer. It can be added (5 pounds) to this mixture for quick clover growth and attraction. However, if you plant berseem clover or oats north of Zone 8, be sure to plant a cold-tolerant variety, which should persist through winter at least as far north as Zone 7. Forage rape (1 pound) also may be added for additional winter grazing as the clovers and chicory are developing (do not add more than 1 pound, or the rape may limit clover establishment).

Management: There is no preemergence herbicide that can be used with this mixture. However, given the rapid germination and early growth of oats and wheat, preemergence herbicides usually are not needed. After the wheat (or oats) have produced seed and died and the clovers have flowered and produced seed, imazethapyr (such as Pursuit) can be sprayed postemergence as necessary to control various weeds. A grass-selective herbicide (such as Clethodim) is recommended to control grass weeds. Spray grasses before they reach 6 inches tall. Top-dressing the appropriate amount of lime and fertilizer in September (one year after planting) with will have this perennial forage plot looking good into October/November. An insecticide treatment for white grubs (Japanese beetles and June bugs) may be necessary in the third or fourth year after establishment. Sevin (water soluble powder at 10 pounds per acre), Dylox (3.75 ounces per 1,000 square feet), Mach II (granular formulation at 1.33 pounds per acre), and Arena (water soluble powder at 10 pounds per acre) have all worked well.

Dry-land perennial mixture (PLS/ac)
    10 pounds alfalfa
     5 pounds red clover
     3 pounds chicory
    40 pounds wheat

Benefit and considerations: This is a perennial mixture that will do well in upland areas prone to becoming fairly dry during summer (but it does best where there is plenty of moisture!). Expect stand thinning to occur during prolonged dry periods; however, with proper management, the stand can be retained and invigorated. Exposed sites that become excessively dry during summer should be planted to an annual cool-season plot or not planted at all, but instead managed as an early successional community with naturally occurring plants. Expect alfalfa and chicory to persist for many years if top-dressed annually according to a soil test and if weeds and weevils are sprayed as necessary. This mixture is not cheap; therefore, it is important to realize the management effort needed to maintain a stand before planting.

Management: Alfalfa is sensitive to acid soils and low fertility, and alfalfa weevils can become problematic. To maintain alfalfa, pH should be raised to 7.0, both macro- (phosphorus and potassium) and micronutrients (especially sulfur and boron) should be applied if needed, and insecticides will be necessary to combat alfalfa weevil infestations in spring (4-8 ounces per acre of Way-Lay 3.2 AG has been successful).

The wheat will produce seedheads in May. Deer will eat the seedheads through summer. If you have amended the soil and planted correctly, and the plot has not been overgrazed, you should not mow until late summer. Incoming weeds can be sprayed with imazethapyr (Pursuit) or imazamox (Raptor) postemergence. Alfalfa, chicory, and red clover flower in midsummer. After they produce seed in late summer, mow the plot to encourage fresh growth for fall. Problem grasses can be controlled with a grass selective herbicide if sprayed before they reach 6 inches in height. Top-dress the plot with lime and fertilizer in August/September as recommended by a soil test.

Fig 8.46; Alfalfa, red clover, and chicory produce highly nutritious forage for whitetails. This perennial mixture is outstanding if you can drive it (requires management), can be maintained for several years, and it will perform well even on relatively dry sites. Picture taken July 5 in Zone 7. Wheat or oats can be drilled into existing plots in late summer/early fall if desired.

Southern bottomland perennial mixture (PLS/ac)
   8 pounds alsike clover
   4 pounds ladino white clover
   12 pounds berseem clover

Benefit and considerations: This perennial mixture is very well-suited for Southern bottomlands that are poorly drained and moist most of the time. Berseem clover is a high-preference annual clover that should be added because it germinates and establishes relatively quickly. Berseem is not cold tolerant. However, there are cold-tolerant varieties of berseem clover, such as Frosty, that persist through winter at least as far north as Zone 7. There are no grasses (such as oats) added to this mixture because this allows a grass-selective herbicide to be applied soon after planting to control ryegrass, which usually infests most Southern bottomland fields.

Fig 8.47a and b; Berseem clover (top photo) is a high-preference non-reseeding annual clover that germinates quickly and grows well in association with ladino and alsike (bottom photo) as they develop.

Management: Imazethapyr (Pursuit) and/or 2,4DB (Butyrac 200) can be sprayed postemergence as needed after the clovers are well-established to control various broadleaf weeds. A grass-selective herbicide can be applied postemergence any time after planting to control grasses. This mixture should be mowed after the clovers have produced seed (usually in August) and as necessary to prevent weeds from flowering if the plot is not managed with the appropriate herbicides. An insecticide treatment for white grubs (Japanese beetles and June bugs) may be necessary in the third or fourth year after establishment. Sevin (water soluble powder at 10 pounds per acre), Dylox (3.75 ounces per 1,000 square feet), Mach II (granular formulation at 1.3 pounds per acre), and Arena (water soluble powder at 10 pounds per acre) have all worked well.

Fig 8.48; Both alsike and berseem clover do well in poorly drained soils, including areas that are periodically flooded for short periods of time. Photo by Ryan Basinger.

Including wheat or oats in perennial mixtures

Oats and wheat are selected forages for white-tailed deer and they germinate and establish quickly. Including wheat or oats as a companion plant when establishing perennial forages provides quicker green-up and increased forage availability. Oats and wheat also help suppress weed growth and prevent soil erosion. Perennial clovers, alfalfa, and chicory are relatively slow to establish. Oats and wheat serve as a "nurse crop" through the first winter after planting, bearing the brunt of grazing pressure and allowing perennial forages to establish by the following spring. Fall-planted oats and wheat produce seedheads the following spring, then die and decompose into the perennial plot through the summer. Barley is not recommended because whitetails did not eat barley in any of our trials. Cereal rye and triticale (wheat/rye hybrid) are not included as companion plants because they grow very tall (5-7 feet) upon maturity and leave a tremendous amount of dead material on top the perennial forages. Brassicas are not good nurse plants for perennial clovers because the leaves are large and can shade-out developing clovers.

Fig 8.49a, b, and c (bottom left); This series of pictures illustrates the typical progression of an annual grain (such as oats) planted in a perennial clover mixture. This plot was sown in September 1999. Soon after planting, the oats germinate, become established, and provide forage for deer. By May 2000 (upper left), the oats have matured and the clover has become well established. While growing, the oats serve as a "nurse crop" for the ladino white clover. By July 2000 (upper right), the oats begin to decompose. By September 2000 (bottom left), the oats "melt" into the clover (the plot was not mowed), leaving a pure clover stand that can be maintained for several years with the appropriate management techniques.

Monitoring food plot success with exclusion cages

Planting success and use of food plots (especially forage plots, such as clovers, chicory, or soybeans) should be monitored using exclusion cages. These cages allow you to observe how much forage is being consumed over time and estimate the success of your planting. Exclusion cages (approximately 4 feet in diameter and 4-5 feet tall) can be made of dog-panel wire wrapped around four stakes driven into the ground. I like the size of the openings in dog-panel wire (2" X 4") better than hog-panel (4" X 4") because rabbits can get through hog-panel, and I like dog-panel better than chicken-wire because chicken-wire is flimsy and is often bent over by deer trying to get to forage inside the cage.

It is easy to assume your planting efforts were futile when you visit a plot and all you see is weeds. Poorly developed, weed-choked food plots are common in areas with high deer densities and little other available food as food plots are ravaged as soon as the plants germinate. Exclusion cages will make this problem evident, as knee-deep lush forage may be found in the cage.

Fig 8.50 a and b; Exclusion cages make it evident when deer density is too great for the forage available. Without an exclusion cage, you may wonder if bad seed, weed pressure, planting technique, soil fertility, and/or weather contributed to crop failure. Excessive deer density may not only prevent soybeans from establishing (left), but perennial clovers and wheat (right) as well.

What about exclusion fences?

Various types of fences and repellents may be used to keep wildlife and livestock out of food plots. A 5-strand barbed-wire fence or a temporary electric fence with a portable solar-powered charger are effective in excluding livestock. An electric fence also may be used to repel wildlife that are damaging plots, such as black bears, raccoons, wild pigs, and deer (number of strands, strength of charger, and design of fence varies according to targeted species). Temporary electric fences with solar-powered chargers are particularly popular and effective. They are commonly used to keep wildlife (especially bears and pigs) out of corn plots until they have matured and to keep deer out of plots managed for doves. Exclusion fences also are sometimes used to keep deer out of soybean plots until they have established and can better withstand grazing pressure.

Do not plant perennial cool-season grasses!

Do not include tall fescue, orchardgrass, bromegrasses, timothy, matuagrass, or bluegrass in any food plot! Perennial grasses are ranked at the bottom in terms of forage preference by white-tailed deer. In our experimental plots used to determine planting recommendations for white-tailed deer, there was virtually no measure of deer grazing these grasses at all, in any year or season. White-tailed deer food habit studies over the past 50 years have noted a lack of perennial grasses in the diet, except in late winter when they have just germinated and there is little other green forage available. Not only are they not selected, perennial grasses are competitive and usually choke-out clovers by the second growing season, leaving nothing but a field of rank grass with relatively high lignin content, providing low palatability, low digestibility, and little nutrition. Even if other desired

Fig 8.51a, b; Here (a, top) is a plot of orchardgrass (foreground) and an adjacent plot of oats, Jan. 24, 2006. Nothing really needs to be said — the picture speaks for itself. Fresh, nutritious oats, or senescent, rank orchardgrass. The choice is yours. Here (b, bottom) is a lush plot of timothy, May 2, 2005. Horses may love this stuff, but deer don't! You won't find a single blade of timothy grazed in the entire plot.

forages were not choked out completely, why would you want a certain percentage of your food plot taken up by non-preferred plants with lower nutritional quality? It doesn't make sense!

Are you interested in wild turkeys or bobwhite on your property? If so, then there are more reasons why you shouldn't plant perennial cool-season grasses (see *Chapter 10*).

Table 8.2

	Forage produced (pounds per acre – dry weight)	Percent eaten by deer
crimson clover	4,050	97
oats	3,676	92
triticale	4,049	89
wheat	3,952	85
orchardgrass	2,212	2

Production of three annual grains, crimson clover, and orchardgrass grown in separate plots in the same field (shown in Figs. 8.51a and b) from October 2005-April 2006. These data were collected at the end of each month, except April, when data were collected prior to flowering for each of the forages. It is clear these annual forages out-produce and are highly preferred over orchardgrass by white-tailed deer. Deer density in this area was approximately 70 per square mile.

Table 8.3

	Forage produced (pounds per acre – dry weight)	Percent eaten by deer
crimson clover	3,726	78
wheat	5,736	65
rye	7,378	50
ryegrass	4,889	10
orchardgrass	2,449	3
timothy	2,486	0

A comparison of three annual grasses, two perennial grasses, and crimson clover grown from October 2004-April 2005 show the same general results as Table 8.2. These data were collected in the same field in the same manner as those from Table 8.2. It is clear the annual small grains and crimson clover are highly preferred by white-tailed deer over ryegrass, orchardgrass, and timothy. Oats (variety not stated) are not included in the comparison, because they were winter-killed in January 2005.

Nine

Chapter 9 — Wild Turkeys

Presence and abundance of wild turkeys are influenced greatly by food availability. Food plots can not only influence wild turkey movements and home range size, but they also can lead to increased body weight and help increase the percentage of eggs that hatch when nutrition is a limiting factor. Food plots also can influence poult survival, depending on what you plant and how you manage your food plots (discussed below). **However, before considering what to plant in a food plot, your primary considerations when managing for wild turkeys should be nest success and brood survival.** Unless these are relatively high, the impact of food plots most likely will be negligible.

Early successional areas with considerable brushy cover provide outstanding nesting cover for wild turkeys. In the woods, improvement cuts that allow at least 20-30 percent sunlight to reach the forest floor enhance nesting cover greatly. These techniques can be combined by felling or killing undesirable trees at least 100 feet into the woods around old-fields managed for high-quality early successional cover. When managing fields, nonnative perennial grasses (such as tall fescue, orchardgrass, bermudagrass, and bahiagrass) should be killed and the seedbank allowed to respond. Eradicating these grasses will improve cover for broods greatly. Succession should be managed with prescribed fire, disking, and selective herbicides. The fire-return interval will depend on plant response and desired vegetation composition and structure, but is 2 to 6 years on most sites. Make no mistake, these are the practices that increase recruitment into the fall population, not food plots. However, after recruitment is increased, food plots can keep birds in an area, influence movements and weight, and help maintain a healthy, vigorous, wild turkey population.

Figs 9.1a, 9.1b, 9.1c, 9.1d, from top; Early successional areas (a and b: May and June) as well as managed woodlands (c: June) and woods (d: July) with a well-developed understory offer optimum nesting and brooding cover for wild turkeys.

Fig 9.2a (above), b (below); Wild turkeys are irresistibly drawn to recently burned areas. Prescribed fire consumes leaf litter and makes seeds and invertebrate parts readily available. Burning woods and old-fields adjacent to food plots is an excellent strategy.

Use of food plots by wild turkeys is increased when food plots are located adjacent to attractive cover. Use in fall, winter, and spring is especially increased when food plots are near favored roosting areas. Use in late winter and early spring also is increased when food plots are in proximity to old-fields and woods that have been burned recently. Burning exposes seeds and charred invertebrates (such as snail shells and beetle parts), which provide an excellent source of calcium for hens about to lay eggs. As hens are attracted to these areas, so are gobblers. Recently burned old-fields and woods are open and provide exceptional strutting areas. Later, use of food plots in summer is increased when favorable brood cover is nearby. Coincidentally, prime brooding cover usually is provided in those old-fields and woods that have been burned recently (refer to *Chapter 10* for discussion on brood cover).

Food plots provide nutrition for wild turkeys via grain (such as corn and wheat) and other seeds (such as soybeans, buckwheat, millets, sunflowers), insects and other invertebrates (such as spiders), green forage (such as clover leaves), and tubers (chufa). Both warm- and cool-season food plots should be considered when managing property for wild turkeys because each provides something useful for turkeys during different seasons.

Warm-season food plots for wild turkeys

The primary benefit of warm-season food plots for wild turkeys is realized during fall and winter. A high-energy food source during cold periods when naturally occurring foods may be limiting can replenish body fat and reduce movements (travel), which can lead to increased winter survival. Corn, soybeans, grain sorghum, and chufa are the primary options. Other warm-season plantings, such as millets and buckwheat, may be used by wild turkeys, but the seed value (energy and fat) is not as great nor are those seeds as available into fall and winter.

Managing corn and grain sorghum plots for wild turkeys

Follow the planting and weed control recommendations for corn and grain sorghum provided in *Chapter 8*. When managing specifically for turkeys, use the widest recommended row spacing (30 inches) and the lowest recommended population seeding rate. This allows more space between plants and makes the structure more favorable for wild turkeys, yet still provides good grain yield. Using relatively short corn varieties that produce corn ears 2-3 feet aboveground is another consideration that can make a corn plot more attractive and usable by wild turkeys. Large varieties with stalks 7-8 feet tall and corn ears 4-5 feet aboveground are not as attractive and the grain not as accessible for turkeys. Likewise, relatively short (3 feet tall) varieties of grain sorghum allow better visibility for turkeys and makes the seed more easily obtainable. If you want grain sorghum to be available to turkeys into winter, choose a variety with higher tannin content that will resist bird depredation. Tannin content drops considerably after a couple of frosts, increasing palatability of red and brown sorghums during late fall and winter. If depredation from blackbirds is not a problem in your area and you want turkeys to get in the grain sorghum soon after maturation, choose a white grain sorghum, such as Hegari, that contains fewer tannins than red or brown grain sorghums (see sidebar *Grain sorghum or milo – which is it?* on page 191).

Although wild turkeys will knock down corn and grain sorghum stalks to get to the grain, some people try to make the grain more accessible by silage-chopping, mowing (bushhogging), or simply knocking the stalks over. If you do this, do not silage chop or mow more than a few rows at a time depending on size of the plot. Corn and grain sorghum seed deteriorate and decompose quickly during winter when cut or knocked over (as opposed to left standing). Silage-chopping and mowing through winter makes corn available for many other bird species, especially nongame birds. Of course, where hunting is a consideration, silage-chopping and mowing can be continued through turkey season (check your state's regulations to make sure this is legal in your state) to help keep birds in the area. Corn fields can be burned in late March/early April to make the corn readily available and to prepare the field for planting another warm-season plot, if that is an objective. Another strategy is to consider paying a local producer to leave a portion of a cornfield unharvested, which could be cheaper than if you planted a plot yourself.

Fig 9.3; Corn plots do not have to be silage-chopped or mowed for wild turkeys. This plot (in March) was left standing through winter and turkeys have been feeding in the plot all winter long. Corn plots at this time of year can be a primary food source and very attractive to wild turkeys.

Fig 9.4; Corn plots allowed to set fallow the year after planting can provide outstanding brooding cover for wild turkeys. Depending on the forbs that respond from the seedbank, the perfect "umbrella" cover develops, protecting broods while they feed upon bugs and various seed below.

Corn or grain sorghum plots may not receive much use when acorns are plentiful. When this occurs, don't be discouraged. Monitor the acorn crop. By mid- to late winter, turkeys are usually looking for an additional food resource and will eat corn and grain sorghum from standing stalks or knock them over. You can silage-chop additional rows, or just leaving them standing. **I leave them standing.** If turkeys want or need the corn, they will get it. You don't have to knock it over or mow it. Leaving it standing provides a good measuring stick for how much you need. If you consistently have standing corn left over, you don't need to plant that much.

Crop rotation is important. One way to rotate corn and grain sorghum plots is to allow them to set fallow through the following summer, then plant a cool-season plot, such as clovers. Fallow corn plots provide outstanding brooding cover as "weeds" establish in the plot. Toward the end of summer (2-3 weeks before dove season), silage-chop or mow the plot. Later, you could plant a cool-season plot. Rotating grass crops with legumes is recommended.

Planting and managing soybeans for wild turkeys

Follow the planting and weed control recommendations for soybeans provided in *Chapter 8*. Soybeans are planted for wild turkeys to provide a high-energy seed during fall and winter. Turkeys may forage for insects in soybean plots, but the primary consideration when growing soybeans for turkeys is seed production. Therefore, they should be allowed to stand through winter, or at least as long as seed are available. The height and structure of vegetation highly influences use of an area by turkeys. When planting soybeans for turkeys, I like to use an MG 4 or 5 soybean with a relatively short bushy canopy and plant on a wider spacing (30-inch rows or broadcast) with lower plant populations (50,000–80,000 seed per acre, or 80–85 pounds per acre broadcast).

Fig 9.5; Standing soybeans are highly attractive to wild turkeys. Soybeans are a high-energy food source that will hold turkeys in an area as long as the beans are available and the birds are not disturbed. Turkeys have been feeding in this plot regularly through winter (seen here in February).

Planting and managing chufa for wild turkeys

Chufa is a variety of yellow nutsedge and a very popular planting for wild turkeys. Turkeys scratch-up and eat the nutlike tubers produced among the roots of chufa. Chufa grows well in a variety of soil types. However, it is more difficult for turkeys to scratch down into heavy clays and turn up the roots and get to the tubers than when chufa is planted in loamy soils. I do not recommend planting chufa in clay soils. If you do, you will need to disk the plot in the fall to make the tubers available. Maximum tuber production usually occurs in sandy loam soils and bottomlands with full sunlight where moisture is not limiting. Drought conditions severely reduce tuber production and plant survival.

Chufa can be planted with a drill or planter, or the tubers can be broadcast and covered 1-2 inches by disking. Planting rate is approximately 50 pounds per acre. Chufa does best in fertile soils; therefore, P and K should be raised accordingly (50 and 160 pounds available per acre, respectively). Top-dress with a nitrogen fertilizer (60 pounds N per acre) when plants reach approximately 6 inches in height and rain is in the forecast. Chufa matures approximately 100 days after germination. "Clean" chufa plots typically produce greater yields than weedy plots. Planting chufa in rows allows cultivation for weed control. Combining cultivation with the appropriate herbicide applications is the best strategy for clean, productive chufa plots. Many forb weeds can be controlled postemergence with 2,4-DB (use Butyrac 200, not 2,4-D as it may kill chufa, especially at 2- to 3-quarts per acre), Banvel, or Clarity. Problem grasses can be controlled postemergence with Clethodim or Poast. When growing chufa, it is important to rotate the crop each year, which will encourage healthier plants and help manage plant density. Chufa plots typically volunteer the year after planting.

Other warm-season plantings for wild turkeys

Wild turkeys will use and benefit from most warm-season plantings, whether for the seed produced, invertebrates available, or for brood cover. The structure of the planting,

Fig 9.6a, b, c, and d, from top; A field of chufa can be a magnet for wild turkeys. Note the sandy-loam soil that chufa is best adapted to and in which turkeys can easily scratch-up the tubers. Plants develop through summer then die in fall when tubers are well-developed and most attractive to turkeys. Planting chufa with a planter enables cultivation for weed control. The energy-rich tubers should be available throughout fall and winter.

Fig 9.7; A relatively light rate of American jointvetch along with buckwheat provides food with excellent cover for broods through summer.

however, influences use greatly. Turkeys like open structure at ground level and will not use plots with dense, tall vegetation as much as those with a canopy 2-3 feet high with a relatively open structure underneath. Structure is an important consideration when deciding what to plant and planting rate.

Southern brooding plot
10 pounds American jointvetch (*Aeschynomene*)
8 pounds buckwheat

Benefit and considerations: If a relatively low rate of jointvetch is planted (10 pounds), the structure provided by this mixture is suitable for brooding wild turkeys — open underneath with an overhead canopy for poults. Buckwheat germinates and grows fast, maturing and providing seed for turkeys within 60 days. It also attracts lots of insects, which supply protein and calcium for young poults. If planted in April or early May, this mixture coincides well with the food requirements of developing wild turkey poults. Insects attracted to the plants are available soon after planting and buckwheat seed is available by the time most poults are 1 month old, when poults are transitioning from a diet composed primarily of insects to seed and other plant parts. Later, growth of jointvetch provides good overhead canopy structure for foraging broods.

Management: Kill existing vegetation with a glyphosate herbicide in the spring prior to seeding. Grass-selective herbicides can be applied postemergence to control grass weeds if sprayed when weeds are young.

Cool-season food plots for wild turkeys
Annual cool-season plots may provide foraging opportunities through fall and winter and, depending on what is planted, may provide a seed source (such as wheat) the following summer. Perennial cool-season plots may provide forage in fall, spring, and summer. Clovers and alfalfa are most readily eaten and the insects associated with these plots are a critical source of protein and calcium for wild turkeys. Chicory leaves also are readily eaten by wild turkeys and if allowed to bolt and produce seed, the structure of flowering chicory is desirable for brood cover.

Management and arrangement of cool-season plots greatly influence use by wild turkeys. Perennial cool-season plots should not necessarily be managed as those for white-tailed deer. The presence of certain weeds, such as ragweed, pokeweed, crotons, goldenrods, and horseweed, can make clover/alfalfa plots more attractive for broods

150

during summer. Instead of keeping perennial plots "clean," you should tolerate a little coverage of certain weeds for the structure they provide. Enhanced structure enables perennial cool-season plots to provide cover as well as forage and insects, which is very important for broods. Mowing should be limited. Mowing once in late summer

Fig 9.8; Clover plots are highly attractive to wild turkeys, both for the forage and associated insects. Note the additional habitat management for wild turkeys on this property (burning woods in the background and a standing "weedy" cornfield).

Fig 9.9; This woods road planted to ladino clover is highly attractive to wild turkeys. Foraging and bugging opportunities are perfectly juxtaposed with a well-developed understory in the adjacent woods, which are burned every 2-3 years.

Fig 9.10a and b; It's OK to have a few "weeds" in a perennial plot for turkeys. In fact, it's good! Plants such as common ragweed and horseweed provide structure that make the plot more attractive to broods. These annual forbs can be mowed late in the growing season before they go to seed. The picture on the top was taken in early August and the plot had two broods in it. We mowed the plot in late August. The following spring (bottom), the perennial clovers are looking great.

(August) will prevent many of the annual weeds, such as ragweed and horseweed, from seeding and becoming too dense. Perennial weeds, such as pokeweed or goldenrod, can be reduced if they are too dense by spot-spraying with a selective herbicide, such as Pursuit or 2,4-DB in late spring. Annual cool-season plots should be allowed to go fallow through summer to provide good brood cover.

Arrangement of cool-seasons plots is important. Relatively large blocks (greater than 2 acres) are not as attractive as smaller plots or those that are more linear or irregularly shaped, especially for broods. Broods younger than 2 weeks old cannot fly. Linear plots adjacent to or within early successional cover will receive much more use than plots adjacent to areas less attractive to broods. Planting firebreaks around and through early successional areas is an excellent strategy. Do not overlook the importance of high-quality early successional cover for wild turkey broods. Food plots are not a surrogate for old-fields and other early successional plant communities! Although high-quality foods (forage and insects) may be provided in cool-season plots, the structure at ground level is relatively dense. That is why it is typical to see older broods (4 to 12 weeks old) frequenting perennial clover and alfalfa plots, whereas younger broods are most often associated with areas that provide good overhead cover with an open structure at ground level.

Cool-season mixtures

Clovers and alfalfa should represent the backbone of any cool-season mixture planted for wild turkeys. Not only do they provide outstanding forage, but they also attract grasshoppers and other insects that turkeys relish. I also add wheat or oats when planting cool-season mixtures for turkeys. The seed is highly attractive to turkeys and the height of wheat and oats is perfect for brooding cover. Rye and triticale grow too tall (5-7 feet) and the resulting thatch creates an undesirable structure and may smother perennial clovers and alfalfa underneath. The height of oats is good, but oats are more susceptible to winter kill, especially above Zone 7.

Perennial bugging plot
  10 pounds alfalfa
   5 pounds red clover
   3 pounds chicory
  30-40 pounds wheat or oats

Benefits and considerations: There is no other plant that attracts grasshoppers and other insects as much as alfalfa. Wild turkeys love it. Both alfalfa and red clover provide excellent forage that turkeys readily eat. Adding red clover in the mixture helps diversify the structure, making it more attractive for turkey poults. The larger and taller stems of alfalfa and red clover (as opposed to ladino white clovers) make them perfectly suited for brooding. About 30 pounds (PLS per acre) of wheat or oats when planting this mixture for turkeys works well. The wheat not only provides forage and seed production but also provides a "nurse" crop to help the alfalfa and red clover establish.

Fig. 9.11; It is early August and wild turkey broods have used this plot of alfalfa, red clover, chicory, and wheat extensively. It may look "unkempt" to some, but this mixture provides exactly what wild turkey broods require. Insects, good forage, and favorable structure make this mixture very attractive to wild turkeys.

Management: Be sure to plant this mixture relatively early (August up North; September down South) so the alfalfa and red clover can become well-established before cold weather. Before committing to planting this mixture, be sure to realize the effort required to maintain alfalfa. It is sensitive to acid soils and low fertility, and alfalfa weevils can become problematic. You will need to raise soil pH to 7.0 and adjust macro- (phosphorus and potassium) and micronutrients (especially sulfur and boron) according to soil test. Insecticides will be necessary to combat alfalfa weevil infestations in spring (4-8 ounces per acre of Way-Lay 3.2 AG has been successful). If you don't think you want to work with alfalfa, you can replace the alfalfa with 5 pounds of ladino white clover.

Most undesirable broadleaf weeds can be controlled with imazethapyr or imazamox postemergence in spring. Problem grasses can be controlled with a grass-selective herbicide if sprayed before they reach 6 inches in height. Grass-selective herbicides can be tank mixed with imazethapyr or imazamox. Top-dress the plot with lime and fertilizer in August or September as recommended by a soil test. Do not mow this plot until late summer! Allow the chicory to bolt and flower and remember, certain "weeds" can make this plot more attractive for wild turkeys because of the structure they provide.

Fig 9.12a, and b; Relatively small plots of crimson clover and wheat are used extensively by wild turkeys. There were turkeys feeding and strutting in this plot (top) just before the picture was taken (April). Relatively light coverage of weeds that are tall in structure, such as ragweed and horseweed (bottom), can be left standing through the brooding season to make the plot more attractive to broods (picture taken in October). Both of these plots were in their third season with no additional seed sown.

Strutting and brooding plot
40-50 pounds wheat or oats
15 pounds crimson clover

Benefits and considerations: If you are south of Zone 5 and want a "green plot" for turkeys to feed, strut, and display in during late winter/early spring, this is it! No other clover will attract turkeys more during spring than crimson. After the crimson clover dies (May), the plot will be attractive through midsummer as the turkeys feed on the wheat or oat seedheads. Be sure to use an awnless (or beardless) variety of wheat. Varieties with long awns on the seedheads are not eaten as readily by turkeys or other wildlife species. If undesirable weeds are not problematic and you want to maintain clover in the plot through the summer, add 5 pounds PLS of red clover to the mixture.

Management: This mixture establishes very quickly and crimson clover is an excellent re-seeder. You can retain crimson clover for years without replanting if it is not overgrazed and it flowers and produces seed. When managing specifically for turkeys, I manage this mixture a little differently than the *Best Annual Forage Mixture* for deer. After the wheat or oats mature in early June, monitor incoming weeds. If desirable weeds are present (such as ragweed, pokeweed, and crotons), sit back and allow them to grow and provide the best brooding cover you could ask for! If undesirable weeds (such as horsenettle, curly dock, morningglories, and sericea lespedeza) are problematic, you have some herbicide options. Of course, glyphosate can be used to control forb and grass weeds. Forb-selective herbicides (such as Harmony Extra, Clarity, and 2,4-D) can be used to control broadleaf weeds. If both desirable and undesirable broadleaf weeds are present, spot-spray the undesirable ones and allow the desirable ones to continue to grow. Undesirable grasses can be controlled with a grass-selective herbicide. In late summer, mow the plot. Mowing will "release" the re-seeding crimson clover. **Note:** if undesirable broadleaf weeds are present, spray them in summer before the crimson clover germinates. Top-dress with lime and fertilizer (if needed) as recommended from a soil test. Additional wheat or oats can be top-sown before mowing or drilled into the plot after mowing. Typically, a fair amount of wheat or oats will re-seed into the plot and do not need to be drilled into the plot until year three.

154

Using wheat for summer attraction and weed control

If you have fields with tough-to-control broadleaf weeds, you should consider planting wheat by itself. Not only does wheat provide an attractive forage through fall and winter, but cool-season broadleaf weeds are easily controlled with Harmony Extra, Clarity, or 2,4-D. After the wheat matures, the wheat seedheads are readily eaten by turkeys and other wildlife (be sure to plant an awnless variety of wheat). Warm-season weeds can be controlled after the wheat matures with glyphosate. If glyphosate-resistant weeds are present, you can use any forb-selective herbicide labeled for fallow areas. Thus, you have lots of options and are able to provide a high-quality food source and "clean-up" the field at the same time. If allowed to continue to set fallow, these fields can provide excellent brooding cover the second summer after planting as more desirable forbs become established from the seedbank.

If you plant wheat specifically for wild turkeys, use a lighter seeding rate (80 pounds per acre) as opposed to a heavier seeding rate for deer forage production (120-200 pounds per acre). A relatively light rate will encourage better growth from the seedbank for brooding cover after the wheat matures and will not have as much thatch at ground level.

Fig 9.13; Do not overlook the value of wheat and oat seed when managing for wild turkeys. This gobbler's crop was full of oat seed, which was growing in a mixture of clovers. Photo by Ryan Basinger.

Fig 9.14; Wheat plots allowed to set fallow can provide outstanding cover for wild turkey broods. Forbs, especially ragweed, provide superb umbrella cover for broods moving about underneath, while they feed on seeds and invertebrates.

Using prescribed fire for wild turkeys

There is no other management practice that will enhance habitat for wild turkeys as much as prescribed fire. Prescribed fire is commonly used in Southern pine stands (Fig 9.15a), the Great Plains (Fig 9.15b), and in early successional communities (Fig 9.15c) to improve cover for nesting and brooding and increase food availability for wild turkeys and several other wildlife species. Fire also can be used for the same effect in upland hardwoods (Fig 9.15d). Before burning, it is important to prepare a burn plan and review regulations with your state forestry agency. In the plan, outline your objectives clearly, which is especially important when timber management also is a consideration with wildlife management objectives.

Fig 9.15a-d (clockwise from top left); Southern pine stands (a), the Great Plains (b), early successional communities (c), and upland hardwoods (d) managed for wild turkeys. Great Plains photo by Dwayne Elmore.

When planned and implemented accordingly, prescribed fire can be used to influence the composition of regeneration, improve understory structure for nesting and brooding wild turkeys, and protect the stand from intense wildfire.

Disturbing nests often is a consideration when using prescribed fire in spring (see nest in lower right of Fig 9.16a). My graduate students and I have found the effect of prescribed fire on vegetation composition is similar when implemented during the dormant season or during the early growing season (April – May). That is, top-killed woody stems resprout, and there is no appreciable difference in composition of herbaceous plant response. Therefore, when managing specifically for wild turkeys, we recommend burning prior to the nesting season or during the latter portion of the growing season (July – October).

Fig 9.16a; Note turkey nest and eggs in right corner; nest was abandoned after fire.

Fig 9.16b; Burning during the late growing season (July-October) sets-back succession, limits encroachment of young trees, stimulates forbs for brooding cover, and does not disrupt nesting.

Just as when burning a field, it is critical to establish firebreaks when burning woods. According to terrain and woody stem density, this may be accomplished with a tractor and disk (Fig 9.17a, top right), a dozer (Fig 9.17b, second from top), or just a leaf-blower (Fig 9.17c, third from top). With forethought, firebreaks can be planned so they provide access throughout your property. Of course, firebreaks must be cleared of debris before burning. Here (9.17d, bottom), we used a leaf blower to clear leaves from the firebreak before burning. If wide enough, firebreaks can be planted and serve as a woods road (see Chapters 11 and 14).

Fig 9.17a-e (from top).

Fig 9.18a-c (left to right); Removing slash and other debris from the base of hardwoods after harvest or thinning helps protect them from prolonged heat when burning that can damage the cambium layer inside the bark and scar the tree, such as this white oak (a). Fire scars allow fungus to enter the tree and lead to tree mortality (b). However, many fire-scarred trees heal and are fine (c).

Figs 9.19a, b; Low-intensity prescribed fire in upland hardwoods (a, left) should not damage overstory trees (b, right) unless there is relatively large debris adjacent to the trunk.

Ten

Chapter 10 — Northern Bobwhite and Cottontail Rabbits

Northern bobwhite

Northern bobwhite (a.k.a. bobwhite or quail) eat a wide variety of seed from naturally occurring plants, as well as planted forbs and grasses. Food, however, is rarely a limiting factor for bobwhite. A lack of suitable cover usually is what limits bobwhite populations. Housing developments, shopping centers, closed canopy woods, and tall fescue (or bermudagrass, dallisgrass, bahiagrass) hayfields and pastures do not represent suitable cover for bobwhite! Cover also is the limiting factor for quail on farms with large crop fields (greater than 10 acres) and "clean" fencerows, creekbanks, and ditches. Moreover, it is important to realize you can have the best quail cover in the county and still not see any bobwhite if your property is only a few acres and you are surrounded by nonhabitat. Food plots will never help quail in these situations, even if you trapped every coon and shot every hawk you saw!

Bobwhite require all of their habitat needs in close proximity. That means brooding cover, loafing cover, and escape cover all must be close together (generally, within 40-80 acres). Therefore, to increase bobwhite populations, you should concentrate on enhancing cover and the proximity of required cover to meet the year-round needs of these game birds.

Bobwhite nest in a variety of grasses and other vegetation. Native grasses, especially broomsedge bluestem, little bluestem, and sideoats grama, have been promoted for nesting cover. However, recent research indicates various grasses are not required for bobwhite to nest successfully. Quail broods frequent "weedy" fields and areas of sparse brush/shrubs as they search for insects and other invertebrates. Areas managed for brood cover should be relatively open at ground level with a canopy of forbs overhead. This type of environment enables

Fig 10.1a,b,c; These scenes are typical across the northern bobwhite's distribution in the eastern US, and people wonder why they don't see quail anymore. Hello?!?! Closed-canopy woods, housing developments, large double-cropped fields, nonnative pasture grazed to the ground, and a landscape maturing to forest do not support quail! Food plots do not help quail in these situations. A change in land-use practices and successional stages is necessary to reverse the trend in declining bobwhite populations.

Fig 10.2a,b; Plant communities dominated by native forbs and grasses with scattered brambles and shrubs provide high-quality habitat for northern bobwhite. Population response to management is greatest in open landscapes dominated by such cover, not forests or other nonhabitat. This is true whether you are in the South (Kyker Bottom Refuge, Tennessee; top) or Midwest (Black Kettle WMA, Oklahoma; bottom), but proper habitat management is a tremendous limitation in the eastern US where annual precipitation exceeds 30 inches per year.

quail chicks in search of invertebrates to move about easily while protected by the "umbrella cover." Forbs that should be encouraged for bobwhite include ragweeds, pokeweed, old-field asters, partridgepea, beggar's-lice, native lespedezas, geraniums, milk pea, butterfly pea, perennial sunflowers, smartweeds, 3-seeded mercury, and crotons (see the *Plant Identification Guide*). Grasses that provide seed for bobwhite include annual panicgrasses and foxtail grasses. Brambles and shrubs, such as blackberry, sumac, wild plum, and elderberry, scattered throughout an area provide protective cover for loafing and escaping predators. Thickets of dense brush and blackberry provide critical winter cover bobwhite need to escape predators and harsh weather. Shrub cover must be adjacent to or within potential loafing and feeding areas.

As you can see, having the right plant composition is only one piece of the habitat management puzzle for quail. *Arranging the appropriate successional stages and vegetation types correctly* within a relatively small area is equally important. It is essential to understand how desirable vegetation types provide food for bobwhite. Food is usually provided **by default** when desirable cover is present.

Desirable cover for quail can be created by killing nonnative perennial grass cover. Perennial cool-season grasses are killed most effectively by spraying a glyphosate herbicide (1.5-2 quarts per acre) in the fall after a couple frosts as the grasses prepare for winter dormancy. Most desirable native vegetation will not be harmed by this application, because warm-season annual plants will be dead and most perennial warm-season plants will be dormant. Selective herbicides (such as Plateau) also can be used to kill tall fescue when desirable forbs and grasses tolerant of imazapic (see herbicide label to identify forbs tolerant of imazapic) are present. Techniques used for eradicating bermudagrass and other undesirable warm-season grasses are described under *Preparing the site — controlling existing vegetation before planting* on page 40.

Fig 10.3; Availability of early successional plant communities with desirable plant species is a major limiting factor for bobwhite throughout the South. This 25-acre field used to be covered with tall fescue and orchardgrass. At that time, bobwhite did not use the field. After killing the tall fescue and orchardgrass with Roundup and Plateau, native grasses and forbs were established. The field now supports two 10-15-bird coveys. Note how various sections of the field are divided with firebreaks (see aerial photo on page 175). This arrangement allows burning and/or disking various sections every two to four years and provides nesting cover, brooding cover, loafing cover, roosting cover, escape cover, and food all within the same field.

 After undesirable plant cover has been killed, desirable plants in the seedbank usually can be stimulated to germinate by disking November through February in the Deep South, through March in the Mid-South, and through April farther north. Disking in April (and later in the spring) is not recommended in the South, as undesirable species, such as broadleaf signalgrass, johnsongrass, crabgrass, goosegrass, sicklepod, and thistles, often are stimulated by disking at this time and can dominate the area *if they are present in the seedbank.* Disking is not an important management practice in arid regions where there is adequate bare ground and succession is much slower. Cover around crop fields can be enhanced by allowing a 50-foot-wide border of grasses, forbs, brambles, and shrubs/brush to develop. Tree invasion into buffers can be controlled with an application of Arsenal AC or Garlon 3-A (see *Appendix 2*). Winter cover for bobwhite can be enhanced along the field/woods interface by killing undesirable trees and allowing additional brush to develop approximately 100 feet into the woods around the field.

Fig 10.4; Bobwhite populations have increased in many areas throughout the South and Midwest after establishing/promoting field borders of native grasses and forbs around crop fields. The concept is simple: bobwhite populations increase when additional habitat is provided. Here, a border of broomsedge, brambles, and forbs has been allowed to establish around a soybean field, providing cover and food for quail.

Once you have established suitable cover, the best way to maintain it is with prescribed fire. Disking also can be used to influence plant composition and structure in many areas. **However, prescribed fire is strongly recommended to manage habitat for bobwhite.** Burning every two to four years in areas with abundant precipitation consumes dead vegetation, recycles nutrients, stimulates fresh plant growth, creates an open structure at ground level, makes seed and invertebrates more available, and maintains an early-successional plant community.

Before burning, establishing a firebreak is necessary to keep fire out of areas not intended to burn. Firebreaks can serve two purposes. In addition to restricting fire from spreading into unwanted areas, firebreaks can be planted to provide additional food resources (see *Chapter 11*). Thus, *linear strips of food* surrounding and within *blocks of cover* represent good habitat arrangement for bobwhite.

Whatever you do, **do not plant tall fescue, orchardgrass, bromegrasses, timothy, or bluegrass in your firebreaks!** These nonnative perennial cool-season grasses can be detrimental to bobwhite and other wildlife for several reasons. To begin, they displace good brood-rearing cover described previously. Because of their sod-forming nature, a dense "carpet" is created at ground level. This dense structure at ground level limits mobility of quail chicks, as well as wild turkey poults. Also, fields dominated by these nonnative grasses typically do not harbor as many invertebrates as fields predominately composed of forbs, providing less food for young game birds. Broods have to spend more time searching for food when invertebrates are less abundant, which leads to increased energy expended and increased exposure. Later in life, when seeds become more important in the diet, the thatch produced by nonnative sod grasses limits seed availability (if any is present). Furthermore, consumption of tall fescue seed by bobwhites may lead to weight loss, cloacal swelling, and ultimately, increased mortality. For these reasons, it is obvious that establishing nonnative sod grasses degrades habitat for bobwhite significantly and, over time, can lead to increased mortality and reduced recruitment into the fall population.

Fig 10.5a,b (from top); Prescribed fire is the preferred technique to set-back succession and maintain bobwhite habitat. Here, we burned during the late growing season (a; early October 2011) to set-back succession where trees were becoming the dominant cover (b; April 2012).

Fig 10.6a,b; This is the structure presented by tall fescue (left) and orchardgrass (right). These grasses inhibit travel for young quail and wild turkeys, reduce seed and invertebrate availability, and inhibit the seedbank from germinating. These grasses are not "wildlife friendly" and should be eradicated.

A word about native warm-season grasses…

Native warm-season grasses — particularly big bluestem, little bluestem, broomsedge bluestem, indiangrass, sideoats grama, switchgrass, and eastern gamagrass — have been promoted to provide cover for wildlife. Bobwhite frequently nest at the base of various native grasses, especially broomsedge bluestem, little bluestem and sideoats grama. However, these grasses are not necessary for successful nesting as bobwhite nest successfully in a wide variety of grasses and forbs. Relatively sparse native warm-season grass cover (20-50 percent grass coverage) allows a variety of forbs and shrubs to develop throughout the field (50-80 percent forb coverage with shrub cover interspersed throughout), providing outstanding cover, forage, and seed production. A field of grass is **not** desirable when managing for northern bobwhite. In fact, research has shown bobwhite avoid areas of dense planted native warm-season grasses. Sparse grass cover with an open structure at ground level provides bobwhite space to travel, dust, and feed throughout the field, not just along the edge. White-tailed deer select old-fields and other early successional communities for fawning sites. Deer also graze various forbs throughout spring and summer. Native warm-season grasses are not an important forage for most wildlife species. They are promoted because of the cover and structure they provide, not food. However, with the exception of grassland songbirds, no wildlife species in the eastern US requires more than 30% coverage of grass. To learn more about early successional plant communities and associated management, refer to UTIA publication *Managing early successional plant communities for wildlife in the eastern US* (bit.ly/earlysuccession).

Fig 10.7a (top), b (below); Bobwhite and many other birds may nest within native grasses. However, to provide cover and food for a wide variety of species, including bobwhite, rabbits, wild turkey, and white-tailed deer, a field of native grass is not desirable. Rather, the field may include native grasses, as well as forbs and shrubs. It is most important to realize that where favorable native grasses, such as broomsedge bluestem, and forbs already occur (whether standing or in the seedbank), planting native grasses and forbs is not necessary! Burning and/or disking every two to five years in most areas promotes desirable plant composition, with an open structure at ground level and an umbrella canopy overhead. Neither of these fields was planted. Native vegetation was promoted through selective herbicides and fire after tall fescue was eradicated.

Want more quail?

If so, then you first should understand how quail **require** an open landscape for healthy populations to persist. **The northern bobwhite is not a woods bird!** If you find bobwhite using closed-canopy woods, that should be an immediate indication of poor-quality habitat. Look around your property. Do you see lots of woods? What about on the properties surrounding you? If you are located in a forested landscape, there may be nothing you can do to increase wild, naturally occurring bobwhite. In general, a minimum of 500 acres should be considered for bobwhite management.

Photo by Dale Rollins

And on that 500 acres, optimally, there should be no closed-canopy woods. Forested areas, including hardwoods and pines, should be cleared or thinned to allow at least 50 – 80% sunlight into the stand. Then, the understory must be managed with frequent prescribed fire to maintain a plant community desirable for bobwhite and not allow a young forest to regenerate. (**Note:** are you really interested in managing your property for quail? If so, read this paragraph again and take a look on Google Earth and see what your property and the surrounding properties look like.)

Next, consider your open areas. Do you see tall fescue, orchardgrass, timothy, bromegrasses, dallisgrass, bahiagrass, and/or bermudagrass? If so, then your next step toward enhancing bobwhite habitat and increasing the bobwhite population should be eradicating these grasses and converting those areas into suitable cover. Don't be afraid if some brambles and shrubs establish. Bobwhite need them for loafing and escape cover, as well as the food they provide. Do you have areas of dense native grasses? Perhaps you planted them. Regardless, you need to reduce grass density in those areas by using selective herbicides or disking. Bobwhite don't have to have grass! If you have more than 30% coverage of grass in open areas, you can make those areas much more productive for bobwhite by reducing the grass and promoting more forb cover. And make sure protective shrub cover is not more than 100 yards apart.

Perhaps this seems extreme. Well, to see a rebound in bobwhite populations in areas where they used to be abundant but now absent, extreme measures are needed. Not planting food plots. And there is no need to worry about eyeworms, diseases, fire ants, raccoons, and hawks. None of that matters if there is no habitat! People quickly discover that while it may not be possible to control predators, they can control predation to an acceptable level for wild bobwhite by establishing and maintaining the type of cover bobwhite require. And finally, yes, releasing pen-reared birds can provide some fun shooting, but they cannot be used to re-establish quail populations. If so, quail would be everywhere! Their survival rate is simply too low.

Cottontail rabbits

There are four species of rabbits that are commonly managed for in the eastern U.S.: the eastern cottontail, Appalachian cottontail, swamp rabbit, and marsh rabbit. The eastern cottontail is by far the most prevalent and is found throughout the eastern U.S. Each species uses slightly different types of cover, but early-successional vegetation managed with periodic disturbance is important for all four. Rabbits primarily eat green forbs and grasses when available, but they also eat bark from saplings and browse when necessary, especially fall through winter. Forage food plots can benefit rabbits greatly at this time (similar to deer in some regions).

Photo by Mark Cunningham

Rabbits are prolific breeders. A female rabbit may raise three to four litters of three to four young per year. This reproductive rate is quite necessary, as they are preyed upon heavily by a wide variety of predators. Less than 25 percent of all rabbits born each year survive until fall. Therefore, good escape cover is absolutely necessary to have a healthy rabbit population. You may wonder why so much space is being devoted to cover in a book about food plots. The reason is that you can have rabbits and quail without food plots *if you have good cover*, but you cannot have rabbits or quail without

Fig 10.8a,b; Planting firebreaks and other strips of food for rabbits is a useless exercise unless there is adequate cover year-round. The top photo shows the type of cover needed to support good rabbit populations. This firebreak (below) was planted to the Annual cool-season firebreak mixture *listed on page 176. High-quality cool-season forages (wheat and clover) adjacent to good cover and a soft edge of brambles represent the perfect arrangement of food and cover for rabbits. A field of cover with a strip of food will support many more rabbits than a field of food and a strip of cover. Again, good cover also provides food. And no, woods do not provide good habitat for rabbits. If rabbits are forced to use woods, they are dying quickly and the population is declining.*

good cover *even if you have food plots*. Again, managing food plots is only one habitat management practice, and it is not as important as developing suitable cover.

High-quality habitat can support at least three rabbits per acre. The lifetime home range of a rabbit is less than 25 acres in good habitat. Areas with considerable early successional cover, including abundant brushy thickets less than head high, are magnets for rabbits. Good early successional cover for rabbits can be created by eradicating nonnative perennial grasses, such as tall fescue and bermudagrass, and allowing various native forbs and grasses with scattered shrubs to establish. Fields of tall fescue can be especially detrimental to rabbits as the endophyte fungus associated with tall fescue is toxic to rabbits, leading to lower weights and smaller litters, not to mention the poor structure at ground level and lack of overhead cover afforded by tall fescue.

If you are interested in maximum rabbit production, manage succession by burning every three to five years (in most areas). Depending on the size of the area, you can manage it in several sections, 2-5 acres each. By burning sections on rotation (at least one-fifth of the acreage each year), you can provide different successional stages and vegetation types within a relatively small area.

Burning requires firebreaks, which can be planted to provide additional forage adjacent to cover. Both warm- and cool-season plots are beneficial, but high-quality cool-season forages are especially important for rabbits in fall and winter. Rabbits will eat the green forage of clovers, chicory, alfalfa, wheat, and oats, as well as peas, beans, and grain produced by warm-season plantings. Mixtures that can be planted in firebreaks to benefit rabbits are provided in the next chapter. Food plots that are square, or rectangular and larger than 1/4 acre are not as beneficial to rabbits as planted lanes (such as firebreaks 10-20 feet wide) adjacent to and within dense cover. Rabbits like to feed adjacent to cover, else they are exposed and overly vulnerable to predation.

Fig 10.9; Fallow corn fields provide outstanding cover for rabbits.

Eleven

Chapter 11 — Managing firebreaks for Quail, Rabbits, and other wildlife

Prescribed fire is the most effective and efficient management technique to set-back succession and enhance and maintain early successional plant communities. Prescribed fire also is recommended to improve habitat for many wildlife species in forested areas, such as upland hardwoods, Southern pines, and Ponderosa pine. Whenever controlled burning is conducted, it is critical to establish a firebreak around the area to help keep fire from spreading into areas where burning is not intended. Natural firebreaks include creeks and roads. However, most firebreaks around fields are created with a tractor and disk. Bulldozers may be used to establish firebreaks within forest stands (some of which may be used as woods roads). Another option for firebreaks in woods is a backpack leaf blower. Clearing a strip approximately 10 feet wide of leaves and other debris around the area to be burned can be an effective firebreak, whether burning woods or a field surrounded by woods.

Firebreaks in fields or other open areas can be managed in several ways. You can plant them or leave them fallow to allow naturally occurring legumes and other forbs to establish from the seedbank. Various sections of firebreaks can be planted with different forages to provide year-round food resources to benefit several wildlife species. An example of managing firebreaks with various plantings is shown in Fig 11.4.

Firebreaks around fields should be established well ahead of the projected burning date. If you plan to burn in March, you can establish your firebreak the previous August or September and plant cool-season forages, if desired. However,

Fig 11.1; Firebreaks are easily established around fields with a tractor and disk.

Fig 11.2; Firebreaks in woods can be established with a leaf blower. Small trees and debris can be cleared from the site. If completed in late summer, the firebreak can be sown to cool-season annuals, such as crimson clover and winter wheat. As the leaves drop, it is necessary to blow them off the planted lane or they will shade-out the clover and wheat.

if the firebreak is located adjacent to woods and you intend to burn the following March or April, you likely will need to disk after the leaves fall because fire will move across a firebreak filled with dead leaves and/or pine needles. In this scenario, it is best to plant a warm-season strip in spring after the field has been burned or allow the strip to remain fallow until the following fall when it can be planted to cool-season forages or left fallow. Regardless, when establishing firebreaks adjacent to woods or a tree line, **position the firebreak outside of the drip line of the trees** — at least 30, if not 50 feet, from the edge of the trees. Distance from the treeline is a major consideration overlooked by many people. Moving the firebreak away from trees will allow much better production from your planting, as there will be reduced competition for sunlight, water, and nutrients, and it will reduce leaf-litter buildup. It also will allow a soft edge to establish between the trees and the firebreak, which will provide desirable cover for quail, rabbits, and many songbirds.

Fig 11.3a and b (from top); Firebreaks can be planted or they can be left fallow to stimulate plants from the seedbank, such as the common ragweed growing in this firebreak, which provides outstanding cover, an open structure underneath, and a good seed source in late summer/fall. Sometimes managers add small amounts of seed to complement growth from the seedbank, such as a light rate of sunflowers or grain sorghum. The intent is not to have a pure planting of sunflowers or sorghum, but merely to add a little to the beneficial plants arising from the seedbank, such as the common ragweed shown here.

Firebreak width is another consideration. Most firebreaks are 1-2 tractor-widths or 1-2 disk-widths wide. Management may be much easier if you establish firebreaks the width of your sprayer. Firebreaks require spraying periodically, just like food plots. It is difficult and troublesome to spray a firebreak that does not "fit" the sprayer. Establishing firebreaks 1-2 sprayer-widths wide (depending on type and width of sprayer) enables you to spray without damaging nozzles or booms in tall vegetation. Disking a firebreak that is 1½ times as wide as a disk is much easier than spraying a firebreak 1½ times as wide as a sprayer.

Firebreaks often are planted to annuals. However, perennial mixtures are applicable if the field (or portion of) is not going to be burned for three to four years.

Otherwise, leaves and dead plant material accumulate in perennial firebreaks over time. **Thus, perennial mixtures usually need disking before burning. The notion of a "perennial green firebreak" is a myth — it will need re-disking before burning.**

This 25-acre field is managed for northern bobwhite and eastern cottontail (see Fig 10.3 on page 163). The plant community consists of various native grasses and forbs with scattered brambles and shrubs throughout the field. Brambles and shrubs include sumac, blackberry, wild plum, and elderberry. The field is managed in at least five sections, separated by firebreaks with 4 sections maintained by burning and/ or disking on a 2- to 4-year fire-return interval, and another section burned every 5 – 7 years to allow more brushy cover. Firebreaks are planted in various forages and grains to provide a supplemental food source throughout the year, juxtaposed to good cover. Warm-season annuals (not all in the same mixture) include

Fig 11.4

grain sorghum, Egyptian wheat, millets, iron-clay cowpeas, and soybeans. Cool-season forages include wheat, oats, crimson clover, arrowleaf clover, red clover, ladino clover, chicory, and alfalfa. The various sections of firebreaks are not always maintained in these plantings, but rotated around the field, depending upon which foods may be needed more than others, timing of management, weed pressure, and landowner objectives.

It is important to maintain some fallow firebreaks. Planned crop rotation and weed pressure determine which sections of firebreak remain fallow around this field. Although two coveys of northern bobwhite and scores of cottontails can be found regularly in this field, white-tailed deer, groundhog, raccoon, striped skunk, coyote, red and gray fox, evening bat, red bat, deer mouse, meadow vole, wild turkey, indigo bunting, field sparrow, yellow-breasted chat, loggerhead shrike, American kestrel, great horned owl, eastern kingbird, eastern bluebird, common yellowthroat, blue grosbeak, dickcissel, American goldfinch, box turtle, fence lizard, garter snake, rat snake, kingsnake, copperhead, and American toad also can be found using this field and benefit from the management efforts of the landowner.

Fig 11.5; Firebreaks that are to be planted can be established the width of your sprayer if weed control is a consideration. This firebreak was sprayed with Prowl and preplant incorporated prior to sowing iron-clay cowpeas and peredovik sunflowers.

Firebreak plantings
The "Best" annual cool-season firebreak mixture (PLS / ac)
15 pounds crimson clover
5 pounds arrowleaf clover
40 pounds wheat

Benefit and considerations: This mixture provides high-quality forage for rabbits, deer, wild turkey, groundhog, and other species. You can plant it in late summer or early fall after a late summer burn, or it can be planted in late winter in the South after burning in February or early March. If you plant in late summer and intend to burn the following winter, be sure to check for debris in the firebreak before burning. Never trust a "green" firebreak without checking for debris. It is not necessary to include the arrowleaf clover, but rabbits will use the overhead cover of arrowleaf clover as it grows over the wheat in June and July.

Management: After the arrowleaf clover matures and dies in July, you can spray the firebreak with glyphosate to kill weeds, or you can just mow in early August and allow

Fig 11.6; A mixture of crimson and arrowleaf clover and wheat works extremely well in firebreaks. Not only does it provide outstanding forage and a high-quality seed source, but it also provides a relatively safe and reliable "green" firebreak through early spring, if planted the previous fall. This firebreak was planted in late August. It provided forage through winter and spring. It is now late July. One strip (left side of picture) was mowed in June after the wheat had matured and dried. As you can see, it reseeded very well. There were more rabbit droppings in this firebreak than anywhere I have ever seen.

the crimson and arrowleaf clovers to reseed. Another option is to allow the firebreak to go fallow. The clovers will reseed, but better clover growth will be realized if weeds are controlled and the firebreak is mowed in early August.

"Just wheat" firebreak (PLS / ac)
   60-150 pounds wheat

Benefits and considerations: There is nothing wrong with planting wheat by itself in a firebreak. High-quality forage is provided through winter and a good seed source is available for many species the following summer. I like to use a higher rate (120-150 pounds) when managing the firebreak for deer or rabbits, and a lower rate (60 pounds) when managing for northern bobwhite or wild turkey. Nongame birds use both, but more grain will be available with the heavier seeding rate. The lower rate provides better structure at ground level for birds. Wheat is a better choice than oats in firebreaks because of the value of the wheat seedheads and because wheat is less susceptible to winter-kill from Zone 7 northward. Winter-killed leaves of oats burn readily and a fire is likely to creep across a firebreak with dead leaves (either from dead grass or fallen leaves from trees). Be sure to plant an awnless variety of wheat.

Fig 11.7; Firebreaks planted to wheat provide excellent forage through winter, and later the seedheads are eaten by many wildlife species. Such areas left fallow the following summer provide optimal brooding structure for northern bobwhite and wild turkey.

Management: Depending on seedbank composition, broadleaf weeds can be controlled with Harmony Extra, Clarity, or 2,4-D. If desirable forbs arise from the seedbank, do not spray. Increased growth can be achieved with a midwinter application of 30-60 pounds of N per acre. Allow the wheat to stand after it matures. Many wildlife species will eat the grain, especially if you plant an awnless variety of wheat. Wheat will re-seed fairly well. When it sprouts, top-dress with 30-60 pounds N per acre if the wheat is pale green. You can allow the firebreak to go fallow, or you can plant a cool-season forage the following late summer/fall.

Perennial forage firebreak mixture (PLS/ac)
    10 pounds alfalfa
    5 pounds red clover
    3 pounds chicory
    40 pounds wheat

Benefits and considerations: This perennial mixture is great to use where wild turkeys and white-tailed deer are focal species. The three perennial components provide outstanding forage for at least 3 – 4 years, making this mixture applicable for firebreaks around fields that will not be burned for several years. The wheat also provides grain that deer and turkeys eat, and there is probably no other plant that attracts grasshoppers as much as alfalfa. This mixture performs very well when planted around upland fields that are relatively dry. If you plant on a site that is more moist, or if you do not want to manage alfalfa, you can replace alfalfa with 5 pounds of ladino clover.

Fig 11.8; This perennial firebreak mixture provides high-quality forage and also offers favorable structure for game bird broods. Undesirable grasses (such as johnsongrass and crabgrass) were sprayed with Clethodim earlier in the season. Rabbits, deer, and wild turkey broods were using this section of the firebreak regularly.

Management: Grasses can be sprayed with a grass-selective herbicide after the wheat matures. Many undesirable forbs can be controlled with imazethapyr (Pursuit), imazamox (Raptor), or 2,4-DB. Broadleaf weeds should be sprayed when young. When managing firebreaks with this mixture, a few desirable forbs, such as ragweed, pokeweed, and goldenrod may appear. The structure provided by these forbs makes the firebreak more attractive for turkey broods. Unless necessary to keep undesirable weeds from producing seed, do not mow this mixture until August. Top-dress with lime and fertilizer in late summer as recommended by a soil test.

Vining legume firebreak mixture (PLS/ac)
  30 pounds Quail Haven soybeans
  5 – 10 pounds peredovik sunflowers

Benefit and considerations: Quail Haven re-seeding soybeans produce forage as well as a seed source for bobwhite and others birds. If you cannot find QH soybeans, use 75 pounds of iron-clay cowpeas instead. Sunflowers are included at a relatively light rate to provide substrate for the soybeans/cowpeas to vine up on and produce more forage than if there was nothing for them to climb. You can adjust the amount of sunflower seed based on your objectives. If birds are more of an objective, you can add up to 10 pounds of sunflower seed. If deer are more of an objective, I reduce the amount to 5 pounds.

Management: Prowl, Dual Magnum, and Treflan can be applied preplant incorporated for weed control. A grass-selective herbicide can be applied postemergence for additional grass weed control. Quail Haven soybeans have excellent re-seeding capabilities. If allowed to mature and produce seed, another stand can be stimulated by light disking. After the planting dies in the fall, there will be a lot of dead plant material in the plot. Just allow it to decompose through winter. Spray cool-season annual weeds with a glyphosate herbicide before they flower and produce seed. In March or April (according to your location), spray the firebreak with Prowl, Dual Magnum, or Treflan, apply fertilizer and/or lime as recommended by a soil test, and then disk the firebreak. You have effectively re-seeded your QH beans. You can use a grass-selective herbicide later (postemergence) if a grass weed problem develops. If you want to add sunflowers or grain sorghum to the firebreak, just sow them before disking or drill them after you have disked the QH beans. If you add grain sorghum, do not apply herbicides. If you choose to use cowpeas, you should reseed them the following year if you want to plant the same mixture because cowpeas do not reseed very well.

Annual "Upland Bird" firebreak mixture (PLS/ac)
  7 pounds Egyptian wheat
  7 pounds white proso millet
  3 pounds grain sorghum

Benefit and considerations: If you are looking to provide additional seed for birds in an area with a lot of deer, you should consider this one. Deer eat very little grass during summer; therefore, this mixture does well even where deer density is high. Seed production is excellent for a variety of game birds — including northern bobwhite, mourning dove, and wild turkey — as well as many songbirds, including northern cardinal, several species of sparrows, juncos, northern flicker, and others.

Management: Undesirable broadleaf weeds can be controlled with 2,4-D, Aim, Banvel, or Clarity.

Fig 11.9; Egyptian wheat grows tall and provides good cover for ground-feeding birds, such as northern bobwhite, northern cardinal, chipping sparrow, and white-throated sparrow. Mourning dove will forage where the structure is open at ground level with bare ground. Grain sorghum and various millets can provide additional seed. Several plants germinating from the seedbank (such as partridge pea) also may provide additional seed.

Annual "Bobwhite" firebreak mixture (PLS/ac)
  10 pounds Kobe and/or Korean lespedeza
  0.5 pound partridge pea

Benefit and considerations: Northern bobwhite relish seed from these annual lespedezas and partridge pea, which are available through winter, making areas planted to this mixture primary feeding spots from December through February. If not already present in the seedbank, partridge pea is a good addition to this mixture. Although they are annual species, these lespedezas and partridge pea reseed readily. Contrary to what you might have read or heard, these lespedezas do not provide high-quality forage for deer and are not selected forages by deer. However, partridge pea is a high-quality forage that is moderately selected by deer.

Fig 11.10; Seed from Kobe and Korean lespedeza and partridge pea are readily eaten by northern bobwhite through fall and winter.

Management: The best time to plant (inoculate seed, disk, top-sow, cultipack) is mid-February through early April, depending on your location, which coincides with when fields are typically burned to maintain early successional plant communities. In years after the lespedezas and partridge pea have established, they can be stimulated to reseed by light disking December-February. Weed-free lespedeza/partridge pea firebreaks are possible by spraying Pursuit preemergence. Postemergence applications of Pursuit over established firebreaks also will be effective if weeds are sprayed while they are young. Problem grasses can be controlled postemergence with a grass-selective herbicide.

Twelve

Chapter 12 — Mourning Dove

Mourning doves are attracted to many different seeds and grains, including sunflowers, millets, grain sorghum, corn, buckwheat, and sesame. Relatively large fields (5-20 acres or more) are recommended to attract large numbers of doves. Freshly cut grain fields are attractive areas for doves to feed. Doves do not scratch and are "weak-beaked." Therefore, they prefer loose grain (as opposed to corn still attached to the cob) or other seed in open areas with plenty of bare ground. Creating this environment is best accomplished by silage-chopping, burning, mowing, or disking to make seed available and feeding sites attractive. Burning is an excellent way to provide an open structure at ground level and make wheat and millet seed readily available. You cannot burn, however, unless there are adequate fine fuels (dead or dry grass and other debris) to carry the fire. A lack of fine fuels is not a problem in a mature millet or wheat plot.

In weed-free plots, such as corn fields without thatch, burning is not needed to enhance conditions at ground level. Silage-chopping or mowing will scatter seed and make them available. Silage chopping provides most desirable results.

Fig 12.1; Millet seed can be made more accessible for mourning doves by mowing or burning. Photo by Marion Barnes.

Doves prefer fields with some structure nearby, such as trees or powerlines, that allow them to perch and loaf near the field. An excellent way to provide perching sites is to plant dove fields along or adjacent to powerlines. If a powerline is not nearby, you can erect poles in a line through the middle of the field and string a cable from the top of each pole. Another way to provide perching sites is to kill trees (such as elms, yellow-poplar, sweetgum, and maples) along the edge of the field and leave them standing. Doves are attracted to dead trees around a good dove field like magnets. Dove hunters are usually "pulled" toward them as well! In addition to perching sites, a source of water and grit (such as a nearby gravel road) will make the area more attractive for doves.

Fig 12.2; Doves are strongly attracted to loafing lines within a dove field or dead trees along the edge of a dove field. Photo by Bill Smith.

Weed control can be a problem in dove fields, but be aware many naturally occurring "weeds" (such as pokeweed, ragweed, barnyardgrass, tropic croton, redroot amaranth [pigweed], geraniums, and foxtail grasses) produce seed readily eaten by doves and actually can make a food plot more attractive to these birds. Preemergence herbicides are recommended where weed control is necessary. Planting single species, instead of mixtures, can make weed control easier, depending on weed composition.

Consider planting grass crops (such as millets, grain sorghum, or corn) where certain broadleaf weeds (such as cocklebur and sicklepod) are problematic, and planting forbs (such as sunflowers, buckwheat, or sesame) where grass weeds (such as goosegrass and johnsongrass) are especially problematic. This strategy will allow you to use postemergence grass-selective or forb-selective herbicides for optimum control of problem plants (see *Appendix 2*).

Juxtaposing single-species plantings for doves

Many planting mixtures are possible for attracting doves. However, planting a section of a field in a single species adjacent to another section(s) with a different species is highly recommended when managing fields for doves. As mentioned above, there are real advantages in weed control when you plant single species, but the biggest advantage is timing maturation dates of different species so that a fresh seed source is available from early summer through winter. When planned properly, your management opportunities increase dramatically.

Buckwheat and browntop millet mature and provide dry seed within 60-65 days, sesame requires 100 days, sunflowers approximately 110 days, and grain sorghum and corn will need 100-145 days, depending on variety. When several species with different maturation dates are mixed together, the field cannot be managed (silage chopped, burned, mowed, disked) until the latest maturing species is ready. If single species are planned in separate sections of a field, and planted accordingly, an area may provide seed in various sections June through the end of winter. For example, wheat planted in the fall provides seed in June. Millets or buckwheat planted in the spring

Fig 12.3; High-tensile powerline rights-of-ways can be great areas for dove plantings. Here, sunflowers and corn have been planted in separate sections for mourning doves.

184

provide seed July through August. Millets and buckwheat planted in mid- to late June mature in August or September, and sesame, grain sorghum, and corn planted in the spring can provide seed from late August through winter, depending upon management. Management strategies include burning or mowing millet and wheat, and silage-chopping or mowing sunflowers, sesame, grain sorghum, and corn when the seed/grain has matured and is dry.

Management should not stop when the crop is silage-chopped, burned, or mowed. After doves consume the wheat crop, buckwheat or browntop millet can be planted to mature in late summer. If buckwheat or millets were planted in spring, the crop can be managed in midsummer. After doves consume the buckwheat or millet in late summer, the field or section can be disked to prepare a seedbed and, where legal, winter wheat planted by top-sowing and cultipacking in late August or September (See *Planting and managing wheat fields for mourning doves* on page 194). Before planting wheat, have soil samples tested and add lime and fertilizer as recommended, to ensure a productive crop and encourage a better seed yield the following June. Wheat also can be planted following management of sunflowers.

Fig 12.4; Single-species plantings, such as these sections of sunflowers, browntop millet, and corn, allow several management options. The separate crops are cut at different times because of varying maturation dates. This strategy also allows for better weed control (as there is no labeled herbicide for many seed mixtures) and can enable double-cropping in some cases if desirable. Photo by Marion Barnes.

When a field is planted in separate sections with different species, each entire section may be managed when the seed is mature. Other sections are managed as they become ready. If, however, you only plant one species, do not manage the entire field at once, but manage sections or strips to provide seed as long as possible.

Planting and managing millets for mourning doves
Many people will argue millets are the preferred seed of mourning doves. Others argue for sunflowers. Regardless, both should be planted in separate sections of a field to attract large numbers of doves. Browntop millet may be preferable in some areas because it is not as prone to chinch bug damage as proso millets. Browntop millet requires approximately 60 days for dry seed as compared to 70-80 days for proso millets.

Top-sow 25 pounds PLS of browntop millet or 30 pounds PLS of dove proso millet per acre on a well-prepared seedbed and cultipack. Seed also can be drilled at approximately 50 percent reduced rate.

Fig 12.5; Millet seed, such as this giant foxtail millet, is a favorite food of mourning doves.

There are no labeled preemergence herbicides for millets. Therefore, you are at the mercy of the seedbank for grass weeds, especially crabgrass, johnsongrass, and broadleaf signalgrass. If you have severe grass weed problems in a particular field, plant a forb crop (such as sunflowers, buckwheat, or sesame) so you can use grass-selective herbicides. Broadleaf weeds should not be a problem when planting millets because several forb-selective herbicides can be used, such as 2,4-D, dicamba, thifensulfuron methyl, and tribinuron methyl. Spray broadleaf weeds when they are young! Usually, this is about two to three weeks after you plant when millets are 4- to 8-inches tall.

Mature millet fields can be burned or mowed to make seed more available. Unlike wheat, most millets (browntop, proso, foxtail) do not stand up long after the seed matures (pearl millet is the exception). When these grasses fall, they create thatch on the ground and mowing usually makes the problem worse. Using a hay rake after mowing will remove thatch and make seed readily available. Burning is the best way to provide a clean, open structure at ground level that mourning doves prefer.

Planting and managing sunflowers for mourning doves

Sunflower seeds contain lots of energy and are a preferred food source for mourning doves and a number of other bird species. Peredovik sunflowers (black-oil type) can be top-sown over a well-prepared seedbed at 25 pounds PLS per acre and covered by

Fig 12.6; It's hard to beat a field of sunflowers for dove hunting. Here, two sections of this field have been planted to sunflowers, but one section (foreground) was planted later than the other to prolong availability of sunflower seed in the field. A section of corn also has been planted specifically for doves adjacent to one of the sunflower sections. Various sections will be managed by silage chopping to make seed available throughout fall and winter. Note also the loafing line that was erected specifically for doves.

Fig 12.7; A 6-foot tall multi-strand electric fence may be necessary to establish dove fields where deer density is too high.

disking approximately 1 inch. Or they may be planted or drilled at approximately 8 pounds PLS per acre in 38-inch rows. Planting in rows allows considerable weed control via cultivation. Prowl, Dual Magnum, or Treflan should be applied preplant incorporated or Prowl, Spartan, or Dual Magnum may be applied preemergence. Clethodim or Poast may be used for postemergence grass control. Beyond can be applied postemergence to Clearfield sunflower varieties only. Applications should be made before weeds exceed 3-5 inches in height and when sunflowers are in the 2- to 8-leaf stage. Do not apply Beyond to sunflower varieties that are not resistant/tolerant to imidazolinone herbicides. Preplant incorporated applications of Prowl and postemergence applications of Beyond can be used with Clearfield sunflower varieties for optimum weed control. Sunflower fields are best managed by silage chopping sections or strips before and during dove season. If you don't have access to a silage chopper, mowing will help make seed available, but seed are not scattered as well as with a silage chopper. Sunflowers may be defoliated with glyphosate or gramoxone to facilitate silage chopping. Defoliation is commonly done a couple weeks before opening of dove season to make fields as attractive to feeding doves as possible.

Fig 12.8; Dove fields may be defoliated with glyphosate (requires couple weeks) or gramoxone (requires a few days) prior to silage chopping to provide optimal conditions for feeding doves (note the bare ground).

Deer may eat developing sunflower seedheads. If deer are problematic, you need to shoot more deer! Concentrate on shooting does — that is the only way you will lower the population. Contact your state wildlife agency for depredation permits if necessary, and use them. If your property is relatively small, try to get your neighbors to join you in your management efforts. Deer meat may be donated to food banks if you can't use it all or give it away to friends and neighbors. Where doves are a focal species, electric poly-tape fencing may be used to repel deer. Nonetheless, management should address overabundant deer herds by lowering the population and implementing additional habitat management. Where deer populations cannot be managed appropriately and fencing is not an option, grass crops or sesame, instead of sunflowers, should be used for dove management.

Managing corn and grain sorghum for mourning doves

Grain sorghum and corn both are readily eaten by doves. If given a choice, doves usually prefer grain sorghum over corn if a low-tannin grain sorghum (such as Hegari) is available or after tannin levels drop following frost. If you want grain sorghum available for doves into winter, choose a variety with higher tannin content that will resist bird depredation. Tannin content drops considerably after a couple of frosts, increasing palatability of red and brown sorghums during late fall and winter. If depredation from blackbirds is not a problem and you want grain sorghum to be as attractive as possible soon after maturation, choose a white grain sorghum, such as Hegari, that contains fewer tannins than red or brown grain sorghums.

Fig 12.9; White grain sorghums, such as Hegari, contain fewer tannins than red or brown grain sorghums and are thus less bitter.

Follow the planting recommendations for corn and grain sorghum provided in *Chapter 8*. When managing specifically for doves, use the planting strategy that will provide the greatest yield per acre (as opposed to managing specifically for wild turkeys). Weed control is important when managing corn fields specifically for doves (as opposed to managing specifically for white-tailed deer). Dense grass weeds can be especially problematic for doves because of the thatch produced. Doves like to feed on bare ground. Corn fields are best managed for doves by silage chopping. Burning corn fields is not recommended when managing for doves because burning does not scatter corn seed and doves do not eat corn off the cob (even if they are on the ground after burning). Therefore, weed control that maintains bare ground along with silage chopping is recommended when managing corn food plots for doves.

Manage fields according to need. If you are managing multiple fields of various plantings, you can manage an entire corn or grain sorghum field at once, depending on field size. However, be aware, corn and grain sorghum seed deteriorate and decompose fairly rapidly once it is cut. Do not expect corn or grain sorghum cut in late summer to be good into winter. If you only have one field and want to spread usage out over a longer time period, do not manage the entire field at once, but leave some to cut later in the season. Try to manage grain sorghum and corn so grain

Fig 12.10; Silage chopping portions of the field through late summer and fall can continually attract doves and provide good late-season hunting opportunities.

is available from late summer through winter. Always leave some grain standing until winter to provide a high-energy source during the cold months, even if dove season is not in!

Crop rotation is important. One way to rotate corn and grain sorghum plots when managing for doves is to follow grass crops with a broadleaf crop, such as buckwheat, sunflowers, or sesame. You also can plant an annual cool-season legume, such as crimson clover, to help increase nitrogen availability. Fallow rotation also can be used if undesirable weeds are controlled by spot-spraying or disking and not allowed to go to seed.

A variety of herbicide applications are possible when planting corn and grain sorghum. Glyphosate can be sprayed postemergence over varieties of Roundup-Ready corn. Postemergence applications may be applied anytime from emergence until the V8 stage (8 leaves with collars) for corn or until the corn reaches 30 inches in height, whichever comes first. An initial application of 1-2 pints should be applied when weeds are 2- to 8-inches tall. A sequential application may be necessary if a new flush of weeds appear. Roundup-Ready technology is a great strategy to use when tough-to-handle weeds are present. However, after growing Roundup-Ready crops, there can be problems with volunteer sprouting if another crop is planted. Residual Roundup-Ready corn can be killed with a grass-selective herbicide. Roundup-Ready technology is especially effective when one of the preplant/preemergence herbicides labeled for corn also is used.

If you are planting **non**-Roundup-Ready corn and grain sorghum, atrazine, Bicep II Magnum, or Dual Magnum may be applied preplant incorporated or preemergence to control a wide variety of grass and forb weeds. However, realize grain sorghum seed must be treated with Concep seed safener before applying Bicep II Magnum or Dual Magnum. Basagran or Permit can be applied postemergence to control several forb weeds and yellow nutsedge. 2,4-D, Aim, atrazine, Banvel, and Clarity are other herbicides that can be applied postemergence to control broadleaf weeds. Remember to read and follow all label directions when applying herbicides. It is important to identify the weeds you have before spraying postemergence when managing for doves. Common ragweed, pokeweed, crotons, and many others (see *Plant Identification Guide*) can provide additional seed relished by doves. If undesirable noxious broadleaf weeds such as cocklebur, curly dock, horsenettle, jimsonweed, and sicklepod are numerous, you should spray to prevent them from becoming problematic.

In addition to the herbicides listed above, Python may be applied preplant incorporated or preemergence for corn, but not grain sorghum. Pursuit can be applied preplant incorporated, preemergence, or postemergence in Clearfield varieties of corn only (do not apply Pursuit to non-imidazolinone varieties of corn).

Grain sorghum or milo – which is it?

The terms "grain sorghum" and "milo" often are used interchangeably. Technically, this is incorrect. Milo is a grain sorghum, but there are several other grain sorghums besides milo.

The sorghums include many species, but they are classified into four broad groups: broomcorn, sorgos, grass sorghums, and grain sorghums.

Fig 12.11; Not all grain sorghums are milo. Hybrids and varieties of grain sorghum grown in the U.S. today were derived from crosses of milo and kafir.

Broomcorn plants have woody stalks with very long panicle branches, which are used by some cultures as brooms. The sorgos generally have tall, juicy, sweet stalks and are grown primarily for livestock forage, silage, or syrup. Grass sorghums, such as sudangrass and johnsongrass, have been produced for livestock forage. The grain sorghums, of course, are grown primarily for cereal grain.

Grain sorghums were introduced into the U.S. from different areas of Africa and Asia and are classed into seven groups: durra, feterita, hegari, kafir, kaoliang, milo, and shallu. The durras have bearded, fuzzy heads, with large, flat seeds, and dry stalks. The feteritas have few leaves, dry, slender stalks, and compact seedheads with large chalky-white seeds. The hegaris have nearly oval heads with chalky white seeds on plants that tiller abundantly. The kafirs have thick, juicy stalks, large leaves, and cylindrical heads with white, pink, or red seed. The kaoliangs have dry, woody, sparsely-leaved stalks that produce bitter brown seeds. The milos have fairly juicy stalks and wavy leaf blades with a yellow midrib. The heads of true milos are compact and fairly oval and the seeds are large and creamy white or salmon-colored.

The vast majority of the grain sorghums grown in the U.S. are hybrids and varieties derived from crosses of milo and kafir. Few, if any, are fully typical of the parent group. The other grain sorghum sometimes grown in wildlife food plots is hegari. The white seed of hegari grain sorghum (see Fig 12.16 on page 197) are quite attractive and readily consumed by wildlife.

Planting and managing buckwheat for mourning doves

Buckwheat is easy to grow, matures quickly (within 60 days), and doves readily eat the seed. Buckwheat most often is top-sown at 40 pounds PLS per acre over a well-prepared seedbed. Seed may be covered by light disking, but it is not necessary. Cultipacking after top-sowing is important. Buckwheat also can be drilled at approximately 25-30 pounds PLS per acre.

Buckwheat is a vigorous reseeder. If planted relatively early (late April in Zones 7-9), you should get three crops from one seeding even without disking. As seed mature and drop, doves will be attracted and begin feeding. Much of the seed will sprout and begin growing. The same scenario will play-out for the reseeded crop. If the structure at ground level is too dense with dead vegetation or if undesirable weeds are establishing, lightly disk the plot. That will create more bare ground space, make seed available, and effectively reseed another crop of buckwheat.

If broadleaf weeds are a problem in the field you are planting, do not plant buckwheat. There are no labeled herbicides to control broadleaf weeds in buckwheat. Grass weeds can be controlled with a grass-selective herbicide.

Fig 12.12; Buckwheat germinates fast, grows fast, and produces seed fast.

Planting and managing sesame for mourning doves

Sesame (also known as benne) is another good seed producer that is often overlooked for doves. Like buckwheat, sesame should be considered where broadleaf weeds are not severe. Sesame is quite drought tolerant and does well in a variety of soils, including sandy soils.

Sesame is best planted in 38-inch rows at about 6 pounds PLS per acre. Planting in 38-inch rows is highly recommended to allow cultivation for weed control and to create bare ground that doves prefer to feed on. Liming and fertilization should follow soil test results, but generally P and K should be raised to medium with about 30 pounds of N at planting and another 30 pounds of N when sesame is about 12-18-inches tall.

Prowl (1 pint per acre), Treflan (1 pint per acre), or Dual II Magnum (1 pint per acre) may be applied preplant incorporated to help control various grass and forb weeds. Refer to herbicide labels to make sure applications are legal in your area or state. Grass weeds can be controlled with Clethodim or Poast. Where broadleaf weeds are severe, you should consider a grass crop, such as millets or grain sorghum.

Some varieties of sesame shatter on their own and do not need to be silage chopped or mowed to make seed available for doves. Sesame also is a good crop to use where deer depredation on other crops is a problem. It is rare for deer to damage sesame crops.

Fig 12.13a,b,c; Sesame can be a great attraction for late season doves (12.13a). Sesame seed is high in oil content and crude protein. Here, sesame was planted adjacent to sunflowers (12.13b). After the sunflower seeds are eaten, the sesame will be mature and the doves will move into the sesame. Weeds within sesame are primarily controlled with cultivation between rows (12.13c). Cultivation makes the field more attractive for doves when feeding.

Fig 12.14a and b; It's late June and this wheat is ready to mow or burn to make seed available for doves. If many weeds are present, spraying the field with 1 quart per acre of a glyphosate herbicide two weeks prior will facilitate burning.

Planting and managing wheat fields for mourning doves

Wheat should be a "backbone" planting for anyone managing property specifically for doves. Not only is wheat readily eaten by doves, but it matures in early June, can provide seed throughout summer, can be managed in a number of ways, and (in some states) can be top-sown and hunted over during the early dove season. There is no other planting for doves that provides that flexibility. Seed availability in summer provides food for nesting doves and attracts doves to your property earlier in the season.

Wheat can be top-sown over a well-prepared seedbed or it can be drilled. Normally, wheat is planted for doves and winter cover at 120 pounds PLS (2 bushels) per acre. Broadleaf weeds can be controlled postemergence with 2,4-D, Aim, Banvel, Clarity, and Harmony Extra. Various weeds (including some grasses, forbs, and nutsedge) can be controlled in Clearfield varieties of wheat with a postemergence application of Beyond. If you have a problem with annual ryegrass, Achieve and Osprey may be applied postemergence. However, if you have a problem with ryegrass, I recommend you plant a forb so you can use grass-selective herbicides to control the ryegrass over a 2- to 3-year period. Be persistent and do not let ryegrass go to seed!

Table 4. Recommended rates for top-sowing winter wheat on a prepared seedbed (UT Extension)[1].

Use	*Seeding rate[2]*	*Seeding dates*
Winter cover, wildlife enhancement, or fall grazing	100-150 pounds (2-3 bushels) per acre	Aug 15-Nov 1

[1]These planting rates are applicable to Tennessee. If you do not live or hunt in Tennessee, check with your state wildlife agency and Cooperative Extension Service to determine what is legal in your area.
[2]Seeding rate may be increased 50 percent if using combine-run seed.

When managing wheat fields for doves, spray the plot as needed with a glyphosate herbicide after the seedheads have matured and dried (usually in June). After the weeds die, burn or mow the wheat to make the seed readily available to doves. If weed coverage was dense prior to spraying, burning is recommended. If weeds were not dense and the plot was relatively clean at ground level, mowing can produce desirable results. If mowed, cut the wheat just below the seedhead and not down low near the ground. Mowing at this height will reduce thatch buildup, scatter the seed better, and provide better feeding conditions for doves.

Managing wheat fields for mourning dove hunting

Hunting over baited fields is always a concern for dove hunters (at least most of them!). Because mourning doves are migratory, the U.S. Fish and Wildlife Service regulates the restrictions placed on dove hunting. The U.S. Fish and Wildlife Service allows dove hunting over mowed or cut grain fields as well as fields grown specifically for doves and other wildlife. In addition, "Lands planted by means of top-sowing or aerial seeding can be hunted [for doves] where seeds are present solely as the result of a normal agricultural planting or normal soil stabilization practice. Normal agricultural planting, harvesting, or post-harvest manipulation means a planting or harvesting undertaken for the purpose of producing and gathering a crop, or manipulation after such harvest and removal of grain, that is conducted in accordance with official recommendations of State Extension Specialists of the Cooperative State Research, Education, and Extension Service of the U.S. Department of Agriculture." (Note: By policy, the Service does not make a distinction between agricultural fields planted with the intent to harvest and those planted without such intent as long as the planting is in accordance with Cooperative Extension Service recommendations.) (Excerpts from Title 50, Code of Federal Regulations, Parts 20.11 and 20.21i)

Therefore, dove hunting over harvested and unharvested agricultural fields is legal, **providing your state wildlife agency allows it**. Furthermore, it is legal to hunt doves over fields of top-sown or drilled winter wheat where the seedbed has been prepared and as long as the seeding rate does not exceed that recommended by the State Agricultural or Cooperative Extension Service, providing the state wildlife agency has not implemented additional restrictions or regulations on managing fields for doves. **Always check current federal, state, and local laws before manipulating and hunting fields for doves.**

Planting and managing mixtures for mourning doves

If you are just planting one field and had rather plant a mixture than single species for doves, there are several options. However, it is important to keep weed control in mind if undesirable seed is present in the seedbank. Two of the three mixtures listed below have herbicide options.

Fig 12.15; A variety of millets and grain sorghum readily attracts doves. Photo by Ryan Basinger.

Millet mixture (PLS/ac)
   10 pounds white proso millet
   10 pounds dove proso millet
   10 pounds browntop millet
    3 pounds grain sorghum

Benefit and considerations: Research has shown white proso millet is the preferred seed for mourning doves, followed closely by dove proso millet and browntop millet. Regardless of type, if a good crop of millet is established, doves will come. The propensity for doves to feed on grain sorghum seed is no secret; however, non-target birds (such as house sparrows) may be a problem in some areas.

Management: Undesirable broadleaf weeds can be controlled with 2,4-D, Aim, Banvel, or Clarity.

Grain mixture (PLS/ac)
   6 pounds corn
   5 pounds grain sorghum
   5 pounds peredovik sunflowers

Benefit and considerations: A variety of grains will attract doves and other wildlife species. This planting rate will leave the corn plants not more than about 4 feet apart, ensuring adequate pollination. Be aware, there are no labeled herbicides for this mixture.

Fig 12.16; A blend of corn, grain sorghum, and sunflowers will attract large numbers of doves. It is important to realize not all weeds are bad when managing a dove field. Seed from common ragweed and giant foxtail in this field were eaten by doves and several other birds. The cover made available by several weeds also is desirable for brooding wild turkeys and bobwhite. Note the white seedheads of the Hegari grain sorghum.

Management: If weeds are likely to be problematic, you should consider another mixture or a single-species planting. Learn to identify the plants in your field so you can identify "good weeds" as opposed to "bad weeds" (refer to the *Plant Identification Guide*). There are several "weeds" (as mentioned above) that are quite beneficial in a field managed for doves. This mixture is ready to silage chop once the corn has matured and dried. If cut in sections or strips, this plot can provide seed well into winter.

Dove/deer mixture (PLS/ac)
   40 pounds iron-clay cowpeas
   6 pounds corn
   5 pounds peredovik sunflowers

Benefit and considerations: This mixture may be considered if you are interested in providing high-quality warm-season forage for deer and want to cut it for doves at the end of summer.

Fig 12.17; A mixture of cowpeas, corn, and sunflowers can double as a warm-season plot for deer and a dove field.

Management: Dual Magnum may be applied preplant incorporated to control annual grass and several forb weeds. Prowl or Pendimethalin may be applied preemergence, but not preplant incorporated (or you may kill the corn as it germinates). The field may be cut for doves after the corn matures (usually in September), or you can allow deer to forage on the cowpeas until first frost. When the plot dies after frost, use a silage chopper to make grain available for doves.

197

Thirteen

Chapter 13 — Waterfowl

Most people think about upland species when they think about food plots. However, relatively flat areas that can be flooded 6-18 inches can be planted for ducks and geese. If you can construct a shallow dike with a water-control structure (such as a flashboard riser), you can flood the field using rainfall or pumped water and attract waterfowl.

Flooding should occur by late August if you are down South trying to attract early-migrating blue-winged teal or wood ducks that hatched in the area. Otherwise, flooding should be initiated in November with full flood by early December (according to your region and objectives). Drawdowns should be initiated in February or March (according to your region) and completed by late April, if another crop is to be grown.

It is important to realize food plots for ducks cannot be manipulated like those for doves. Federal regulations allow waterfowl hunting over standing crops (food plots) and harvested crops (flooded or not). However, crops can be manipulated only by standard agricultural practices used to establish, manage, and harvest the crop. Grain or other seed inadvertently scattered during harvest operations or as you enter or exit the field while hunting, placing decoys, or retrieving birds is **not** considered bait. **Nonetheless, you should always check the current U.S. Fish and Wildlife Service and state hunting regulations before hunting.**

Insert 13.1 and 13.2; Constructing a shallow dike across a flat field can enable you to provide excellent feeding and loafing areas for waterfowl. A water control structure allows the water level to be manipulated when fields are flooded. Top photo by Ryan Basinger; bottom photo by USDA-NRCS.

Facing page: Flooded impoundment planted to corn (left).

The main consideration when establishing a food source for waterfowl is seed availability, primarily after inundation (flooding). Waste grain availability in a harvested corn field (that is, a production agricultural crop, not a food plot) averages 150-200 pounds per acre. That is not much food for wintering waterfowl. Availability is reduced even further when you consider **40-60 percent is lost every month post-harvest** prior to inundation through decomposition, depredation, and/or germination. Therefore, timing of crop harvest (production agricultural crops, not food plots) and initial flooding is critical. After flooding, the deterioration rate of waste corn after harvesting is 35-50 percent after 90 days. Thus, the overall loss rate is greater prior to flooding a harvested field than post flooding. Obviously, if the field is planted specifically for ducks and the corn is not harvested (a food plot), an abundance of grain should be available.

As mentioned above, harvested grain fields (such as corn) provide some waste grain, but the amount is miniscule compared to an unharvested field (5,600 pounds per acre in a field that only produces 100 bushels of corn per acre). To increase grain availability in production agricultural fields, leave as much of the field unharvested as possible. Even if only a few rows are left unharvested, there will be substantial increases in grain availability. If you own land, but do not have the means to plant, you can allow a producer to lease the field for row cropping and arrange for a section of the field to be left unharvested to increase available food for migrating and wintering waterfowl. If you lease fields to hunt, consider paying the producer the same price per acre he is getting for the harvested grain to leave a few acres unharvested. This will be a bargain for the producer because he will not have to spend time and fuel harvesting the grain off that area, and the increase in the number of ducks using the field will be worth the extra expense for you.

An approach to facilitate hunting and avoid potential problems associated with manipulating unharvested corn is to plant the area in front of your blind to low-growing plants, such as millets. Buckwheat and millets (with the exception of pearl millet) fall to the ground soon after a frost. Therefore, an open "hole" is provided for landing and feeding in front of the blind within a larger field of corn. Decoys can be placed around the edge of the corn, where they are visible and attract ducks to land in the open hole. [Note: Some people may find this practice distasteful — to manipulate a food source to facilitate shooting an animal. What they fail to understand or acknowledge is that these management efforts provide food and resting areas to support a surplus of ducks, whose populations are carefully managed by the U.S. Fish and Wildlife Service and state wildlife agencies. Also, a myriad of nongame wildlife species including shorebirds and other water birds and amphibians, receive benefit from fields flooded primarily for duck hunting. Obviously, many more ducks and geese benefit from these management efforts than are hunted. The last thing duck hunters want to see is declining populations of waterfowl. Indeed, this is true for other hunters and game species as well.]

Another important consideration, because you actually care about wildlife, is the crop planted. Soybeans, for example, is **not** a good crop to flood for ducks for several reasons. Soybeans decompose rapidly after inundation (70 to nearly 100 percent over 90 days), thus food availability is low. Even when soybeans are available, waterfowl do not select for them. Some studies have noted a relative lack of soybean consumption by waterfowl, even when acreage of planted soybeans within a particular region has increased. This is difficult for some people to understand, especially when they see ducks using a harvested soybean field that has been flooded. Most often, the ducks are not actually eating soybeans, but primarily loafing in the flooded field and eating weed seed and invertebrates. Certainly, ducks may eat a few soybeans, but decreases in body mass and fat reserves have been recorded among mallards after feeding upon flooded soybeans. Furthermore, soybeans may cause food impaction in a duck's crop, which can be fatal.

Regardless of crop, agricultural grains do not represent a complete diet for waterfowl. As with doves, it is a good idea to provide a variety of foods to attract and hold waterfowl through the winter. Although many grains contain lots of energy that can be metabolized, they do not contain the amino acids, vitamins, and minerals

Fig 13.2; Areas containing naturally occurring moist-soil plants, such as this smartweed, provide highly attractive seed that contain vitamins and minerals that are critical for migrating and wintering waterfowl.

needed for migrating and wintering waterfowl to maintain body mass. Therefore, naturally occurring moist-soil plants should be encouraged as well, either in the same flooded unit or in an adjacent flooded unit, to provide optimum feeding conditions for waterfowl. An added benefit of naturally occurring moist-soil plants is they persist longer than agricultural crops. Many moist-soil plants (such as smartweeds, sedges, panicgrasses) experience only 20-25 percent deterioration after 90 days of flooding.

Planting and managing millets for ducks

Corn may be "king of the hot crops," but in terms of being economical and efficient, it's hard to beat millets when planting a food plot for ducks. All of the millets are annual and none are native, but a couple are naturalized and occur commonly in bottomlands and riparian areas.

Wild millet (or duck millet)

Wild millet may include Japanese millet (*Echinochola frumentacea*) as well as barnyardgrass (*E. crusgalli*; see page 405). Wild millet produces an abundance of seed relished by ducks. It is commonly planted by top-sowing seed on mud flats in spring in and around shallow wetland areas, including beaver ponds. Wild millet also can be planted in fields with shallow dikes that can be flooded using water control structures.

*Fig 13.3; Wild millet (*Echinochola frumentacea, *shown here) whether planted or naturalized provides a good seed source for migrating and wintering waterfowl. The U.S. Fish and Wildlife Service allows you to manipulate (burn, mow, disk) moist-soil vegetation for duck hunting. However, it is illegal to manipulate planted crops, whether grown for ducks or agriculture.*

Seed may be top-sown (25 pounds PLS per acre) on a well-prepared seedbed and cultipacked, or it can be drilled after competing vegetation has been killed, usually with an application of glyphosate. Top-dress with 30-60 pounds N per acre when the wild millet reaches 4-6 inches in height.

Fig 13.4; Wild millets, such as this Japanese millet, can be shallowly flooded after establishment to help control weeds. This field was shallowly flooded to help control cocklebur. Photo by Ryan Basinger.

Wild millet can be flooded shallowly, but not inundated, soon after establishment to help provide weed control. Wild millet has a maturation date of approximately 55 days after germination and a deterioration rate of 57 percent after 90 days of inundation. Although wild millet will mature relatively quickly, increased seed yields greater than 1,000 pounds per acre are commonly realized if wild millet is planted prior to July 1. Later plantings may yield less than 500 pounds per acre. *Chiwappa* is a variety of *E. frumentacea* that is taller and produces larger heads and additional seed per plant. It should be planted earlier (by June 1) because it requires 120 days for maturity; 20 pounds PLS per acre is recommended.

Wild millet (especially *E. frumentacea*) usually reseeds naturally in wetlands where previously planted if it is allowed to mature and produce seed. "Naturalized" wild millet (that coming up naturally, at least one growing season after it was planted) can be manipulated legally, as can other naturally occurring, moist-soil plants. What this means is, it is legal to mow, disk, or burn a naturalized stand of wild millet if nothing else has been planted with the naturally occurring millet. [Note: This manipulation is allowed by the U.S. Fish and Wildlife Service. However, state wildlife agencies may place additional restrictions that do not allow manipulation. Always check both federal and state laws before managing areas for waterfowl to make sure you are legal.]

Fig 13.5; Chiwappa millet produces larger seedheads than standard Japanese millet, but it needs to be planted earlier because it requires 120 days to mature. Photo by Ryan Basinger.

Fig 13.6a; Control of broadleaf weeds and insects, such as armyworms, is possible with a variety of herbicides and insecticides when managing millet food plots. Photo by Ryan Basinger. Fig 13.6b; Browntop and proso millets are easy to grow and readily eaten by ducks. If broadleaf weeds are not a problem in the field where you are planting, buckwheat can be added to the millets for additional seed.

Undesirable broadleaf weeds can be controlled with 2,4-D, Aim, or Banvel. Armyworms sometimes can become problematic, especially when wild millet is planted relatively early (June). They can be controlled with an insecticide application, such as 4-8 ounces per acre of Intrepid (See *Appendix 4*).

Tame millets

A variety of "tame" millets can be planted and the seed of all of them are eaten by ducks. The millets are easy to grow, have low fertility requirements, and mature relatively quickly. Follow soil-test recommendations for liming and fertilization. Seed can be top-sown on a well-prepared seedbed. Cultipack before and after top-sowing. Apply 30-60 pounds N per acre when millets are 4-6 inches tall. There is nothing magic about a millet mixture for ducks. However, you might need to consider timing of maturation. If planting relatively early in the growing season (June), you can plant any of the millets you want. If you are planting relatively late in the growing season, you might want to plant browntop millet because it matures more quickly than the other tame millets (see *Appendix 1*). My favorite tame millet mixture includes white proso (20 pounds PLS per acre) and browntop (10 pounds PLS per acre). If taller structure is desirable, replace browntop with pearl millet, which will grow 4-6 feet tall. White proso millet has a maturation date of 70 days and a low deterioration rate after 90 days of inundation. Pearl millet remains standing for a while after flooding, which lengthens its availability and helps retain seed quality longer because its deterioration rate is approximately 70 percent after 90 days of inundation. Foxtail millet can be used in place of browntop if desired.

To create a "soft edge" within a duck food plot, I like to plant the outer edges (width depends on size of field and where blind is located) with corn or grain sorghum. Just inside that, I'll plant the white proso/pearl millet mixture or *Chiwappa* wild millet.

Then, in the center, or just in front of the blind, I'll only have white proso or perhaps chufa (see below), neither of which will be above water, providing an open hole. Decoys are placed according to wind direction along the edge of the corn and within the tall millet (pearl or *Chiwappa*) mixture. The ducks are left with an open hole just in front of the blind.

A big advantage to planting millets for ducks is broadleaf weed control. Without a forb included in the mixture, it is easy to control cocklebur, morningglories, and other

Fig 13.7; *This diagram shows how various plantings can be arranged to influence where ducks land in the field. Taller structure is presented on the outside and shorter structure is presented on the inside, within 40 yards of the blind. The black squares are duck blinds, the yellow around the edge is corn, the red represents chufa or moist-soil management, the blue is a mixture of millets and buckwheat, and the green is winter wheat.*

problematic broadleaf weeds with several forb-selective herbicides, such as 2,4-D, Aim, or Banvel. If broadleaf weeds are not problematic and if you are planting relatively late in the growing season, you can include buckwheat (10 pounds PLS per acre) with the browntop millet. If you add buckwheat, you have no herbicide option.

Planting and managing corn and grain sorghum for ducks

Corn is the energy king for migrating and wintering waterfowl. Standing corn plots are particularly important during mid- to late winter when other foods may be scarce. Realize, however, *it is illegal to hunt over corn that has been manipulated* (other than harvesting for grain). That is not a problem, however, because ducks will knock the stalks over if they cannot reach the corn from the water, particularly if relatively short varieties are planted. Amazingly, ducks (at least the larger species) can shuck an ear of corn to get to the grain. Flooded standing corn also provides cover ducks use for loafing and protection from wind. If the corn has been harvested, expect approximately 50 percent of the available waste grain to deteriorate within 90 days after inundation.

Grain sorghum can be planted for ducks, but it is not as preferable to ducks as corn, rice, or millets, and grain sorghum is highly susceptible to red-winged blackbird damage in wetlands. A relatively light seeding rate (8 pounds PLS per acre, as opposed to 10) is recommended because the lighter rate will still provide thermoregulatory cover and not be too dense for ducks to feed amongst. Depending on water depth, short varieties of grain sorghum (such as W.G.F. or wild game food) may be used to make seed more available for ducks. Grain sorghum has a deterioration rate of 42 percent after 90 days. White proso (10 pounds PLS per acre) or pearl millet (10 pounds PLS per acre) and/or buckwheat (20 pounds PLS per acre) can be added if desired. If so, decrease grain sorghum rate to 4 pounds PLS per acre.

Follow the planting recommendations for corn and grain sorghum provided in *Chapter 8*. Weed control is not as important when managing corn fields that are going to be flooded for waterfowl as it is for doves. Remember, you cannot manipulate (silage-chop, burn) the crop prior to hunting, the field will be flooded (so structure at ground level is relatively unimportant), and ducks will be eating corn from the cob, not on the ground. Nonetheless, where undesirable weeds, such as cocklebur, are expected, you might consider the appropriate herbicide applications.

Corn requires a lot of nitrogen. It is important to manage soil fertility as recommended after soil testing when trying to maximize grain production. The

Fig 13.8; Flooded corn provides much-needed energy and attractive loafing cover for migrating and wintering waterfowl. When the water level is near the corn ears, ducks have easy access to the grain. Photo by Bill Smith.

amount of N is highly correlated to the number of bushels of corn grain produced. In most areas, a good yield of corn is about 160 bushels per acre (or about 9,000 pounds). This yield will require about 160 pounds of actual N, or about 10 bags of ammonium nitrate (34-0-0) per acre, as well as considerable amounts of phosphate and potash, as recommended from a soil test. Weeds also respond to N application. If extreme weed pressure is expected where you are planting, use the appropriate herbicide applications.

As with upland food plots, crop rotation is important. Agricultural producers commonly rotate corn and soybeans. However, soybeans are not recommended for ducks. Corn and grain sorghum plots planted for ducks can be rotated with millets/buckwheat, chufa, or left fallow. Do not overlook the value and attractiveness of moist-soil management for ducks (analogous to early successional vegetation for quail, rabbits, white-tailed deer, and wild turkeys).

A variety of herbicide applications are possible when planting corn and grain sorghum. Glyphosate can be sprayed postemergence over varieties of Roundup-Ready corn. Postemergence applications may be applied anytime from emergence until the V8 stage (8 leaves with collars) for corn or until the corn reaches 30 inches in height, whichever comes first. An initial application of 1-2 pints should be applied when weeds are 2- to 8-inches tall. For best results on several perennial weeds, allow them to grow to 6 inches before spraying. A sequential application may be necessary if a new flush of weeds appears, which should be expected in low-lying, moist areas typically planted for ducks.

If you are planting **non**-Roundup-Ready corn and grain sorghum, Bicep II Magnum or Dual Magnum may be applied preplant incorporated to control a wide variety of grass and forb weeds. However, grain sorghum seed must be treated with Concep seed safener before applying Bicep II Magnum or Dual Magnum. Basagran or Permit can be applied postemergence to control several forb weeds and yellow nutsedge. 2,4-D, Aim, Banvel, and Clarity are other herbicides that can be applied postemergence to control forb weeds only. Python may be applied preplant incorporated or preemergence for corn, but not grain sorghum. Pursuit can be applied preplant incorporated, preemergence, or postemergence in Clearfield varieties of corn only (do not apply Pursuit to non-imidazolinone varieties of corn).

There are various insect pests that can reduce corn yields considerably. The most efficient and effective approach in managing insect pests and diseases is to plant corn varieties resistant to various pests and diseases and to plant seed already treated for insect/disease problems.

Planting and managing rice for ducks

Rice is a "Cadillac" duck food. Managed correctly, rice provides large seed yields with lots of digestible energy that ducks feed upon readily. Optimally, rice should be grown on flat ground that can be flooded for weed control. However, this is not absolutely critical when growing rice specifically for ducks, as opposed to grain production. Planting success generally is best when rice is sown (100 pounds PLS per acre) on a prepared seedbed (that can be flooded later) by disking or drilling about 1 inch deep. Rice also may do well if not flooded while growing, but it does require considerable moisture, such as a bottomland field that doesn't drain very well. Don't expect to grow rice in dry conditions. Rice also can be top-sown on mud flats or in shallow water (no more than 2- to 3-inches deep), but it does best when and where the soil can be amended. Medium-grain rice varieties generally have better seedling vigor and produce more seed than long-grain varieties. Medium-grain varieties also seem to be preferred by ducks. Early maturing (110-115 days) medium-grain varieties include *Alan, Jackson, Lagrue,* and *Kaybonnet.* Late-maturing (130-140 days) medium-grain varieties include *Bengal, Cypress, Lemont, Newbonnet,* and *Orion.*

Rice responds well to N. Soil pH, as well as P and K, should be amended according to soil test, and 90 pounds of N should be applied when the rice is approximately 2-3 inches in height and another 90 pounds of N 60 days later. Rice will respond best if flooded shallowly

Fig 13.9a and b; Rice is grown primarily along the coast and major rivers. However, it also can be grown farther inland within impoundments. The ability to shallowly flood rice for weed control is desirable, but not absolutely necessary. Rice can be grown in bottomland fields that remain fairly moist through the summer.

(2-4 inches) when it is about 6-8 inches in height for seven to 10 days, and then drained. Obviously, flooding provides adequate moisture, but it also helps with weed control. It is best to apply N just prior to rain if irrigation or controlled flooding are not possible. Facet or Prowl may be applied preemergence or postemergence, and Basagran or Permit may be applied postemergence to control various forbs and grasses. 2,4-D or Storm can be applied postemergence to control broadleaf weeds. Be sure to check herbicide labels before any application because there are several restrictions and limitations when spraying rice. Also, be sure the "weeds" you are trying to control aren't actually desirable plants for ducks. Wild millet, smartweeds, and various panicgrasses complement rice when grown for ducks by providing a diversity of food.

Planting and managing chufa for ducks

Chufa can be planted for ducks, just as for wild turkeys. Ducks "dig" the tubers out with their beaks. Waterfowl have an amazing ability to detect underground tubers, even though the field is flooded. And whereas wild turkeys may have a problem scratching up tubers in clay soils, this is not as big of a problem for ducks because flooded soils are relatively soft, enabling ducks to get to the tubers.

Chufa can be planted with a drill or planter, or the tubers can be broadcast and covered 1-2 inches by disking. Planting rate should be approximately 50 pounds per acre. Chufa does best in fertile soils; therefore, P and K should be raised accordingly

Fig 13.10; The tubers produced among the roots of chufa are a strong attractant for waterfowl. When shallowly flooded (2 – 10 inches), ducks can grub up the tubers quite well. Here, chufa has been planted in a river bottom between 2 corn patches. Guess where the blind will be?!?

(50 and 160 pounds available per acre, respectively). Top-dress with a nitrogen fertilizer (60 pounds N per acre) when plants reach about 6 inches in height and rain is in the forecast. Chufa matures approximately 100 days after germination. "Clean" chufa plots typically produce greater yields than weedy plots. Planting chufa in rows allows cultivation for weed control. Combining cultivation with the appropriate herbicide applications is the best strategy for clean, productive chufa plots. Broadleaf weeds can be controlled postemergence with 2,4-DB (use Butyrac 200, **not** 2,4-D; 2,4-D may kill chufa, especially at 2 to 3 quarts per acre), Banvel, or Clarity, and problem grasses can be controlled postemergence with Clethodim or Poast.

Just as when you grow chufa for wild turkeys, it is important to rotate the crop each year, at least every other year. Crop rotation will encourage healthier plants and help manage plant density. Planting corn, millets, or a fallow rotation following chufa allows several weed control options. It is good to wait at least three years before planting chufa in the same plot again.

Planting and managing wheat for waterfowl

Some species of ducks, especially American wigeon and green-winged teal, and Canada geese readily eat green growing vegetation. Winter wheat can complement warm-season food plots (such as corn and millets) or moist-soil vegetation by providing green vegetation outside the edge of the flooded field.

Fig 13.11; Geese and wigeon readily feed on flooded wheat. Here, wheat has been planted along the outer edge of a moist-soil management field (right) several weeks prior to flooding.

Wheat sown in late summer/early fall can be flooded after it gets about 6 inches tall. I like to plant wheat along the edge of the flooded area. As it rains through the duck season, the water level will fluctuate along this zone and ducks will often concentrate their feeding along this zone, consuming both the green wheat and invertebrates.

Therefore, the wheat may extend 30-50 feet beyond the high-water level, and 30 feet or so within the flooded area. Wigeon, teal, mallards, and Canada geese will "grub" along the flooded zone and "dry feed" above the water line. Providing a diversity of foods, such as a grain/seed plot as well as a green plot, is highly attractive for a variety of duck species.

Wheat may be top-sown and lightly disked-in, or drilled at 120 pounds PLS per acre. Fertilize according to soil test and apply another 30-60 pounds of N before flooding. Broadleaf weeds are controlled easily in wheat plots with numerous broadleaf-selective herbicides, such as Harmony Extra, Aim, Banvel, and Clarity. Weed control is especially useful when planting wheat in these moist areas where cool-season forbs such as henbit, purple deadnettle, and chickweeds can be problematic, actually competing with the wheat and limiting growth. Spray weeds before they are 3-4 inches tall.

Fig 13.12 ; Let's be honest—the reason the vast majority of you are flooding fields for ducks is to shoot them. And that's OK! However, keep in mind, you are responsible for bringing on the next generation, and that includes your daughters! Ethical hunting is one of the most wholesome activities you can take part in. Share it with others.

Fourteen

Chapter 14 — Managing woods roads for wildlife

Planting and maintaining woods roads (or logging roads) can do more than prevent erosion, it also can provide high-quality forage for various wildlife species in areas where forage availability is often limited. Planted woods roads can become linear wildlife openings. Linear openings are particularly important to many species of wildlife in forested areas where early successional vegetation and high-quality forage are limiting. Planting woods roads can increase overall food plot acreage without clearing additional openings and help disperse forage throughout a property. Planted woods roads can impact more animals per planted acre than food plots when roads traverse and wind through an area, encompassing the home range of more animals.

All of the principles and procedures outlined in *Chapters 3, 4,* and *5* also apply when planting woods roads. However, plowing and heavy disking are not possible or desirable on roads where large roots may be just under the soil surface and where vehicle travel will continue. Light disking, no-till top-sowing, and drilling seed are recommended techniques when planting woods roads.

Soil pH often is a limiting factor along woods roads, especially where acidic leaves and needles have fallen and accumulated over the years and where topsoil has been removed. Another limitation when planting woods roads is the amount of light reaching the road. At least four hours of direct sunlight is desirable to maintain planted forages. Unless the adjacent forest stand has been thinned or

Fig 14.1; Woods roads can provide much-needed nutrition when planted to high-quality forages (as opposed to perennial cool-season grasses). This road in the mountains of NC was planted to clovers and has received heavy use by white-tailed deer, wild turkeys, black bears, and ruffed grouse.

regenerated recently, the road will need "daylighting" – that is, some trees will need to be killed and/or felled along at least one side of the road to allow sufficient sunlight to reach the road for desirable forage production. You do not have to remove all the trees. You can leave residual trees with mast-bearing potential and remove the others.

Roads can be managed for wildlife in a variety of ways if the road is closed to traffic. If the road is gated, yet still receives considerable traffic from land managers or hunters, it probably should be graveled. These roads still can be managed for wildlife by clearing and planting the sides of the roads.

Fig 14.2; This road has been daylighted on one side to produce additional browse for deer and nesting cover for wild turkeys. The road is graveled because it is traveled frequently.

Fig 14.3; Another option for frequently traveled roads is to plant the sides of the road after daylighting. The sides of this road were planted to crimson clover and wheat and receive tremendous use by white-tailed deer and wild turkeys.

Roadsides also can be left fallow. Woody growth can be suppressed by spraying selective herbicides, such as imazapyr and triclopyr. Light disking every other year just before spring green-up will stimulate and encourage additional herbaceous growth from the seedbank. If the road is gated and does not get too much traffic, the road itself can be planted.

Many of the same forages used in food plots can be planted on woods roads; however, some are much better suited than others. For example, crimson clover, subterranean clover, and white clovers are relatively shade tolerant. Ladino white clover persists well on roads traversing through bottomlands and on hillsides with an eastern or northern exposure. Ladino white clover may not, however, do well on southern or western exposures, especially from the Mid-South southward. Red clover and alfalfa do not respond to traffic as well as white clovers. Taller forages, such as arrowleaf clover, are not desirable on roads and do not stand up to traffic well. If the road has been closed and is essentially no longer used by vehicular traffic, it truly can be treated as a linear opening and planted to anything you might consider for your other food plots, including warm-season forages (see Fig. 14.5).

Soil erosion and siltation often are associated with woods roads and logging decks after logging. In fact, research from the Coweeta Hydrologic Lab near Otto, NC has shown more than 95 percent of the erosion and siltation into creeks following logging comes from improperly constructed and planted roads, not the logging

Fig 14.4; When woods roads do not receive much traffic, the road itself can be planted. This road was initially planted to ladino white clover and oats. After three years, a solid stand of clover remains. Encroaching japangrass (Microstegium vimineum) has been sprayed along the sides. Can't you just see deer feeding or a gobbler strutting along this road?

itself. Because erosion and siltation is such an important factor, many land managers have been led to the false assumption that it is necessary to include perennial cool-season grasses, such as tall fescue or orchardgrass, in a mixture sown on woods roads. This assumption is not true and certainly counterproductive for wildlife! Remember, there is no planting that corrects poor road construction, especially on those roads that receive vehicle traffic. Excessive slope is corrected with a bulldozer and gravel, not seed.

Germination and growth of annual cool-season grasses (such as wheat) are considerably faster than perennial cool-season grasses, which is important to prevent erosion from winter rains. The preference for oats and wheat as forage over perennial cool-season grasses, was discussed on page 140 and illustrated in *Tables 8.2* and *8.3* on page 141. The value of wheat seed and the resulting cover for brooding wild turkeys and northern bobwhite was covered in *Chapters 9* and *11*. This practice also benefits ruffed grouse in the same manner when implemented on woods roads where grouse occur.

It also should be made clear that, for a number of reasons, native warm-season grasses are not suited for planting woods roads either. These grasses are established for wildlife to provide cover, not forage (see *A word about native warm-season grasses…*, on page 166). Deer, rabbits, groundhogs, and other wildlife rarely, if ever, eat native warm-season grasses. In fact, all perennial grasses, whether native or not, simply are not preferred forages. Cover along old logging roads is hardly ever a limiting factor, as brambles, forbs, and other "weedy" growth occur along woods road edges

Fig14.5; *This woods road, or linear wildlife opening, was a skid trail that was cleared, amended with the appropriate amount of lime and fertilizers, and planted to soybeans. Woods roads that are not traveled with vehicles during summer can be planted as a warm-season food plot.*

Fig 14.6; *This road was planted to orchardgrass in September after the logging operation was completed. Orchardgrass is a perennial cool-season grass and is relatively slow to establish. Precipitation through the fall and winter eroded this road before the orchardgrass could become established. The problem with this erosion was not related to orchardgrass, but improper road construction. No planting will hold soil together on a road coming straight down the hill. The bottom line is orchardgrass and tall fescue are not cure-alls to prevent erosion of forest roads. Sensible road construction and use of annual and wildlife-friendly perennials, as well as the seedbank, is a much better solution for managing woods roads after logging.*

Fig 14.7; High-quality forage can be limiting in winter for several wildlife species. This old skid trail was cleared and now serves as a linear wildlife opening and allows vehicular access into areas of the property previously not accessible. Photo by Jim Phillips.

Fig 14.8; Planting woods roads is not always necessary. If the seedbank contains desirable species, such as the common ragweed growing along this woods road, that suit your objectives, and slope is not an issue, planting may not be needed.

along these roadsides. The primary limitation for many species of wildlife on vastly forested tracts is forage. Therefore, high-quality forages should be planted along roads in these areas, not warm-season grasses.

Logging roads usually are planted soon after the logging operation is finished. However, planting logging roads November through February or in July is a waste of time and money in most areas, during most years. Nothing is going to germinate and establish a root system anytime soon if planted in November through January. Frost-seeding clovers often is conducted in February, but cool-season plants in general are best sown in March or April and August through October, depending on your location. Warm-season plantings should be sown mid-April through June. If a road needs to be sown in late spring or early summer, a warm-season annual planting should be considered. If planting can wait until late summer, plant a cool-season forage.

Below are some planting mixtures recommended for woods roads. Before deciding what to plant, remember the critical limitations associated with planting woods roads: 1) available sunlight and 2) amount of traffic. These factors limit your options. Amend soils as recommended by soil test, don't plant outside the recommended planting window, do not plant perennial grasses, and manage weed pressure as you would with food plots. If you do these things, you will have a productive linear wildlife opening that will positively affect available nutrition and movements of various wildlife species.

Ruffed grouse don't like orchardgrass either!

We collected 53 ruffed grouse during March 2000-2002 in western North Carolina to determine the physiological condition of grouse and see what grouse were eating during late winter. The effort was part of a regional project — the Appalachian Cooperative Grouse Research Project — that studied the ecology and management of ruffed grouse in the central and southern Appalachians. Crop contents from all birds killed were identified, weighed, and preserved. All of the grouse were killed from gated woods roads initially planted in an orchardgrass and white-dutch clover mixture. Leaves and flowers of herbaceous plants were found in 92 percent of the 53 crops examined and comprised 40 percent of the material in the crops over the three-year period. Other foods included evergreen and deciduous leaves, acorns, ferns, soft fruits, buds, and twigs. Of the herbaceous material eaten, cinquefoil, clover, and wild strawberry represented the vast majority, followed by avens and ragwort.

The interesting thing was that orchardgrass, which was the dominant cover on most of the roads, was not present in any of the grouse crops. In fact, the graduate student who sorted through crops of 326 grouse from North Carolina, Virginia, West Virginia, Kentucky, Maryland, and Pennsylvania reported, "Grasses were not eaten much at all at any site in any year. I did get a few grasses in crops, but their quantities usually were not measurable and were classified as 'trace' (<0.1 gram dry mass). It seemed like grouse only ate grass incidentally while foraging on the forbs in between the grasses." (Bob Long, 2007, M.S. Thesis, West Virginia University).

Figs 14.9a and b; This woods road was limed, fertilized, and sown with a mixture of white-dutch clover and orchardgrass in the fall of 1993. By July 1995, the clover was out-competed and disappeared from the site, resulting in a road of orchardgrass, which offered poor structure for poults and fewer invertebrates. Nonnative perennial cool-season grasses should never be included in a planting mixture where wildlife is a consideration.

Plantings for woods roads

Perennial woods road mixture (PLS/ac)
    8 pounds ladino and/or intermediate white clover
    50 pounds wheat

Benefit and considerations: This mixture is an excellent one to use on woods roads. No need to try and get fancy. Wheat germinates and grows relatively quickly, helping prevent soil erosion while providing high-quality forage for wildlife. The amount of wheat planted on a woods road might be increased a little over that for a food plot, especially on areas of the road where slope is a little steeper. Choose a variety of ladino and/or intermediate white clover that is adapted to your site. The clovers will be retained longer on sites that are not overly dry during summer. Chicory and red clover can be added if desired. If so, add 2 pounds chicory and 5 pounds red clover and reduce white clover to 4 pounds and wheat to 40 pounds.

Management: Pursuit and/or 2,4-DB can be sprayed postemergence to control various weeds after the wheat matures. Clethodim or Poast can be sprayed postemergence to control problem grasses (such as *Microstegium vimineum*; see *Dealing with japangrass* on pages 220-221 and the *Plant Identification Guide*) after the wheat matures. The road can be mowed after the clovers have produced seed (late summer) and as necessary to prevent weeds from flowering if the road is not managed with the appropriate herbicides. Or, desirable incoming forbs and grasses from the seedbank can be allowed to pioneer into the road and provide good brood-rearing cover for wild turkeys and ruffed grouse.

Annual cool-season woods road mixture (PLS/ac)
    15 pounds crimson clover
    50 pounds wheat

Benefit and considerations: This annual mixture will provide lots of high-quality forage, especially during winter when green forage is limited.

Management: Spray incoming weeds after the wheat matures in July with a glyphosate herbicide. Mow a couple weeks after weeds die. The crimson clover will reseed naturally and you can retain good clover cover on the road for many years. Another option is to allow the weeds to grow, especially if there are desirable forbs and grasses. Succession will provide sufficient cover along the road and provide good brood-rearing cover. Where japangrass (*Microstegium vimineum*) is problematic, the glyphosate application mentioned above will kill it, or you can use a grass-selective herbicide, or you can use 8-12 ounces per acre of Plateau.

Annual cool-season woods road planting (PLS/ac)
  150 pounds wheat

Benefit and considerations: If all you need is a temporary cover and you intend to allow the seedbank to establish cover along the road, just sowing wheat is an excellent strategy. Sowing a road to wheat protects it from erosion. In addition, forage and seed for wildlife are provided (be sure to plant an awnless variety of wheat). Naturally occurring annual grasses and forbs in the seedbank will germinate and provide excellent cover for various wildlife species after the wheat matures and dies. Wild strawberry, annual panicgrasses, beggar's lice, cinquefoil, asters, and blackberry provide forage, soft mast, and seed for deer, turkey, grouse, black bear, and songbirds, while the perfect structure for brood cover is created.

Management: Several forb-selective herbicides, such as Harmony Extra, 2,4-D, and Clarity, can be used to control broadleaf weeds, such as chickweeds, henbit, and purple deadnettle, in fall/winter while the wheat is establishing.

Annual warm-season woods road mixture (PLS/ac)
  12 pounds browntop millet
  12 pounds foxtail millet
  20 pounds buckwheat

Benefit and considerations: According to your location, logging operations may finish in spring or early summer. A warm-season planting mixture may be the best option. Planting a road with cool-season forages during this time of year usually is a waste of time and money, especially south of Zone 5 (see Figure 2.3). Buckwheat and millets germinate very quickly and provide excellent cover along a woods road. Buckwheat provides forage for deer and rabbits, whereas the millets and buckwheat provide seed for a number of bird species. Some reseeding may occur, but the real benefit is having a quick-establishing annual planting that grows through summer, provides forage and seed, and helps prevent erosion. Unless the road was cut deep and the seedbank removed, expect grasses and forbs germinating from the seedbank to dominate the following year.

Management: Once the warm-season mixture produces seed, a cool-season mixture can be sown in late summer, especially if additional cool-season forage is needed or if possible erosion through fall or winter is a concern. The road can be disked lightly to prepare a seedbed and a cool-season mixture top-sown, then cultipacked, or the cool-season mixture can be drilled. Another option is no-till top-sowing white clover over the existing warm-season mixture after it has produced seed.

Dealing with japangrass (*Microstegium vimineum*)

Japangrass (also called Japanese grass, Japanese stiltgrass, Nepalese browntop) is a nonnative, annual, warm-season grass that is extremely invasive and can quickly outcompete native groundcover and degrade habitat quality for many wildlife species. It is shade tolerant and flourishes along riparian areas and other moist sites, along roadsides, ditches, and forest edges. Japangrass also may grow extensively throughout upland hardwood areas, often forming mats that exclude other vegetation. It has no known wildlife value. Japangrass seed are small and transported by wind and water, and on wildlife, equipment, and clothing. Therefore, it is very important to clean equipment (such as a tractor and rotary mower) of japangrass seed (and other undesirable seed) before moving equipment into other areas. Japangrass can quickly spread into a forested area, especially when already present and when additional sunlight is made available in the stand through thinning, cutting, and burning.

Aggressive action should be taken when japangrass is found because it can spread so quickly. Fortunately, japangrass is easy to control—you just have to do it! Imazapic, glyphosate, and clethodim may be used to control japangrass. My graduate students and I evaluated several herbicides at different rates to control japangrass. Our recommendation is 8-12 oz/ac of Plateau or 12 oz/ac of Clethodim (add nonionic surfactant with either postemergence application) because they selectively kill japangrass and do not injure various nontarget plants that may be important to wildlife. If nontarget plants are not present, glyphosate (2 qts/ac) is very effective.

We usually spray japangrass via ATV sprayer along woods roads and field edges. We drive ATVs in the woods and use a spray gun where possible, and we commonly use backpack sprayers in woods. We have found a single herbicide application in spring is very effective in controlling japangrass. However, be sure to check the area 3 – 4 weeks after spraying and spray any plants that were missed with the initial application. Japangrass coverage should be reduced significantly the following year, but subsequent herbicide applications will be necessary. Japangrass seed may persist in the soil for up to 3 years. This technique and effort is very effective, but again, you have to get out there and do it. You can't wish japangrass away!

Fig 14.10; Japangrass can quickly outcompete and exclude native groundcover important to wildlife. You should aggressively treat this nonnative invasive wherever it occurs on your property. This woods road has been overtaken by japangrass. However, it is very easy to control with the appropriate herbicides.

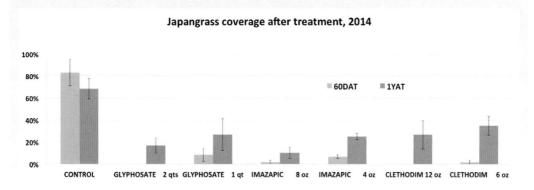

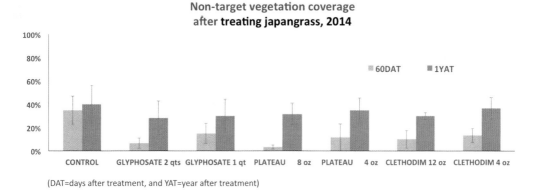

(DAT=days after treatment, and YAT=year after treatment)

Figs 14.11a and b; A variety of herbicides can be used to control japangrass. We found broad-spectrum (glyphosate; 41% active ingredient), broad-spectrum selective (imazapic), and grass-selective (clethodim) herbicides all worked well to control japangrass during the growing-season immediately following application. However, best control after one year was achieved with imazapic. Interestingly, non-target vegetation coverage was similar among herbicides used one year after treatment. Of course, non-target vegetation response will vary with site. Rates are per acre. (Brooke et al. 2015)

Fifteen

Chapter 15 — Final thoughts

Wow. If you just finished reading this book, you now realize there is a lot of information to digest (pun intended) when considering food plots. Planting and managing food plots is an involved process, and it can be complex. However, if you take it step by step and follow the recommendations provided, you will be successful and should have fun doing it. Without question, planting and managing food plots can be very rewarding. To work with the land and watch wildlife respond and benefit from your efforts can be intoxicating. Working with the land instills a land ethic, which unfortunately is quickly disappearing in our society today.

It is my hope that after reading this book, you have learned and appreciate the fact that food plots are a relatively small component of habitat management and that food plots are not a surrogate for or replace other practices, such as forest and old-field management and adjusting and arranging coverage of successional stages and vegetation types to meet the requirements of focal species. Hopefully, your interest in food plots will encourage you to do more with your property and begin to focus on holistic habitat management. If you haven't already, I highly recommend you spend some time with a Certified Wildlife Biologist® and a Registered Forester and develop a comprehensive management plan for your property.

As you are now aware, a successful food plot program requires planning and dedication, and it can be expensive. It often requires year-round effort, especially if you hope to actually increase the nutritional carrying capacity of your property for a particular species. It also requires you to carefully consider exactly what you are planting. I hope you have learned how important it is to consider plant phenology when creating planting mixtures. Each species in the blend should have a specific purpose for a specific time period.

Be aware of marketing when considering food plots. Marketing is a powerful tool. Many of the commercial blends available are sensible and will work well for a specific purpose. However, many do not represent the best planting for your objective. Carefully consider your objectives and how a particular blend may help you meet your objectives. Don't be misled or fooled by advertisement claims that sound too good to be true. There is no trophy in the bag. High-quality food plots and healthy wildlife populations are the result of planning, persistence, and hard work.

Finally, stay focused on the big picture and have fun with your efforts! You are planting food plots and managing your property because you have a passion for wildlife. Realize your limitations, whether related to your location or economics, and consider your success relative to your limitations. If you do this, you should continue to have fun and realize success as you continually improve existing habitat conditions for the focal species on your property.

Psalm 34:3
Mark 16:15-16

Appendix 1

Planting information for various crops.

Crop species[1]	Seeding rate[2] (lbs/ac)	lbs/ bushel	Approximate days to dry seed	Planting date[3]	Planting depth (inches)	Optimum pH	USDA Plant Hardiness Zone	Preferred soil type
Cool-season legumes[4]								
Alsike clover (perennial)	10	60		Aug 15-Oct 15; Feb 15-May 1	1/4	5.8-6.5	3-8	Adapted to cool climate; tolerates wet bottomland soils
Arrowleaf clover (annual)	10	60		Aug 15-Oct 15	1/4	6.0-6.5	6-10	Fertile, well-drained sandy loams and light clay; good re-seeder
Balansa clover (annual)	8	60		Aug 15-Oct 15	1/4	5.8-7.5	5-9	Grows on wide range of soils, including wet bottoms
Ball clover (annual)	6	60		Aug 15-Oct 15	1/4	5.8-7.0	7-10	Sandy loams and clay loams; tolerates poor drainage and relatively low fertility; good re-seeder
Berseem clover (annual)	20	60		Aug 15-Oct 15	1/4	5.8-7.5	5-9	Tolerates poor drainage; high fertility requirements; not cold tolerant, but winter hardy variety available; poor re-seeder; not shade tolerant
Crimson clover (annual)	20	60		Aug 15-Oct 15	1/4	5.8-7.0	5-10	Well-drained sandy loams to heavy clays; moderately shade tolerant, but winter hardy variety available; good re-seeder
Red clover (biennial)	12	60		Aug 15-Oct 15; Feb 15-May 1	1/4	6.0-7.0	4-9	Sandy loam to clay; wide range of moisture regimes; fairly drought tolerant
Rose clover (annual)	20	60		Aug 15-Oct 15	1/4	6.0-7.0	7-10	Well drained sandy loam to clay; tolerant to low soil fertility and drought; good re-seeder
Subterranean clover (annual)	25	60		Aug 15-Oct 15	1/4	5.5-7.0	7-10	Sandy loam to clay; moderately shade tolerant; tolerates both dry and moist sites as well as low fertility; fair re-seeder
White (including ladino and intermediate) clover (perennial)	6	60		Aug 15-Oct 15; Feb 15-May 1	1/4	6.0-6.5	3-10	Sandy loam to clay; moderate fertility requirements; mildly shade tolerant; tolerant to poor drainage
Alfalfa (perennial)	15	60		Aug 15-Oct 15; Feb 15-May 1	1/4	6.5-7.5	3-9	Well-drained loams; high P,K,S,B requirements; moderately drought tolerant
Austrian winter pea (annual)	50	55		Aug 15-Oct 15	1-2	6.0-7.0	3-8	Loam to heavy clay; moderate fertility requirements
Birdsfoot trefoil (perennial)	10	60		Aug 15-Oct 15; Feb 15-May 1	1/4	5.8-7.0	3-9	Widely adapted; moderately tolerant to drought and poor soil drainage

Crop species[1]	Seeding rate[2] (lbs/ac)	lbs/ bushel	Approximate days to dry seed	Planting date[3]	Planting depth (inches)	Optimum pH	USDA Plant Hardiness Zone	Preferred soil type
Cool-season grasses								
Oats (annual)	100-150	32	170	Aug 15-Oct 15; Feb 15-Mar 15	1-2	6.0-6.5	5-10	Sandy loam to clay; well-drained
Rye (annual)	100-150	56	180	Aug 15-Oct 15	1-2	5.8-6.5	3-10	Sandy loam to clay; well-drained
Triticale (annual)	100-150	48	180	Aug 15-Oct 15	1-2	5.8-6.5	3-10	Sandy loam to clay; well-drained
Wheat (annual)	100-150	60	180	Aug 15-Oct 15	1-2	6.0-7.0	3-10	Light-textured soils; not in poorly drained or heavy clay
Warm-season legumes[4]								
Alyceclover (annual)	20	60		Apr 1-June 15	1/4	6.5-7.0	4-10	Sandy loam to clay
American jointvetch (annual)	20	60		Apr 1-June 15	1/4-1	5.8-6.5	4-10	Sandy loam to clay
Catjang cowpeas (annual)	30			Apr 1-June 15	1/2-1	5.8-7.5	4-10	Widely adapted; well-drained soils
Iron-clay cowpeas (annual)	75-100	60	110	Apr 1-June 15	1/2-1	5.8-7.5	4-10	Well-drained soils; drought tolerant; tolerates relatively low fertility
Lablab (annual)	25-40			Apr 1-June 15	1	5.8-7.5	5-12	Well-drained soils; drought tolerant; tolerates relatively low fertility
Soybeans (annual)	130,000-200,000 seeds per acre or 75-150 pounds	60	125	Apr 1-June 15	1-2	5.8-6.5	4-10	Widely adapted; well-drained soils
Re-seeding soybeans (annual)	40			Apr 1-June 15	1/2-1	5.8-6.5	4-10	Well-drained soils
Perennial peanut	50-100 bushels of sprigs per acre			Jan-Mar	1 1/2-2	5.5-6.0	lower 8-10	Sandy and sandy loam soils
Florida beggarweed (annual)	10			Apr 1-June 15	1/4	5.8-6.5	8-10	Sandy loam to clay
Korean lespedeza (annual)	15	45		Feb 15-June 15	1/2-1	5.8-6.5	7-10	Widely adapted; tolerates relatively low fertility; not wet soils
Partridge pea (annual)	1-4		110	Feb 15-Jun 15	1/2-1	6.0-6.5	3-9	Sandy loam to clay
Warm-season grasses								
Corn (annual)	16,000-30,000 seeds or 10-15 pounds	56	145	Apr 1-May 15	1-2	5.8-7.0	3-10	Widely adapted, well-drained soils; high fertility requirements
Grain sorghum (annual)	8	50	120	Apr 15-June 15	1	5.8-7.0	3-10	Widely adapted, well-drained soils; moderate fertility requirements
Egyptian wheat (annual)	15		110	Apr 15-June 15	1/2	6.0-7.0	3-10	Widely adapted, well-drained soils; moderate fertility requirements
Browntop millet (annual)	25	50	65	Apr 15-July 15	1/4-1/2	5.8-7.0	3-10	Well-drained soils; moderate fertility requirements

Crop species[1]	Seeding rate[2] (lbs/ac)	lbs/ bushel	Approximate days to dry seed	Planting date[3]	Planting depth (inches)	Optimum pH	USDA Plant Hardiness Zone	Preferred soil type
Foxtail millet (annual)	25	40	80	Apr 15-June 15	1/4-1/2	5.8-7.0	3-10	Well-drained soils; moderate fertility requirements
Wild millet (annual)	25	35	55	May 1-Aug 15	1/4-1/2	5.8-7.5	3-10	Loams and clays; tolerates shallow flooding after establishment; moderate fertility requirements
Pearl millet (annual)	25	48	100	Apr 15-June 15	1/4-1/2	5.8-7.0	3-10	Well-drained soils; moderate fertility requirements
Dove proso millet (annual)	30	50	80	Apr 15-June 15	1/4-1/2	5.8-7.0	3-10	Well-drained soils; moderate fertility requirements
White proso millet (annual)	30	50	80	Apr 15-June 15	1/4-1/2	6.0-7.0	3-10	Well-drained soils; tolerates dry sites; moderate fertility requirements
Rice (annual)	100	45	150	Apr 15-June 1	1	6.0-7.0	3-10	Poorly drained soils; not on sandy soils; prefers shallow inundation

Other plantings

Crop species[1]	Seeding rate[2] (lbs/ac)	lbs/ bushel	Approximate days to dry seed	Planting date[3]	Planting depth (inches)	Optimum pH	USDA Plant Hardiness Zone	Preferred soil type
Buckwheat (warm-season annual)	40	40	60	Apr 15-Aug 15	1/2-1	5.8-7.0	3-10	Widely adapted; tolerates relatively low fertility
Burnet, small (cool-season perennial)	20			Aug 15-Oct 15; Mar 1-May 1	1/4	5.8-7.0	3-10	Widely adapted; moderate fertility requirements; drought tolerant; does not tolerate poor drainage
Chicory (cool-season perennial)	10			Aug 15-Oct 15; Mar 1-May 1	1/4	5.8-7.0	3-10	Widely adapted; drought tolerant
Chufa (warm-season perennial)	50		110	Apr 15-June 1	1-2	5.8-7.0	3-10	Sandy or loam soils; avoid clay soils; moderate fertility requirements
Radish (cool-season biennial)	10-12			Aug 1-Oct 15	1/2-1	6.5-7.0	3-10	Widely adapted; does not tolerate wet soils
Rape and kale (cool-season annual)	6-8	50		Aug 1-Oct 15	1/2-1	5.8-7.0	3-10	Widely adapted; high fertility requirements
Sesame (warm-season annual)	12	46	100	Apr 15-June 1	1/2	6.0-7.0	3-10	Well-drained loams and clay; medium P and K
Sugarbeet (cool-season biennial)	8	23		Apr 1-May 31; Aug 1-Sept 15	1/4-1/2	6.5-7.5	4-6	Well-drained loams (avoid heavy clay or sand); high fertility requirements
Sunflower (warm-season annual)	25	32	110	Apr 15-May 15	1-2	5.8-7.0	3-10	Well-drained soils; high P and K requirements
Turnip (cool-season biennial)	4-5	55		Aug 1-Oct 15	1/4	5.8-7.0	3-10	Widely adapted; high fertility requirements

[1]Annual plantings complete their life cycle in one growing season and — depending on the plant, variety, and management strategy — may or may not reseed. Biennials normally require two growing seasons to complete their life cycle. Perennials continue living after flowering and producing seed and, depending on management, may be present for many years.

[2]All seeding rates in this chart are for a single-species broadcast planting. When planting mixtures, the seeding rate for each species included should be reduced according to the number of species in the mixture, the composition preferred, and the growth form and desired structure of the resulting stand. Drilled plantings typically require approximately 25 percent less seed, depending on plant species. All seeding rates are based on Pure Live Seed (see page 52).

[3]Appropriate planting dates vary with location. The approximate dates shown in this chart are for USDA Plant Hardiness Zones 5-9 (see Fig. 2.3). For Zones 3 and 4, many cool-season species are best planted in the spring, as opposed to the fall. There are even specific varieties of some species, such as spring wheat and spring oats, for these regions.

[4]All legume seed should be inoculated with species-specific inoculant prior to planting unless the seed was purchased preinoculated.

Appendix 2

Various herbicides[1] and applications for wildlife food plots. This chart should be used as a reference only. Always read the herbicide label prior to application. Refer to herbicide labels prior to purchase by visiting http://www.cdms.net.

Primary active ingredient	Trade name (% active ingredient)	Suggested rate per acre[2]	Application[3]	Selected labeled crop / application	Rate per crop
Broad-spectrum herbicides (may kill any type of plant that photosynthesizes)					
glyphosate	Roundup Weather Max (48.8); Gly-4 Plus (41); Accord (53.8); several others	1 – 5 quarts	postemergence	Roundup Ready™ corn, Roundup Ready™ soybeans; many other applications	
glufosinate	Ignite 280 SL (24.5), Finale (11.32), Liberty 280 SL	22 – 43 oz	postemergence	LibertyLink™ Corn, LibertyLink™ Soybeans, many other applications	
Grass-selective herbicides (only kills grasses)					
clethodim	Clethodim 2 EC (26.4); Select Max (12.6); Tapout (12.6)	9 – 32 ounces (Select Max); 6 – 16 ounces (Clethodim 2 EC)	postemergence	alfalfa, birdsfoot trefoil, brassicas, cowpeas peanuts, soybeans, sunflowers, non-crop areas	(Clethodim 2 EC) alfalfa 6 – 16 oz; cowpeas 6 – 8 oz; brassicas 6 – 8 oz; clover 6 – 16 oz; peanut 6 – 16 oz; soybean 6 – 16 oz; sunflower 6 – 16 oz
sethoxydim	Poast (18); Poast Plus (13)	0.5 – 2.5 pints (Poast)	postemergence	alfalfa, clovers, birdsfoot trefoil, brassicas, Austrian winter peas, soybeans	(Poast) alfalfa <2.5 pts; trefoil <2.5 pts; clovers <2.5 pts; brassicas <1.5 pts; peas <2.5 pts.; soybeans <2.5 pts.
fluazifop	Fusilade DX (24.5)	8 – 24 ounces	postemergence	soybeans	soybeans 8 – 24 oz
quizalofop	Assure II (10.3)	5 – 12 ounces	postemergence	brassicas, soybeans	brassicas 5 – 12 oz; soybeans 5 – 12 oz
tralkoxydim	Achieve Liquid (35)	7 – 9 ounces	postemergence	wheat, triticale	wheat and triticale 7 – 9 oz

Manufacturer	Residual soil activity	Herbicide mode of action (site of action group), purpose for spraying, comments[4]
several	No	EPSP synthase inhibitor (Group 9); controls wide variety of weeds in Roundup Ready™ crops; controls existing vegetation for new plots; kills weeds in dormant plots; minimum time from application to rainfall vaires from 0.5 – 4 hours depending on the product
Bayer CropScience	No	Glutamine synthetase inhibitor (Group 10); controls wide variety of weeds in Liberty Link™ crops; apply to glyphosate-resistant pigweeds and horseweed before they are 6 inches tall; rain-fall within 4 hours of application may reduce weed control
Valent and others	No	Lipid biosynthesis (ACC'ase) inhibitor (Group 1); Select Max contains surfactant; controls various grass weeds; does not harm yellow nutsedge, including chufa; use higher rates and/or multiple applications for perennial grasses (such as johnsongrass, tall fescue, orchardgrass, and bermudagrass); rainfall within 1 hour of application may reduce weed control
BASF	No	Lipid biosynthesis (ACC'ase) inhibitor (Group 1); controls various grass weeds; does not harm yellow nutsedge, including chufa; apply before annual grasses exceed 4 inches; use higher rates and/or multiple applications for perennial grasses (such as tall fescue, orchardgrass, and bermudagrass) but use caution to not exceed maximum rate per season; rainfall within 1 hour of application may reduce weed control
Syngenta	No	Lipid biosynthesis (ACC'ase) inhibitor (Group 1); controls various grass weeds; split applications (10 and 8 ounces) are recommended for rhizome johnsongrass; rainfall within 1 hour of application may reduce weed control
DuPont	No	Lipid biosynthesis (ACC'ase) inhibitor (Group 1); controls various grass weeds; does not harm yellow nutsedge, including chufa; two 5-ounce split applications are recommended for rhizome johnsongrass; higher rates are needed for perennial grasses; rainfall within 1 hour after application may reduce weed control
Syngenta	Yes	Lipid biosynthesis (ACC'ase) inhibitor (Group 1); controls annual ryegrass and foxtail grasses; will kill oats; spray when ryegrass is in 1 – 4-leaf stage; dieback may take 4 weeks; does not control forbs; crop rotation minimum (days) following application: cereal grains and leaf crops (30), all other crops (106); rainfall within 1 hour of application may reduce weed control

Primary active ingredient	Trade name (% active ingredient)	Suggested rate per acre[2]	Application[3]	Selected labeled crop / application	Rate per crop
Forb-selective herbicides (or broadleaf-selective; kills forbs, brambles, vines, and some trees/shrubs)					
2,4-D	2,4-D Amine (46.8)	0.5 – 3 pints (refer to label for specific rate)	postemergence	field corn, grain sorghum, wheat, oats, rye, millets, rice	wheat, millet, rye 0.5 – 2 pts; oats 0.5 – 1.5 pts; corn (PRE) 1 – 2 pts; corn (POST) 0.5 – 1 pts; rice (PPI) 0.5 – 2 pts; rice (POST) 1 – 2 pts.
2,4-DB	Butyrac 200 (25.9); 2,4-DB 200 (26.2)	11-26 ounces; 1-3 quarts	postemergence	soybeans, alfalfa, clovers, birdsfoot trefoil, peanuts	soybean (PRE) 11 – 15 oz; soybean (POST) 11 – 15 oz; peanuts 13 – 26 oz; forage legumes 1 – 3 qrts
clopyralid	Stinger (40.9)	4 - 11 ounces	postemergence	brassicas, corn, oats, wheat, sugarbeets, turnips	brassicas 4 – 8 oz; corn 4 – 11 oz oats, wheat 4 – 5 oz; sugarbeets 4 – 11 oz; turnips 5 – 8 oz
cloransulam-methyl	FirstRate (84)	0.3 – 0.75 ounces	preplant incorporated, preemergence, postemergence	soybeans	0.75 oz PPI or PRE; 0.3 – 0.6 oz POST
dicamba	Banvel (48.2), Rifle (48.2)	2 – 16 ounces	postemergence	field corn, grain sorghum, wheat, oats, rye	wheat, oats, rye 2 – 4 oz; grain sorghum 8 oz; corn 8 – 16 oz
dicamba	Clarity (56.1), Vanquish (56.8), Clash (56.8)	2 – 16 ounces	postemergence	field corn, grain sorghum, oats, triticale, wheat	(Clarity) corn (PPI or PRE) 8 or 16 oz; corn (POST) 8 or 16 oz; wheat, oats, triticale 2 – 4 oz; grain sorghum 8 oz.
dicamba and 2,4-D	Weedmaster (12.4, 35.7), Outlaw (12.2, 24.28), Latigo (18.3, 24.6)	1 – 2 pints	postemergence	grain sorghum, wheat	(Weedmaster) grain sorghum 1 pt; wheat 0.5 – 1 pts.
saflufenacil	Sharpen (29.7)	1 – 6 ounces	preplant incorporated, preemergence, postemergence	alfalfa, corn, rice	dormant alfalfa (POST) 1 – 2 oz; corn (PPI or PRE only) 2 – 3.5 oz; rice 1 – 4 oz (PPI, PRE), 1 oz (POST)
thifensulfuron-methyl (33.3)	Harmony Extra SG	0.45 – 0.9 ounces	postemergence	wheat, triticale, oats	wheat, triticale 0.45 – 0.9 oz; oats 0.45 – 0.6 oz
Broad-spectrum selective herbicides (kills various plants of different types)					
acifluoren	Ultra Blazer (20.1)	1 – 1.5 pints	postemergence	soybeans, peanuts, rice	soybeans and peanuts 1 – 1.5 pts; rice 1.5 pts.

Manufacturer	Residual soil activity	Herbicide mode of action (site of action group), purpose for spraying, comments[4]
several	Yes (about 2 weeks)	Growth regulator (Group 4); controls various forb weeds; 2,4-D volatilizes quickly in hot weather and is highly susceptible to spray drift; 2,4-D should not be applied near growing broadleaf agricultural crops (such as tobacco or cotton); applications prior to crop emergence can be made to emerged weeds; oats are less tolerant to 2,4-D than wheat—do not spray over oats during or immediately after cold weather; rainfall within 1 hour of application may reduce weed control
Agri-Star	Yes (about 2 weeks)	Growth regulator (Group 4); controls various forb weeds; good control on sicklepod; does not control chickweed, henbit, plantain or dock; apply before weeds exceed 3 inches and when legume has two or more trifoliate leaves; rainfall within 6 hours of application may reduce weed control
Dow AgroSciences	Yes	Growth regulator (Group 4); rainfall within 6 hours of application may reduce weed control
Dow AgroSciences	Yes	ALS inhibitor (Group 2); controls various broadleaf weeds including horseweed, pigweeds, smartweeds, morningglories, jimsonweed, and cocklebur; partial control of Palmer amaranth and waterhemp when applied preemergence; weak on sicklepod; can be applied postemergence anytime after development of first trifoliate leaf and prior to full bloom; improved control of ALS-resistant weeds when mixed with Dual II Magnum, Prowl, or Treflan; rainfall within in 2 hours of postemergence application can reduce weed control; crop rotation minimum (months) following application: wheat (4), alfalfa, clovers, corn, peanuts, rice, sorghum, oats (9), sunflowers, sugarbeets (30)
Micro Flo	Yes	Growth regulator (Group 4); controls various forb weeds; use caution to prevent drift and injury to sensitive crops; apply to grain sorghum after all sorghum plants have emerged, but before they are 15 inches tall; apply to small grains after seedlings reach 3-leaf stage; crop rotation guidelines following application: corn, grain sorghum and soybeans may be planted in the spring following applications made the previous year; rainfall within 4 hours of application may reduce weed control
BASF	Yes	Growth regulator (Group 4); controls various forb weeds; use caution to prevent drift and injury to sensitive crops; apply to grain sorghum after all sorghum plants have emerged, but before they are 15 inches tall; apply to small grains after seedlings reach 3-leaf stage; no crop rotation restrictions 120 days or more following application; rainfall within 4 hours of application may reduce weed control
NuFarm	Yes	Growth regulator (Group 4); controls various forb weeds; rainfall within 4 hours of application may reduce weed control
BASF	Yes	Damages plant membranes (Group 14 or Group E); used for preemergence control of various forbs, including pigweeds and smartweeds, in various crops; do not apply after crop emergence (except rice); before applying to rice, verify selectivity of Sharpen on your variety with seed company to avoid potential injury to sensitive varieties; rate varies with soil texture (see label); crop rotation minimum for most crops is 4 – 6 months (see label)
DuPont	No	Amino acid synthesis inhibitor (Group 2); controls various forb weeds and wild garlic; apply after wheat and oats reach 2-leaf stage; crop rotation minimum (months) following application: brassicas (2), any other crop (1½); rainfall within 4 hours of application may reduce weed control
UPI	Yes	Cell membrane disrupter (Group 14); controls certain broadleaf weeds, including glyphosate-resistant pigweeds when they are young and sesbania; rice must be past 3-leaf stage; do not apply after rice reaches boot stage; crop rotation minumum following application: small grains (40 days), all other crops (100 days); rainfall within 4 hours after application will reduce effectiveness

231

Primary active ingredient	Trade name (% active ingredient)	Suggested rate per acre[2]	Application[3]	Selected labeled crop / application	Rate per crop
atrazine	Atrazine 4L (restricted use, 42.2); AAtrex 4L (restricted use, 42.6); others	3 – 4 pints	preplant incorporated, preemergence, postemergence	field corn, grain sorghum	(Atrazine 4L) corn 3.5 – 4 pts; Sorghum 3.5 – 4 pts
atrazine (33) and S-metolachlor (26.1)	Bicep II Magnum (restricted use, 33, 26.1); Cinch ATZ (restricted use, 33, 26.1)	1.5 – 2.5 quarts	preplant incorporated, preemergence	field corn, (also grain sorghum if seed treated with Concep™)	(Bicep II Magnum) corn and sorghum (single, PRE) 2.1 – 2.58 qts, (split PPI/PRE) 1.4 qts/0.7 qts, 1.75qts/0.9 qts
benefin	Balan DF	2 – 2.5 pounds	preplant incorporated	alfalfa, clover	alfalfa and clovers 2 – 2.5 lbs
bentazon	Basagran (44)	1 – 2 pints	postemergence	field corn, Southern peas (cowpeas), grain sorghum, soybeans, rice	corn, Southern peas, grain sorghum, soybeans 1 – 2 pts; rice 1.5 – 2 pts
bentazon and acifluorfen	Storm (29.2, 13.4)	1 – 1.5 pints	postemergence	soybeans, peanuts, rice	soybeans and peanuts 1 – 1.5 pts; rice 1.5 pts.
carfentrazone-ethyl	Aim EC (22.3)	0.25 – 6.4 ounces	postemergence	field corn, grain sorghum, soybeans, rice	soybeans 0.25 – 0.5 oz; corn and grain sorghum 0.5 – 1.0 oz; millets, oats, cereal rye, triticale, wheat 0.5 – 2.0 oz; rice 1.6 – 6.4 oz
EPTC	Eptam 7-E (87.8)	2.25 – 4.5 pints	preplant incorporated	alfalfa, birdsfoot trefoil, clovers, lespedeza	alfalfa, birdsfoot trefoil, clovers, lespedeza 2.25 – 4.5 pts;
fomesafen	Flexstar (22.1), Reflex (22.8)	1 – 1.6 pints	postemergence	soybeans	soybeans 1.6 pts (varies by region; see label)
halosulfuron-methyl	Permit (75), Sandea (75)	0.6 – 1.3 ounces	postemergence	field corn, grain sorghum, rice	(Permit) corn 0.6 oz; grain sorghum 0.6 – 1 oz; rice 0.6 – 1.3 oz

Manufacturer	Residual soil activity	Herbicide mode of action (site of action group), purpose for spraying, comments[4]
several	Yes	Photosynthetic inhibitor (Group 5); controls various forb and grass weeds; atrazine can travel through soil and can enter ground water; wastes from atrazine may be toxic; refer to label for restrictions; weed control may be reduced without adequate soil moisture; do not rotate to any crop except corn or sorghum until the following year, or injury may occur; if applied after June 10, do not plant crop other than corn or grain sorghum the following year; do not plant spring-seeded small grains or small-seeded legumes and grasses the year following application or injury may occur; rainfall within 4 hours after postemergence application may reduce weed control
Syngenta	Yes	Photosynthetic inhibitor (Group 5) and shoot growth inhibitor (Group 15); do not use on soils with less than 1% organic matter; controls various forb and grass weeds (including crabgrass); do not rotate to any crop except corn or sorghum until the following year, or injury may occur; if rain does not occur within a few days after application, weed control may be decreased; crop rotation minimum (months) following application: soybeans, peanuts (12), spring-seeded small grains, small-seeded legumes (24)
Loveland	Yes	Seedling root growth inhibitor (Group 3); preemergence control of various grass and broadleaf weeds in alfalfa and clover; must be incorporated thoroughly into top 3 inches of seedbed within 4 – 8 hours of application; within 4 hours during extremely high temperatures and intense sunlight
BASF	No	Photosynthetic inhibitor (Group 6); controls various forb weeds, including smartweeds, and yellow nutsedge; may cause yellowing or speckling in soybeans and cowpeas, but this is temporary and outgrown within 10 days; add 2 oz of Butyrac 200 for improved control of morningglory; rainfall within 4 hours after postemergence application may reduce weed control
UPI	Yes	Photosynthetic inhibitor (Group 6) and cell membrane disrupter (Group 14); controls certain broadleaf weeds; may also provide partial control of some grasses; may cause foliar burn on soybeans, but it is short-lived; rice must be past 3-leaf stage; do not apply to rice when field is flooded or splashing will wash herbicide off leaves; crop rotation restrictions: do not plant small grains within 40 days following treatment; rainfall within 4 hours after postemergence application may reduce weed control
FMC	No	Cell membrane disrupter (Group 14); controls various forb weeds; rainfall within 6 – 8 hours after application may reduce weed control; do not apply to soybeans until 3-leaf stage; up to 0.5 ounce may be applied to later maturing soybeans than Group 3.5; soybeans may show burn, speckling or necrosis, but will quickly outgrow initial effects; apply to rice when rice is at 2-leaf stage or larger, but before internode elongation
Syngenta	Yes	Seedling shoot growth inhibitor (Group 5); incorporate immediately after application; controls annual grasses and winter annual forb weeds, as well as yellow nutsedge; does not control plantains or docks; do not use if small grain nurse crop is planted with legumes; does not control established weeds; may be difficult to find
Syngenta	Yes	Cell membrane disruptor (Group 14); controls various forbs, including glyphosate-resistant pigweeds; weak on grasses; may cause speckling of soybean leaves, but they quickly outgrow it; commonly mixed with glyphosate; rainfall within one hour after application may reduce weed control; crop rotation minimum (months) following application: cereal grains, peanuts (4), corn (10)
Gowan	Yes	Amino acid synthesis inhibitor (Group 2); controls various forb weeds, including pigweeds and horseweed, and nutsedge; weak on sicklepod and morningglory; do not use more than 1 ounce per acre on grain sorghum; make only one application per season; crop rotation minimum (months) following application: field corn (1), oats, proso millet, rice, rye, grain sorghum, spring cereal crops, wheat (2), peanuts (6), alfalfa, clovers, dry beans, field peas, soybeans (9), canola (15), sunflowers (18); rainfall within 4 hours after postemergence application may reduce weed control

Primary active ingredient	Trade name (% active ingredient)	Suggested rate per acre[2]	Application[3]	Selected labeled crop / application	Rate per crop
imazamox	Raptor (12.1)	4 – 6 ounces	postemergence	alfalfa, chicory, Southern peas (cowpeas), soybeans	chicory, cowpeas, and soybeans 4 oz; alfalfa 4 – 6 oz.
imazamox	Beyond (12.1)	4 – 6 ounces	postemergence	Clearfield™ varieties of sunflowers, wheat, rice, or corn	rice 4 – 6 oz; sunflower 4 oz; spring wheat 4 – 5 oz; winter wheat 4 – 6 oz;
imazapic	Plateau (23.6); Cadre (23.6); Panoramic 2SL (23.3)	4 – 12 ounces	preemergence, postemergence	many applications	
imazapic and glyphosate	Journey (8.1, 21.9)	16 – 32 ounces	preemergence, postemergence	many applications	
imazapyr	Arsenal AC (53.1); Arsenal (28.7); Chopper (27.6); Imazapyr 4 SL (52.6); Rotary 2 SL (27.8); Polaris AC Complete (53.1)	6 – 24 ounces	postemergence	many applications	
imazaquin	Scepter 70 DG (70.0)	1.4 – 2.8 ounces	preplant incorporated, preemergence, postemergence	soybeans	soybeans 1.4 – 2.8 oz
imazethapyr	Pursuit (22.9), Thunder (22.9)	3 – 6 ounces	preplant incorporated, preemergence, postemergence	alfalfa, clovers, Austrian winter peas, Southern peas (cowpeas), soybeans, Clearfield™ corn	(Pursuit) alfalfa and clover 3 – 6 oz; Clearfield™ corn 4 oz; soybeans 4 oz; Austrian winter peas and cowpeas 3 – 4 oz
mesosulfuron methyl	Osprey (4.5)	4.75 ounces	postemergence	wheat	wheat 4.75 oz
metsulfuron methyl	Escort XP (60)	0.33 – 2 ounces	postemergence	many applications	

Manufacturer	Residual soil activity	Herbicide mode of action (site of action group), purpose for spraying, comments[4]
BASF	Yes	Amino acid synthesis inhibitor (Group 2); controls various forb and grass weeds; most effective when applied before weeds exceed 5 inches; good control on cocklebur, morningglory, pigweed, velvetleaf, and johnsongrass; weak on sicklepod and 3-seeded mercury; applications must be made at least 1 hour prior to rain; crop rotation minimum (months) following application: Clearfield™ sunflowers, Clearfield™ wheat (0), wheat (3), rye (4), corn (8½), grain sorghum, millets, oats, peanuts, rice, turnips (9); rainfall within 1 hour after postemergence application may reduce weed control
BASF	Yes	Amino acid synthesis inhibitor (Group 2); controls various broadleaf and grass weeds in Clearfield™ varieties of sunflowers, wheat or corn only; good control on cocklebur, morningglory, pigweed, velvetleaf, and johnsongrass; apply to sunflowers during 2- to 8-leaf stage; crop rotation minimum (months) following application: alfalfa, wheat (3), rye (4), field corn (8½), grain sorghum, millets, oats, peanuts, rice, sunflowers, turnips (9); rainfall within 1 hour after postemergence application may reduce weed control
BASF and others	Yes	Amino acid synthesis inhibitor (Group 2); selectively kills tall fescue, crabgrass, johnsongrass, sicklepod, yellow nutsedge, and others to promote nesting and brood-rearing cover for bobwhite and wild turkeys; also can be used to control japangrass (*Microstegium vimineum*) along woods roads, including those planted to crimson clover, and can be used to control sandbur (*Cenchrus*); Plateau™ is available through select government agencies; Panoramic™ can be purchased at local seed/herbicide supply stores; Cadre™ is labeled for peanuts only in select states; crop rotation minimum (months) following application: wheat, rye (4), field corn, cowpeas, soybeans (9), grain sorghum, oats (18); rainfall within 3 hours after postemergence application may reduce weed control
BASF	Yes	Amino acid synthesis inhibitor (Group 2) and EPSP synthase inhibitor (Group 9); kills tall fescue, crabgrass, johnsongrass, and others to promote nesting and brood-rearing cover for bobwhite and wild turkeys; crop rotation minimum (months) following application: wheat, rye (4), cowpeas, soybeans (9), grain sorghum, oats (18); rainfall within 1 hour after postemergence application may reduce weed control
BASF and others	Yes	Amino acid synthesis inhibitor (Group 2); kills bermudagrass in the season prior to planting food plots; good for spot-spraying woody encroachment in perennial plots and broadcast spraying around plots to prevent woody encroachment; rotational crops may be planted 12 months after application at the recommended pasture and rangeland rate; rainfall within 3 hours after postemergence application may reduce weed control
BASF	Yes	Amino acid synthesis inhibitor (Group 2); not for use in many western states (see label); often mixed with other herbicides for more broad-spectrum and complete weed control; crop rotation minimum (months) following application: wheat (3; may be longer for upper Midwest—check label), corn (10), grain sorghum, oats, peanuts (11), sugarbeets (40)
BASF	Yes	Amino acid synthesis inhibitor (Group 2); controls various forb and grass weeds; apply postemergence only on alfalfa or clover plots (2 trifoliate stage or larger); apply to Clearfield™ varieties of corn only; postemergence applications most effective when applied before weeds exceed 3 inches; for optimum control with preemergence applications, sufficient moisture in the top 2 inches of soil is necessary within 7 days after application; preemergence applications for labeled crops are generally more effective and for a broader range of weeds, such as yellow nutsedge, than postemergence applications; crop rotation minimum (months) following application: alfalfa, clover, rye, wheat (4), field corn (8½), oats, grain sorghum, sunflowers (18); rainfall within 1 hour after postemergence application may reduce weed control
Bayer CropScience	Yes	Amino acid synthesis inhibitor (Group 2); controls annual bluegrass, annual ryegrass and a few annual forbs, including wild mustard, chickweed, henbit and redroot pigweed; do not apply to wheat sown with legumes; rainfall within 4 hours after application may reduce weed control; crop rotation minimum (months) following application: sunflowers (1), soybeans, rice peas, peanuts (3), corn (12); rainfall within 4 hours after postemergence application may reduce weed control
DuPont	Yes	Amino acid synthesis inhibitor (Group 2); apply at 0.33-0.5 oz/ac to control bahiagrass, apply at 1 oz/ac to control sericea lespedeza when in flower; rainfall within 4 hours after postemergence application may reduce weed control

Primary active ingredient	Trade name (% active ingredient)	Suggested rate per acre[2]	Application[3]	Selected labeled crop / application	Rate per crop
metasulfuron methyl and chlorsulfuron	Cimarron Plus (48, 15)	0.125 – 1.25 ounces	postemergence	many applications	
nicosulfuron	Accent Q (54.5)	0.9 – 1.8 ounces	postemergence	field corn	corn 0.9 – 1.8 oz
pendimethalin	Prowl 3.3 EC (37.4); Prowl H2O (38.7); Stealth (37.4); Framework 3.3 EC (43)	1 – 4 pints, 1 – 5 quarts (varies by crop and soil type),	preplant incorporated, preemergence, postemergence incorporated	field corn, alfalfa, forage legumes, Southern peas (cowpeas), Austrian winter peas, soybeans, sunflowers, rice, peanuts	(Prowl 3.3 EC) alfalfa 1.2 – 4.8 qrts; cowpeas 1.8 – 3.6 pts; forage legumes 1.2 – 3 pts; peanuts 1.2 – 2.4 pts, rice (PRE) 1.8 – 2.4 pts, (POST) 1.8 – 2.4 pts; soybeans (PPI) 1.8 – 3.6 pts, (PRE) 1.8 – 3.0 pts, sunflower 2.4 – 3.6 pts
pyroxasulfone	Zidua (85)	1 – 4 ounces	preemergence	corn, soybeans, wheat	corn 1.5 – 4 oz; soybeans 1.5 – 3.5 oz; wheat 1 – 2 oz
quinclorac	Facet 75 DF (75)	0.33 – 0.67 pounds	preemergence, postemergence	rice	rice 0.33 – 0.67 lbs
rimsulfuron (16.7) and thifensulfuron-methyl (16.7)	Leadoff	1.5 – 2.0 ounces	preplant incorporated, preemergence	field corn	field corn 1.5 – 2.0 oz
S-metolachlor	Dual Magnum (83.7); Medal (83.7)	1 – 2 pints	preplant incorporated, preemergence	field corn, Southern peas (cowpeas), soybeans, peanuts, sunflowers (also grain sorghum if seed treated with Concep™)	(Dual Magnum) corn 1 – 2 pts; soybeans 1 – 2 pts; peanuts 1 – 2 pts; Southern peas 1 – 2 pts; grain sorghum 1 – 1.67 pts; sunflowers 1 – 2 pts.
S-metolachlor	Dual II Magnum (82.4); Cinch (82.4)	1 – 2 pints	preplant incorporated, preemergence	field corn, Southern peas (cowpeas), soybeans, sunflowers, peanuts, (also grain sorghum if seed treated with Concep™)	(Dual II Magnum) corn 1 – 2 pts; soybeans 1 – 2 pts; peanuts 1 – 2 pts; Southern peas 1 – 2 pts; grain sorghum 1 – 1.67 pts; sunflowers 1 – 2 pts.
saflufenacil (6.24) and dimethenamid (5.04)	Verdict	5 – 18 ounces (varies by crop, application, and soil type)	preplant incorporated, preemergence, postemergence	field corn, grain sorghum (if using treated seed), soybeans	field corn and grain sorghum 10 – 18 oz; soybeans 5 – 10 oz

Manufacturer	Residual soil activity	Herbicide mode of action (site of action group), purpose for spraying, comments[4]
DuPont	Yes	Amino acid synthesis inhibitor (Group 2); apply at 0.375 oz/ac to control bahiagrass, apply at 0.625 oz/ac to control sericea lespedeza when in flower; rainfall within 4 hours after postemergence application may reduce weed control
DuPont	Yes	Amino acid synthesis inhibitor (Group 2); controls several grasses and a few forb weeds; rainfall within 4 hours after application may reduce weed control; crop rotation minimum (months) following application: soybeans (0.5), oats, rye, wheat (4), canola, grain sorghum, peas (10), sunflowers (11), alfalfa, red clover (12)
BASF	Yes	Seedling root growth inhibitor (Group 3); controls various grass and forb weeds; do not apply preplant incorporated in corn plots, but only preemergence; corn should be planted at least 1.5 inches deep and completely covered with soil; use no more than 2.4 pints per acre on rice plots; most effective with adequate rain soon after application; pendimethalin does not control established weeds; wheat may be planted in the fall 4 months after application
BASF	Yes	Seedling root-and-shoot growth inhibitor (Group 15); controls various annual grasses and forbs, including glyphosate-resistant pigweeds, crabgrass, broadleaf signalgrass, and ryegrass; do not apply near creeks, ponds, wetlands, or wells; activated by at least 0.5 inch rainfall; predissolve product in bucket of water before adding to sprayer; crop rotation restriction (months) at 3.5-oz rate: wheat and sunflower (4), field peas (6), alfalfa and grain sorghum (10), small grains other than wheat (11), sugarbeets (15)
BASF	Yes	Cellulose inhibitor/synthetic auxin (Group 4); controls a variety of annual grasses and broadleaf weeds; optimum weed control is dependent upon adequate soil moisture, including flush irrigation; do not plant any crop other than rice for at least 10 months after application; rainfall within 6 hours after postemergence application may reduce weed control
DuPont	Yes	Cell membrane disrupter (Group 2); controls various annual grass and forb weeds (does not control Palmer amaranth, jimsonweed, or sandbur); do not apply to coarse-textured soils (sandy, sandy loam) with less than 1% organic material; may be applied after fall harvest through early spring up to planting; may be tank-mixed with glyphosate; crop rotation minimum (months) following application: cereal grains (4), soybeans (10), sunflowers (10)
Syngenta	Yes	Seedling shoot growth inhibitor (Group 15); controls various grass and forb weeds (including pigweeds) and nutsedge; does not control emerged weeds; if rain does not occur within a few days after application, weed control may be decreased; crop rotation minimum (months) following application: alfalfa (4), oats, rye, wheat (4½), clovers (9), buckwheat, grain sorghum, rice (12)
Syngenta	Yes	Seedling shoot growth inhibitor (Group 15); controls various grass and forb weeds (including pigweeds) and nutsedge; does not control emerged weeds; if rain does not occur within a few days after application, weed control may be decreased; crop rotation minimum (months) following application: alfalfa (4), oats, rye, wheat (4½), clovers (9), buckwheat, rice (12)
BASF	Yes	Cell membrane disrupter (Group 14) and seedling shoot growth inhibitor (Group 15); causes plant cell membrane damage; primarily preemergence herbicide that controls various annual grasses and forbs and some sedges, including yellow nutsedge; controls pigweeds; at least 0.5 inches of rain or irrigation needed prior to weed seed germination; postemergence applications rainfast after one hour; do NOT apply postemergence onto crop leaves; use MSO (not nonionic surfactant) with postemergence applications; crop rotation minimum (months) following application: sugarbeets, sunflowers (9)

Primary active ingredient	Trade name (% active ingredient)	Suggested rate per acre[2]	Application[3]	Selected labeled crop / application	Rate per crop
sulfentrazone (7.55), S-metolachlor (68.25)	BroadAxe	19 – 38.7 ounces	preplant incorporated, preemergence	soybeans, cowpeas, sunflowers	19 – 38.7 oz (varies by soil type; may not be recommended for cowpeas in sandy soils with <1.5% organic matter; see label)
sulfosulfuron	OutRider (75)	1 – 2 ounces	preemergence, postemergence	wheat	wheat 0.66 oz
triclopyr	Garlon 3A (44.4); Element 3A (44.4); Triclopyr 4 EC (44.3)	1 – 8 quarts	postemergence	many applications	
trifluralin	Trifluralin 4EC (43); Trifluralin HF (43); Treflan HFP (43); Treflan 4D (43); Trust (43)	1 – 2.5 pints	preplant incorporated	brassicas, chicory, Southern peas, soybeans, wheat, sunflowers	(Treflan HFP) chicory 1 – 2 pts; brassicas 1 – 2 pts; Southern peas 1 – 2 pts; wheat 1.5 – 2 pts; soybeans 1 – 2.5 pts; sunflowers 1 – 2 pts.

[1]Use of brand, trade, or company names in this publication is for clarity and information; it does not imply approval of the product or company to the exclusion of others, which may be of similar composition or equal value. Always be sure to read, understand, and follow directions, precautions, and restrictions on herbicide labels before use. For optimum long-term herbicide weed control, it may be necessary to use herbicides with different modes of action to reduce potential for weed resistance. As herbicides, herbicide labels, and their availability and recommendations may change, it is best to consult your local Extension agent or farm supply distributor for the latest recommendations on herbicide use.

[2]Various crops labeled for a particular herbicide often require or tolerate different application rates. Application rates may differ depending on soil texture and percent organic material. Always refer to herbicide labels for specific application rates for a given crop.

[3]Various crops labeled for a particular herbicide often require or tolerate different types of applications, such as preplant incorporated, preemergence, or postemergence. A surfactant should be added to all postemergence herbicide applications unless the herbicide already contains surfactant. Refer to the herbicide label as to which surfactant to use, mixing instructions, and recommended rates.

[4]Many herbicides have multiple uses. Read the herbicide label before use. The purposes stated in this table are for general information. Herbicide mode of action describes how a herbicide inhibits plant growth. Moreover, herbicide site of action describes the exact location within the plant where the herbicide binds. Refer to UT Extension publication *PB 1580 Weed Control Manual for Tennessee* (http://weeds.utk.edu) to identify herbicide site of action and other information. Crop rotation restrictions may preclude you from planting specific crops for a given amount of time after applying various herbicides. Refer to herbicide label for additional information concerning crop rotation restrictions. The majority of postemergence herbicides work best when applied to actively growing plants, often before they reach a certain size or height. Refer to herbicide label to identify optimum application effectiveness.

Manufacturer	Residual soil activity	Herbicide mode of action (site of action group), purpose for spraying, comments[4]
Syngenta	Yes	Cell membrane disrupter (Group 14) and seedling shoot growth inhibitor (Group 15); soil-applied herbicide that controls various forb (including pigweeds and morningglories), grass, and sedge weeds; 1/2-1 inch rainfall required within 7 – 10 days after application or shallow incorporation (< 2 inches) is necessary for weed control; not recommended in sandy soils with <1.5% organic matter; crop rotation minumum (months) following application: triticale, peanuts (4), rye, wheat (4.5), corn, grain sorghum, rice (10), alfalfa, buckwheat, millets, oats (12), sugarbeets (36)
Monsanto	Yes	Amino acid synthesis inhibitor (Group 2); may be used for weed control in wheat crops in certain states (see label); controls a variety of grass and broadleaf weeds, including cheatgrass, chess, downy brome, johnsongrass, buttercup, chickweed, mustard, and nutsedge; rainfall within 2 hours of application may reduce weed control; crop rotation minimum following application varies according to region (see label)
Dow AgroSciences and others	No	Growth regulator (Group 4); good for spot-spraying woody encroachment in perennial plots and broadcast spraying around plots to prevent woody encroachment; also very effective in eradicating hard-to-control perennial forb weeds prior to plot establishment; rainfall within 1 hour after postemergence application may reduce weed control
Dow AgroSciences	Yes	Seedling root growth inhibitor (Group 3); controls various grass and forb weeds, including pigweeds; incorporate immediately after application; trifluralin does not control established weeds; crop rotation minimum (months) following application: proso millet, grain sorghum, oats, annual grass crops (12 – 14)

Summary Of Herbicide Mechanism Of Action According To The Weed Science Society Of America (WSSA)

Site of Action Group	Herbicide Mode of Action
1	Acetyl CoA Carboxylase (ACCase) Inhibitors
2	Acetolactate Synthase (ALS) or Acetohydroxy Acid Synthase (AHAS) Inhibitors
3, 15, 23	Mitosis Inhibitors
4	Synthetic Auxins
5, 6, 7	Photosystem II Inhibitors
8, 16	Fatty Acid and Lipid Biosynthesis Inhibitors
9	Enolpyruvyl Shikimate-3-Phosphate (EPSP) Synthase Inhibitors
10	Glutamine Synthetase Inhibitors
11, 12, 13, 27	Carotenoid Biosynthesis Inhibitors
14	Protoporphyrinogen Oxidase (PPG Oxidase or Protox) Inhibitors
17, 25, 26	Potential Nucleic Acid Inhibitors or non-descript mode of action
18	Dihydropteroate Synthetase Inhibitors
19	Auxin Transport Inhibitors
20, 21, 28, 29	Cellulose Inhibitors
22	Photosystem I Inhibitors
24	Oxidative Phosphorylation Uncouplers

wssa.net/wp-content/uploads/WSSA-Herbicide-MOA-20160911.pdf
Adapted from UTExtension publication PB 1580.

Appendix 3

Response[1] of various weeds to specific herbicides[2,3].

	Glyphosate	Liberty 280 SL	Accent Q	Aim	Atrazine (preemergence)	Atrazine (postemergence)	Banvel	Basagran	Select Max	Poast	2,4-D
Grass weeds											
barnyardgrass	9	5	2	0	6	4	0	0	9	9	0
bermudagrass	X			0		1	0	0	X	X	0
bluegrass, annual	5	3							X		0
cheat											0
large crabgrass	9	6	5	0	7	6	0	0	9	9	0
smooth crabgrass	9	6	5	0	3	4	0	0	9	9	0
foxtail grasses	9	7	9	0	6	7	0	0	9	9	0
goosegrass	9	5		0	6	7	0	0	9	9	0
seedling johnsongrass	10	5	9	0	1	0	0	0	9	8	0
rhizome johnsongrass	9	2	9	0	0	0	0	0	9	7	0
fall panicum	9	5	X	0	3	6	0	0	9	9	0
annual ryegrass	X	2	X	0					8	8	
sandburs	X										
broadleaf signalgrass	9	5	8	0	4	6	0	0	9	9	0
sprangletop	X										
Forb weeds											
3-seeded mercury	X										X
catchweed bedstraw							X				
buttercups	X						X				9
carpetweed	X						X				
chickweed	5	9					9			0	7
cocklebur	10	8	6	6	7	7	9	9	0	0	9

2,4-DB (Butyrac)	Clarity	First Rate	Harmony Extra	Dual Magnum	Bicep II Magnum	Pursuit (pre-emergence)	Pursuit (post-emergence)	Prowl 3.3 EC	Treflan	Python	Beyond/Raptor	Eptam
0	0	0	0	X	9	X	7	X	X		X	8
0	0	0	0		0							X
0	0	0	0						X			
0	0	0	0						X		X	
0	0	0	0	9	9	8	7	9	9	5	7	8
0	0	0	0	9	9	8	7	9	9	5	7	8
0	0	0	0	9	9	8	7	9	9	5	7	8
0	0	0	0	9	9	8	7	9	9	5	7	8
0	0	0	0	8	8	8	7	8	9	4	8	8
0	0	0	0		0		6				X	6
0	0	0	0	X	9	X	7	X	X		X	8
0		0	0	X					X		X	8
		0							X	X		
0	0	0	0	8	8	6	7	8	8	4	7	8
									X			
X												
						X	X					
	X		9									
					X			X	X	X		X
2	9		8		X	X	X		X	X	X	9
8	9	X		0	7	8	8	0	0	7	8	2

	Glyphosate	Liberty 280 SL	Accent Q	Aim	Atrazine (preemergence)	Atrazine (postemergence)	Banvel	Basagran	Select Max	Poast	2,4-D
hophornbeam copperleaf	8										
croton	X						X	X			
bur cucumber	9	9	7			4	X	3			3
dandelion	X						X				
purple deadnettle	5	7					5				4
curly dock	4	7					7		0		7
hairy galinsoga	X										
wild garlic	X						X				7
Carolina geranium	1	8					X				9
groundcherry	6	X		7	X	7	9	3	0	0	8
henbit	5	7					5		0		4
horsenettle	X	4	2	4	3	4	6	0	0	0	4
horseweed (marestail)	2	8					8				9
jimsonweed	8	X		7	8	X	X	8	0	0	8
kochia	X										
lambsquarters	8	6	2	8	9	9	9	6	0	0	8
common mallow	X						X	X			
mayweed chamomile	X						X	X			
morningglories	7	9	7	8		8	9	4	0	0	9
wild mustard	X	X				X	X	X			8
black nightshade	6	X		7	X	7	9	3	0	0	8
pennycress	X						X				X
pepperweed	X						X				9
Palmer pigweed	3	8	3	7	9	9	9	7	0	0	8
smooth pigweed	9	8	9	8	9	9	9	9	0	0	9

2,4-DB (Butyrac)	Clarity	First Rate	Harmony Extra	Dual Magnum	Bicep II Magnum	Pursuit (pre-emergence)	Pursuit (post-emergence)	Prowl 3.3 EC	Treflan	Python	Beyond/ Raptor	Eptam
				4				0	0			
			X									
											X	
1	5		7									8
1	7		9									0
					X							
	X		9									
	8		5									
				7	8	8	8	0	0	9		
1	5		7		X	X	X	X	X	X		8
	6				3							
	8	X	6							X		
X	9	X		0	8	8	9	0	0		6	
								X			X	
4	9			6	9	7	5	8	7	X	5	7
					X		X				X	
			X									
8	9			2	8	7	8	6	6	6	7	4
X	X		9		X					X	X	
	X			7	8	8	8	0	0	9	X	
X												
	X					X	X					
6	9			6	9	3	3	7	8	4	3	
6	9	X		8	9	9	9	7	8	9	8	7

	Glyphosate	Liberty 280 SL	Accent Q	Aim	Atrazine (preemergence)	Atrazine (postemergence)	Banvel	Basagran	Select Max	Poast	2,4-D
pineappleweed	X										
plantain							X				7
poorjoe	X						X				
Florida pusley	X						X				
common ragweed	8	9			9	8	9	5	0	0	9
giant ragweed	6	9	2	2	6	6	9	5	0	0	9
sicklepod (coffeeweed)	9	8	6	1	6	6	8		0	0	8
prickly sida	6	X		4		8	8	X	0	0	7
purslane	X						X	X			
sesbania	X						X	X			
shepherd's-purse	X						X	X			
smartweed	8	7		7	9	8	8	7	0	0	6
sowthistles	X						X				
spotted spurge	9	X		6	X			0	0	0	
thistles	X						X	X			8
velvetleaf	7	7	7	9	6	7	8	8	0	0	8
vetch	5	8					9				8
wild turnip											8
common yarrow	X						X				
Sedge weeds											
yellow nutsedge	7	0	4	0	4	6	0	8	0	0	0

[1]Key to response ratings: 0 = no control; 10 = 100% control; --- = data not available; ratings based on application of each herbicide at labeled rates applied at optimum timing for each weed. An 'X' indicates the herbicide label lists the plant as being controlled or at least suppressed. However, we do not have response data for these plants.

2,4-DB (Butyrac)	Clarity	First Rate	Harmony Extra	Dual Magnum	Bicep II Magnum	Pursuit (pre-emergence)	Pursuit (post-emergence)	Prowl 3.3 EC	Treflan	Python	Beyond/Raptor	Eptam
			X									
2												
					X			X	X	X		X
6	9	X		0	9	8	7	0	0		6	2
6	9	X			6	X	6	0			X	
	8			3	7	0	0	0	0	7	0	X
X				0	7	8	6	0	0	9	6	X
					X			X		X		X
			9			X	X			X		
X	9	X	X	0	9	8	6	3	4	9	6	
	X			7		9	8	0	0	9	8	
X											X	
X	8	X		0	6	8	8	2	2	9	8	
	9		7									
1	0			8	7	3	3	0	0		X	8

[2]Table adapted from UT Extension publication *PB 1580 Weed Control Manual for Tennessee* (http://weeds.utk.edu).

[3]Data collected by personnel with UT AgResearch and UT Extension.

Appendix 4
Various insecticides and applications for wildlife food plots[1].

Trade name[2]	Active ingredient	Rate per acre[3]	Labeled crops[3]	Relative insecticide risk level[4]
Baythroid XL 1*	β-cyfluthrin	1.0-2.8 ounces	corn, forage grasses	Low-Medium
Belt SC	flubendiamide	2.0-4.0	alfalfa, corn, cowpeas, grain sorghum, lablab, peas, peanut, soybean, sunflower	Low
Brigade 2E*, Discipline 2E*, Fanfare 2E*, etc.	bifenthrin	3.0-6.4	corn, soybean	Low-Medium
Declare*	gamma-cyhalothrin	0.5-1.5 ounces	alfalfa, brassicas, corn, cowpeas, forage grasses, grain sorghum, oats, peanut, rice, rye, soybean, sunflower, wheat	Low-Medium
Intrepid 2F	methoxyfenozide	4-8 ounces	brassicas, chicory, chufa, corn, cowpeas, forage grasses, nongrass forages, peanut, soybean, sugarbeet	Low
Karate 2.08*, Warrior II*	β-cyhalothrin	1.0-1.9 ounces	corn, forage grasses, soybean	Low-Medium
Lorsban 15G	chlorpyrifos	5-7 pounds	alfalfa, brassicas, corn, grain sorghum, peanut, soybean, sugarbeet, sunflower	Low-Medium
Malathion 57E	malathion	32 ounces	forage grasses	Low
Mustang Max 0.8*	Z-cypermethrin	1.3-4.0 ounces	corn, forage grasses, soybean	Low-Medium
Orthene 90S	acephate	0.3-1.1 pounds	soybean	Low-Medium
Pounce	permethrin	3-12 ounces	alfalfa, brassicas, corn, soybean	Low-Medium
Prevathon	chlorantraniliprole	14-20 ounces	alfalfa, brassicas, clovers, corn, forage grasses, sesame, soybean, sunflower	Low
Sevin XLR Plus	carbaryl	16-48 ounces	alfalfa, brassicas, clovers, corn, cowpeas, forage grasses, grain sorghum, peanut, rice, soybean, sugarbeet, sunflower, wheat, noncrop	Low
Tracer	spinosad	1-3 ounces	buckwheat, corn, grain sorghum, grass forages, millets, oats, peanut, soybean, rye, triticale, wheat	Low

[1]This table is intended to provide general information on insecticides that may be used for pest control in various crops. It is adapted from information provided in *PB 1768 Insect Control Recommendations for Field Crops,* accessible at http://www.utcrops.com. Visit http://www.cdms.net to view or download pesticide labels.
[2]Use of brand, trade, or company names in this publication is for clarity and information; it does not imply approval of the product or company to the exclusion of others, which may be of similar composition or equal value. Always be sure to read, understand, and follow directions, precautions, and restrictions on pesticide labels before use. As pesticides, pesticide labels, and their availability and recommendations may change, it is best to consult your local Extension agent or farm supply distributor for the latest recommendations on pesticide use.
[3]Control of various insect pests requires different application rates. Furthermore, timing of insecticide treatment varies with crop stage of development. Refer to pesticide labels or to *PB 1768 Insect Control Recommendations for Field Crops,* for specific application recommendations for a given crop.
[4]Relative index of acute toxicity for humans, primarily as related to skin exposure. Never enter a field immediately after an insecticide application. Always read insecticide labels prior to use and follow directions.
*Restricted Use pesticides

Appendix 5
Comparative information for various forages for white-tailed deer.

Species[1]	Germination and initial growth rate	Grazing preference[2]	Resistance to grazing	Crude Protein (percent)[3]	Acid Detergent Fiber (percent)[3]	Date forage collected	High-quality forage available[4,5]
Cool-season legumes							
Arrowleaf clover	slow	moderate	excellent	31.0	19.7	April	March–July
Balansa clover	moderate	moderate	excellent	28.6	13.0	April	October–April
Berseem clover	moderate	high	excellent	24.8	16.8	April	October–May
Crimson clover	moderate	high	excellent	28.4	14.9	April	October – April
Alsike clover	slow	high	excellent	23.8	23.5	June	March–July; October–December
Ladino clover	slow	high	excellent	31.5	17.7	April	March–July; October–December
Red clover	slow	moderate	excellent	23.7	26.7	June	March–August; October–December
Sweetclover	slow	moderately low	good	31.1	18.1	April	March–early June
White-dutch clover	slow	moderate	excellent	31.3	17.7	April	March–July; October–December
Alfalfa	slow	moderate	excellent	24.7	35.1	April	March–early August; October–December
Austrian winter peas	moderate	moderately low	fair	28.0	18.9	April	October–April
Birdsfoot trefoil	slow	low	good	28.2	19.9	April	March–July; October–December
Hairy vetch	moderate	moderately low	good				February–April
Crownvetch	extremely slow	no use recorded	n/a	23.7	36.9	July	n/a
Blue lupine	slow	no use recorded	n/a	34.1		Jan	n/a

continued on next page

Species[1]	Germination and initial growth rate	Grazing preference[2]	Resistance to grazing	Crude Protein (percent)[3]	Acid Detergent Fiber (percent)[3]	Date forage collected	High-quality forage available[4,5]
Sainfoin	slow	no use recorded	n/a	20.6	26.5		n/a
Cool-season grasses							
Barley	fast	extremely low	n/a	23.9	23.4	March	n/a
Oats	fast	high	excellent	26.5	17.9	March	September–early April
Wheat	fast	high	excellent	24.9	21.4	March	September–early April
Rye	fast	high	excellent	23.6	23.1	March	September–early April
Triticale	fast	high	excellent	20.5	26.0	March	September–early April
Ryegrass	fast	low	excellent	12.0	23.5	March	September–early April
Orchardgrass	slow	extremely low	n/a	14.6	36.8	early April	n/a
Tall fescue	slow	no use recorded	n/a	16.6	31.5	early April	n/a
Matuagrass	relatively slow	low	excellent	22.0	19.6	March (after planting)	November–April
Timothy	slow	no use recorded	n/a	29.8	34.8	July	n/a
Bluegrass	slow	no use recorded	n/a				n/a
Warm-season legumes							
Alyceclover	slow	moderate	good	25.7	25.5	July	June–October
American jointvetch	slow	moderate	good	25.3	22.6	July	June–October
Iron-clay cowpeas	moderate	high	good	29.7	19.1	July	June–October
Red ripper peas	moderate	high	good	29.4	24.7	July	June–October
Lablab	moderate	moderate	good	25.7	28.8	July	June–October
Soybeans	moderate	extremely high	poor	33.7	25.9	July	June–October
Quail Haven soybeans	moderate	high	good	24.5	30.5	August	June–October
Florida beggerweed	moderate	high	excellent				June–October
Lathco flatpeas	extremely slow	no use recorded	n/a				n/a
Velvetbean	moderate	no use recorded	n/a	28.8	36.2	July	n/a

248

Species[1]	Germination and initial growth rate	Grazing preference[2]	Resistance to grazing	Crude Protein (percent)[3]	Acid Detergent Fiber (percent)[3]	Date forage collected	High-quality forage available[4,5]
Non-legume broadleaf forbs							
Buckwheat	fast	moderately low	good	25.6	27.0	July	May–September
Chicory	relatively slow	high	excellent	23.6	19.5	December	September–December; March–July
Radish (daikon)	moderate	moderate	good	18.0	23.2	November	October–January
Rape (dwarf essex)	moderate	moderately low	good	32.9	13.1	December	October–April
Turnips (pasja)	moderate	moderately low	good	25.8	24.4	July	October–April
Small burnette	slow	low	good	21.8	24.7	July	

[1]This table shows growth, deer selectivity, and nutritional information for various forages as determined after 9 years of experimentation using side-by-side comparisons and collecting monthly data (measuring, clipping, and analyzing forage) inside and outside exclusion cages. Data collected and information gained during these experiments were used to help create various planting mixtures and recommendations for forage food plots for white-tailed deer. Several of the forages listed are not recommended for planting, whether because of low selectivity, difficulty to establish, and/or because of their invasive nature and difficulty to control once established.

[2]Selectivity and preference are relative to what is available. These preference ratings are based on their use within the demonstration/research plots where many other forages also were available. You may find deer differ somewhat in their selections on your property.

[3]Crude protein includes digestible proteins as well as those proteins found in plant cell walls that may be lignified and indigestible. Acid detergent fiber is a measure of the lignified, indigestible portion of the plant. Levels of crude protein and acid detergent fiber vary greatly with respect to plant maturity, soil fertility, and soil moisture. The data presented in this appendix represent nutritional quality of these forages on certain sites at certain times of the year. All of the forage samples collected for this appendix were taken when the plant was actively growing and before stem elongation and flowering had begun.

[4]This column represents the general time period(s) when forage production is best, starting from the time of planting. Forage availability is naturally dependent upon many factors, such as time of planting, region, soil conditions, weather, and weed control.

[5]Perennial cool-season legumes generally do not produce considerable forage during the fall of establishment. Production is best the following spring through midsummer, then increases again in the fall. In addition, perennial clovers normally "wilt down" in the winter following hard frosts and very cold temperatures. In milder winters, midwinter production may be significant.

Appendix 6

Annual production of forages planted for white-tailed deer.

Forage	Average annual production[1,2] (lbs dry matter)	Percent consumed by deer[3]		
		Low density[4]	Medium density	High density
Perennial				
ladino clover	7,500	25	41	83
red clover	6,500	12	38	91
alfalfa	7,500	15	52	91
birdsfoot trefoil	3,500	13	--	73
chicory	4,500	59	66	85
Cool-season annual				
crimson clover	7,000	20	79	73
berseem clover	4,000	27	85	91
arrowleaf clover	6,500	11	46	76
Austrian winter peas	3,000	13	39	97
oats	3,000	16	87	80
wheat	3,000	19	58	57
dwarf essex rape	3,000	10	13	30
Warm-season annual				
iron-clay cowpeas	8,500	5	63	100
lablab	7,500	45	44	100
Quail Haven soybeans	6,000	27	77	100
soybeans	8,000	81	90	100
American jointvetch	7,500	28	46	91
alyceclover	6,000	1	40	93
buckwheat	3,000	--	29	--

[1]Annual production per acre was estimated by clipping forages within exclusion cages at the end of each month and summing the production (not standing crop) through the year. In general, forage production estimates were lower on the high deer-density site and higher on the low deer-density site.

[2]Production data do not include biomass from the stem elongation/bolting/flowering/seed formation stages for chicory, wheat, oats, or dwarf essex rape because that would have inflated production estimates and biased consumption percentages low. Production data for warm-season annuals did include relatively large stems if they were present.

[3]Percent consumed by deer was measured by clipping forages within and outside exclusion cages at the end of each month. Keep in mind, however, a variety of forages were available in each field throughout each year data were collected. Therefore, these values represent

selectivity among the forages present at that time, and not necessarily the amount of forage you should expect deer to eat at a given density if other forages are not present. For example, you should not expect deer to necessarily eat 13 percent of available Austrian winter pea production (as shown in the table for low deer density) when there is little else for the deer to eat. Another important consideration is food availability in the surrounding area. On the low-density site, habitat quality and food availability was excellent. On both the medium- and high-density sites, surrounding habitat and food availability was relatively poor.

[4]Low, medium, and high deer density are relative terms. Here, low deer density represents an average of approximately 30 deer per square mile. Medium deer density represents an average of approximately 50 deer per square mile. High deer density represents an average of approximately 90 deer per square mile. These density estimates are for three sites across Tennessee. Deer density was estimated at each site by using infrared-triggered cameras and by visual observations at each field, all recorded for three to seven years consecutively at each field.

A virtual cafeteria at the high deer-density site.

The next time you are walking around in an old-field or along the edge of the woods, look at the plants closely and see which ones the deer are eating. Various field management practices, such as burning and disking, can be used to maintain these important forages. Clockwise from upper left; beggar's-lice, goldenrod, wildlettuce, and partridge pea.

Appendix 7

Wildlife and nutritional value of various forbs and shrubs.[1]

Common name	Scientific name	CP	ADF	Selectivity by deer[2]	Value as brood cover	Seed value for birds
pokeweed	Phytolacca americana	32.0	12.0	High	High	High
old-field aster	Aster pilosus	23.3	30.7	High	High	None
wildlettuce	Lactuca serriola	21.7	21.2	High	Low	None
blackberry	Rubus spp.	19.3	18.9	Med	High	High
partridgepea	Chamaecrista fasciculata	29.6	36.5	Med	High	High
beggar's-lice	Desmodium spp.	28.2	20.7	Med	High	High
common ragweed	Ambrosia artemisiifolia	17.8	23.9	Med	High	High
sumac	Rhus spp.	23.1	12.5	Med	High	Med
goldenrod	Solidago spp.	16.1	26.2	Med	Med	None
3-seeded mercury	Acalypha virginica	24.7	16.7	Med	Low	Med
honeysuckle	Lonicera japonica	16.2	34.2	Low	Low	Low
Canadian horseweed	Conyza canadensis	32.9	19.8	Low	Med	None
sericea lespedeza	Lespedeza cuneata	22.2	32.6	None	Low	None
passionflower	Passiflora incarnata	36.6	18.9	None	None	Low

[1]All samples were collected in June 2005 from an old-field in McMinn County, TN, that was burned the previous April.

[2]Selectivity by deer was measured by estimating the percentage of individual plants eaten and the frequency that deer fed upon that plant species throughout the field. It is interesting to note that deer did not select forages based on nutritional content. Deer did not browse or graze all of the plants in the chart above. Old-field aster, prickly lettuce, and pokeweed were grazed heavily, whereas blackberry, goldenrod, common ragweed, and 3-seeded mercury were only browsed or grazed occasionally. For other species, such as passionflower and sericea lespedeza, there was no sign of grazing at all, even though crude protein and digestibility ratings were high. It also is important to note deer density in this area was approximately 25 deer per square mile and there was an abundance of soybean fields around the field where these data were collected.

Appendix 8
Plant selectivity by white-tailed deer in forested and woodland environments.[1]

Common name[2]	Scientific name	Selectivity	CP[3]	ADF
Forage selection by deer at Chuck Swan State Forest and Wildlife Management Area, Union County, TN, 2003, 2007-2008				
Hogpeanut	*Amphicarpaea bracteata*	+	20	29
Virginia snakeroot	*Aristolochia serpentaria*	-		
Bluejoint grass	*Calamagrostis canadensis*	-	8	38
Pennsylvania sedge	*Carex pensylvanica*	-	8	40
Spotted wintergreen	*Chimaphila maculata*	-	9	21
Beggar's-lice	*Desmodium spp.*	+	21	36
Low panicgrass	*Dichanthelium spp.*	-	11	33
Wild yam	*Dioscorea villosa*	+	14	32
Burnweed	*Erectites hieracifolia*	-	12	40
Boneset	*Eupatorium spp.*	-	10	36
Strawberrybush	*Euonymus americanus*	+	11	37
Catchweed bedstraw	*Gallium aparine*	=	9	32
Rattlesnake plantain	*Goodyera pubescens*	-		
Little brown jug	*Hexastylis arifolia*	-	10	18
Lespedeza (native species)	*Lespedeza spp.*	-		
Japanese honeysuckle	*Lonicera japonica*	-	13	28
Japangrass	*Microstegium vimineum*	-	15	35
Partridgeberry	*Mitchella repens*	-		
Virginia creeper	*Parthenocissus quinquefolia*	=	14	26
Solomon's seal	*Polygonatum commutatum*	-	8	23
Christmas fern	*Polystichum acrostichoides*	-	11	41
Pokeweed	*Phytolacca americana*	+	30	12
Blackberry	*Rubus spp.*	+	13	25
Greenbriar	*Smilax spp.*	+	13	24
Poison ivy	*Toxicodendron radicans*	-	11	23
Violet	*Viola sororea*	-	12	24
Grape	*Vitis spp.*	+	20	25
Red maple	*Acer rubrum*	-	11	27
Hickory	*Carya spp.*	-	12	30
Flowering dogwood	*Cornus florida*	=	18	24
Green ash	*Fraxinus pennsylvanica*	-		

Common name[2]	Scientific name	Selectivity	CP[3]	ADF
Yellow-poplar	Liriodendron tulipifera	-	13	38
Blackgum	Nyssa sylvatica	+	11	18
Sourwood	Oxydendrum arboreum	-	10	20
Black cherry	Prunus serotina	-	13	35
Oak	Quercus spp.	-	14	34
Sumac	Rhus spp.	-	10	12
Sassafras	Sassafras albidum	-	14	44
Blueberry	Vaccinium spp.	-	9	33
Mapleleaf viburnum	Viburnum acerifolia	+	7	

Forage selection by deer at private property in Sequatchie County, TN, 2004 - 2005

Common name	Scientific name	Selectivity	CP	ADF
Blackberry	Rubus spp	+	14	22
Greenbriar	Smilax spp.	+	13	31
Red maple	Acer rubrum	=	11	28
Hickory	Carya spp.	-	12	28
Blackgum	Nyssa sylvatica	+	12	20
Sourwood	Oxydendrum arboreum	-	12	24
Sassafras	Sassafras albidum	-	15	42
Blueberry	Vaccinium spp.	-	9	33

Forage selection by deer at Ames Plantation, Fayette County, TN, 2002-2005

Common name	Scientific name	Selectivity	CP	ADF
Alabama supplejack	Berchemia scandens	+	13	21
Japanese honeysuckle	Lonicera japonica	-	13	27
Virginia creeper	Parthenocissus quinquefolia	=	10	34
Blackberry	Rubus spp.	=	12	20
Greenbriar	Smilax spp.	+	13	30
Poison ivy	Toxicodendron radicans	=	12	30
Red maple	Acer rubrum	=	10	26
Hickory	Carya spp.	-		
Carolina buckthorn	Frangula caroliniana	-		
Green ash	Fraxinus pennsylvanica	=	12	39
Yellow-poplar	Liriodendron tulipifera	=		
Blackgum	Nyssa sylvatica	+	11	22
Black cherry	Prunus serotina	-		
Oak	Quercus spp.	-		
Sassafras	Sassafras albidum	-		
Winged elm	Ulmus alata	+	13	28

Common name[2]	Scientific name	Selectivity	CP[3]	ADF
Forage selection by deer at North Cumberland Wildlife Management Area, Anderson, Campbell, and Scott Counties, TN, 2013-2015				
Black cohosh	Actaea racemosa	-	13	20
White snakeroot	Ageratina altissima	+	17	25
Hogpeanut	Amphicarpaea bracteata	=	22	28
Southern lady fern	Athyrium filix-femina	-	13	31
Carex spp.	Carex spp.	-	10	34
Blue cohosh	Caulophyllum thalictroides	-	10	21
Clematis spp.	Clematis spp.	=	18	19
Poverty oatgrass	Danthonia spicata	-	7	32
Deer-tongue grass	Dicanthelium clandestinum	-	10	36
White wood aster	Eurybia divaricata	+	16	26
Hairy bedstraw	Galium pilosum	-	10	32
Jewelweed	Impatiens spp.	+	27	22
Japangrass	Microstegium viminuem	-	14	33
Mayapple	Podophylum peltatum	-	13	24
Solomon's seal	Polygonatum biflorum	-	17	25
Christmas fern	Polystichum acrostichoides	-	13	35
Common cinquefoil	Potentilla simplex	-	11	20
Cankerweed	Prenanthes spp.	+	14	22
Hoary mountain mint	Pycnanthemum incanum	-	12	28
Black raspberry	Rubus occidentalis	+	13	20
Cat greenbrier	Smilax glauca	+	11	22
Common greenbrier	Smilax rotundifolia	+	11	22
Canada goldenrod	Solidago canadensis	+	16	20
Poison-ivy	Toxicodendron radicans	=	14	26
Wood violet	Viola sororia	=	12	23
Vitis spp.	Vitis spp.	=	15	22
Striped maple	Acer pensylvanicum	+	13	27
Red maple	Acer rubrum	=	11	24
Sugar maple	Acer saccharum	=	11	26
Yellow buckeye	Aesculus flava	-	10	29
Black birch	Betula lenta	=	12	24
Bitternut hickory	Carya cordiformis	-	13	30
Pignut hickory	Carya glabra	=	13	30
Mockernut hickory	Carya tomentosa	-	13	30

Common name[2]	Scientific name	Selectivity	CP[3]	ADF
Eastern redbud	*Cercis canadensis*	-	10	22
Strawberrybush	*Euonymus americanus*	+	12	34
American beech	*Fagus grandifolia*	-	13	42
Green ash	*Fraxinus pennsylvanica*	=	12	38
Wild hydrangea	*Hydrangea arborescens*	+	15	23
Mountain laurel	*Kalmia latifolia*	-	7	25
Spicebush	*Lindera benzoin*	=	20	30
Yellow-poplar	*Liriodendron tulipifera*	=	13	29
Cucumber magnolia	*Magnolia acuminata*	-	13	30
Blackgum	*Nyssa sylvatica*	+	14	16
Sourwood	*Oxydendrum arboream*	=	13	24
Virginia creeper	*Parthenocissus quenquifolia*	+	14	23
Black cherry	*Prunus serotina*	-	13	22
Buffalo nut	*Pyrularia pubera*	+	19	14
White oak	*Quercus alba*	-	13	27
Chestnut oak	*Quercus montana*	-	14	24
Northern red oak	*Quercus rubra*	-	14	24
Black oak	*Quercus velutina*	-	14	24
Black locust	*Robinia pseudoacacia*	+	21	25
Blackberry	*Rubus argutus*	+	12	20
Sassafras	*Sassafras albidum*	-	14	26
Lowbush blueberry	*Vaccinium angustifolium*	=	9	27

[1] Plant selectivity by deer was determined by recording the amount and frequency of grazing or browsing upon plant species along transects during studies conducted in forested and woodland environments (as opposed to early successional communities). Selectivity was based on use vs. availability analysis. If a plant species was eaten "more than expected" (+), that means the species was eaten more than would be expected based on the availability or occurrence of that species. If a plant species was eaten "as expected" (=), that means the species was eaten as would be expected given the availability or occurrence of that species. If a plant species was eaten "less than expected" (-), that means the species was eaten less than would be expected based on the availability or occurrence of that species. You will notice that some species were selected by deer at some sites, but not at other sites, or vice versa. This discrepency in selection reflects how deer chose to eat those species in relation to what was available at that particular site. Variable selection of different species by deer across different sites is very common. However, there are a few plants that are always selected by deer, no matter where they occur and no matter what else is available. For example, greenbriar, blackgum, winged elm, strawberrybush, and grape are a few species that are almost always selected by deer in forested or woodland environments wherever they occur. Data from Chuck Swan SF and WMA reported in Lashley (2008). Data from Sequatchie County and Ames Plantation reported in Shaw (2007). Data from North Cumberland WMA reported in Nanney (2016).

[2] Plants are arranged alphabetically by latin name; herbaceous species, brambles, and vines are listed prior to shrub and tree species.

[3] Generally, crude protein levels decrease and acid detergent fiber levels increase through the growing season as plants mature. These data represent average forage quality of leaves collected in June-July across years at each site.

Shaw, C.E. 2008. An evaluation of quality deer management programs in Tennessee. M.S. thesis. University of Tennessee, Knoxville, TN. 219 pages.

Lashley, M.A. 2009. Deer forage available following silvicultural treatments in upland hardwood forests and warm-season plantings. M.S. thesis. University of Tennessee, Knoxville, TN. 53 pages.

Nanney, J.S. 2016. Forage availability and nutritional carrying capacity for cervids following prescribed fire and herbicide applications in young mixed-hardwood forest stands in the Cumberland Mountains, Tennessee. M.S. thesis. University of Tennessee, Knoxville, TN. 76 pages.

Appendix 9

Various legumes and associated inoculant groups.

	Inoculant code	Bacterium
Alfalfa group		
Alfalfa	A	*Sinorhizobium meliloti*
Sweetclover		
Clover group		
Alsike clover	B	*Rhizobium leguminosarum biovar trifolii*
Ball clover		
Ladino white clover		
Red clover		
White-dutch clover		
Arrowleaf clover	O	*Rhizobium leguminosarum biovar trifolii*
Crimson clover	R	*Rhizobium leguminosarum biovar trifolii*
Berseem clover		
Rose clover	WR	*Rhizobium leguminosarum biovar trifolii*
Subterranean clover		
Pea and vetch group		
Austrian winter peas	C or "Garden"	*Rhizobium leguminosarum biovar viceae*
Field peas		
Flat peas		
Hairy vetch		
Sweet peas		
Cowpea group		
Alyceclover	EL	*Bradyrhizobium* spp.
American jointvetch		
Cowpeas		
Lablab		
Lespedezas		
Partridgepea		
Peanuts		
Velvet bean		
Lupine group		
Blue lupine	H	*Rhizobium lupini*
White lupine		
Other		
Prairieclover	F	*Rhizobium* spp. (Petalostemum)
Sainfoin		*Rhizobium* spp. (Onobrychis)
Birdsfoot trefoil	K	*Mesorhizobium loti*
Soybeans	S	*Bradyrhizobium japonicum*

References and recommended reading

There is a tremendous list of scientific articles that are pertinent to agriculture, forage, land management, and the biology and ecology of wildlife species. The list below represents scientific articles as well as books and manuals that synthesize information provided in scientific articles in an easy-to-read format that will benefit any serious land manager.

Ball, D.M., C.S. Hoveland, and G.D. Lacefield. 2002. Southern forages. Potash and Phosphate Institute and Foundation for Agronomic Research. Norcross, GA.

Baskett, T.S., M.W. Sayre, R.E. Tomlinson, and R.E. Mirarchi. 1993. Ecology and management of the mourning dove. Stackpole Books. Harrisburg, PA.

Brooke, J.M. and C.A. Harper. 2018. Renovating native warm-season grass stands for wildlife: A land manager's guide. Purdue Extension and UT Extension, PB 1856.

Brooke, J.M., P.S. Basinger, J.S. Nanney, and C.A. Harper. 2015. Effectiveness of three postemergence herbicides in controlling an invasive annual grass, *Microstegium vimineum.* Journal of the Southeastern Association of Fish and Wildlife Agencies 2:262-267.

Brooke, J.M., D.C. Peters, A.M. Unger, E.P. Tanner, C.A. Harper, P.D. Keyser, J.D. Clark, and J.J. Morgan. 2015. Habitat manipulation influences northern bobwhite resource selection on a reclaimed surface mine. Journal of Wildlife Management 79:1264-1276.

Brooke, J.M., E.P. Tanner, D.C. Peters, A.M. Unger, C.A. Harper, P.D. Keyser, J.D. Clark, and J.J. Morgan. 2017. Northern bobwhite breeding season ecology and resource selection on a managed reclaimed surface mine in Kentucky. Journal of Wildlife Management 81:73-85.

Bryson, C.T. and M.S. DeFelice. 2009. Weeds of the South. University of Georgia Press. Athens, GA.

Crawford, H.S., C.L. Kucera, and J.H. Ehrenreich. 1969. Ozark range and wildlife plants. U.S. Department of Agriculture, Forest Service, Agriculture. Handbook Number 356. 236 pages.

Dickson, J.G. 1992. The wild turkey: Biology and management. Stackpole Books. Harrisburg, PA.

Dickson, J.G. 2001. Wildlife of Southern forests: Habitat and management. Hancock House Publishers. Blaine, WA

Donahue, R.L., R.W. Miller, J.C. Shickluna. 1983. Soils: An introduction to soils and plant growth. Fifth edition. Prentice-Hall, Inc. Englewood Cliffs, NJ.

Foster, M.A., M.J. Gray, C.A. Harper, and R.M. Kaminski. 2010. Post-harvest fates of agricultural seeds in Tennessee croplands. Proceedings Annual Conference of Southeastern Association of Fish and Wildlife Agencies 64:81-87.

Foster, M.A., M.J. Gray, C.A. Harper, and J.G. Walls. 2010. Comparison of agricultural seed loss in a flooded and unflooded field on the Tennessee National Wildlife Refuge. Journal of Fish and Wildlife Management 1:43-46.

Gee, K.L., M.D. Porter, Se. Demarais, F.C. Bryant, and G. Van Vreede. 1991. White-tailed deer: Their foods and management in the Cross Timbers. Samuel Roberts Noble Foundation, Ardmore, OK.

Gruchy, J.P. and C.A. Harper. 2014. Effects of management practices on northern bobwhite habitat. Journal of the Southeastern Association of Fish and Wildlife Agencies. 1:133-141.

Halls, L.K. 1984. White-tailed deer: Ecology and management. Stackpole Books. Harrisburg, PA.

Halls, L.K. and T.H. Ripley, editors. 1961. Deer browse plants of Southern forests. U.S. Department of Agriculture, Forest Service, Southern and Southeastern Forest Experiment Stations.

Harlow, R.F. and R.G. Hooper. 1971. Forages eaten by deer in the Southeast. Proceedings of the Annual Conference, Southeastern Association of Game and Fish Commissioners 25:18-46.

Harper, C.A. 2017. Managing early successional plant communities for wildlife in the eastern US. University of Tennessee Institute of Agriculture. Knoxville, TN. 90 pages. ISBN 978-0-692-90231-8.

Harper, C.A., W.M. Ford, M.A. Lashley, C.E. Moorman, and M.C. Stambaugh. 2016. Fire effects on wildlife in the Central Hardwoods and Appalachian regions, USA. Fire Ecology 12(2):127-159.

Hernandez, F. and F.S. Guthery. 2012. Beef, brush, and bobwhites: Quail management in cattle country. Texas A&M Press, College Station, TX.

Hewitt, D.G. 2011. Biology and management of white-tailed deer. CRC Press. Boca Raton, FL.

Horn, D. and T. Cathcart. 2005. Wildflowers of Tennessee, the Ohio Valley, and the Southern Appalachians. Lone Pine Publishing and the Tennessee Native Plant Society. Auburn, WA.

Kammermeyer, K., K.V. Miller, and L. Thomas, Jr. 2006. Quality food plots: Your guide to better deer and better deer hunting. Quality Deer Management Association. Bogart, GA.

Kenyon, I. 2000. Beyond the food patch: A guide to providing bobwhite quail habitat. Virginia Department of Game and Inland Fisheries.

Kilburg, E., C.E. Moorman, C. DePerno, D. Cobb, and C.A. Harper. 2014. Wild turkey nest survival and nest-site selection in the presence of growing-season prescribed fire. Journal of Wildlife Management 78(6):1033-1039.

Koerth, B.H. and J.C. Kroll. 1998. Food plots and supplemental feeding. Institute for White-tailed Deer Management and Research. Stephen F. Austin State University, Nacogdoches, TX.

Kroll, J.C. 1991. A practical guide to producing and harvesting white-tailed deer. Center for Applied Studies in Forestry. Stephen F. Austin State University, Nacogdoches, TX.

Landers, J.L. and A.S. Johnson. 1976. Bobwhite quail food habits in the Southeastern United States with a seed key to important foods. Tall Timbers Research Station, Tallahassee, FL. Miscellaneous Publication 4. 90 pages.

Larson, J.A., T.E. Fulbright, L.A. Brennan, F. Hernandez, and F.C. Bryant. 2010. Texas bobwhites: A guide to their food habits and habitat management. University of Texas Press, Austin, TX. 280 pages.

Lashley, M.A., M.C. Chitwood, C.A. Harper, C.E. Moorman, and C.S. DePerno. 2015. Poor soils and density-mediated body weight in deer: Forage quality or quantity? Wildlife Biology 21:213-219.

Lashley, M.A., M.C. Chitwood, R. Kays, C.A. Harper, C.S. DePerno, and C.E. Moorman. 2015. Prescribed fire affects female white-tailed deer habitat use during summer lactation. Forest Ecology and Management 348:220-225.

Lashley, M.A. and C.A. Harper. 2012. Effects of extreme drought on native forage nutritional quality and white-tailed deer diet selection in the Central Hardwoods. Southeastern Naturalist 11(4):699-710.

Lashley, M.A., C.A. Harper, G.E. Bates, and P.D. Keyser. 2011. Forage availability for white-tailed deer following silvicultural treatments. Journal of Wildlife Management 75:1467-1476.

Lashley, M.A., J.M. McCord, C.H. Greenberg, and C.A. Harper. 2009. Masting characteristics of white oaks: Implications for management. Proceedings Annual Conference of Southeastern Association of Fish and Wildlife Agencies 61:21-26.

Martin, A.C. and F.M. Uhler. 1951. Food of game ducks in the United States and Canada. USDI Research Report 30 (reprint of USDA Technical Bulletin 634). US Government Printing Office. Washington, D.C.

Martin, A.C., H.S. Zim, and A.L. Nelson. 1951. American wildlife and plants: A guide to wildlife food habits. Dover Publications. New York, NY.

McCord, J.M., C.A. Harper, and C.H. Greenberg. 2014. Brood cover and food resources for wild turkeys following silvicultural treatments in mature upland hardwoods. Wildlife Society Bulletin 38:265-272.

Miller, J.I., S.T. Manning, and S.F. Enloe. 2010. A management guide for invasive plants in southern forests. USDA-Forest Service, General Technical Report SRS-131.

Miller, K.V. and R.L. Marchinton. 1995. Quality whitetails: The why and how of quality deer management. Stackpole Books. Mechanicsburg, PA.

Miller, J.H. and K.V. Miller. 1999. Forest plants of the Southeast and their wildlife uses. Southern Weed Science Society.

Lashley, M.A. and C.A. Harper. 2012. Effects of extreme drought on native forage nutritional quality and white-tailed deer diet selection in the Central Hardwoods. Southeastern Naturalist 11(4):699-710.

Nanney, J.S., C.A. Harper, D.A. Buehler, and G.E. Bates. 2018. Nutritional carrying capacity for cervids following disturbance in hardwood forests. Journal of Wildlife Management 82(6):1219-1228.

Nelms, K.D., B. Ballinger, and A. Boyles, editors. 2007. Wetland management for waterfowl handbook. Natural Resources Conservation Service. www.mdwfp.com/media/8838/wetlandmgtforwaterfowl.pdf.

Odom, E.P. and G.W. Barrett. 2005. Fundamentals of ecology. Fifth edition. Thomson Brooks/Cole, Belmont, CA.

Rohnke, A.T. and J.L. Cummins, editors. 2014. Fish and wildlife management: A handbook for Mississippi landowners. University Press of Mississippi, Jackson, MS. 418 pages.

Rosene, W. 1984. The bobwhite quail: Its life and management. The Sun Press. Hartwell, GA.

Rosene, W. and J.D. Freeman. 1988. A guide to and culture of flowering plants and their seed important to bobwhite quail. Morris Communications Corporation. Augusta, GA.

Shaw, C.E., C.A. Harper, M.W. Black, and A.E. Houston. 2010. The effects of prescribed burning and understory fertilization on browse production in closed-canopy hardwood stands. Journal of Fish and Wildlife Management 1:64-72.

Southern Weed Science Society. 1993. Weed identification guide. Sets I-VI. Edited by C.D. Elmore. Southern Weed Science Society, Champaign, IL. 300 pages.

Stauffer, D.E., editor. 2011. Ecology and management of Appalachian ruffed grouse. Hancock House Publishers, Blaine, WA. 176 pages.

Steckel, L. 2006. Weed control manual for Tennessee. Edited by T.C. Mueller. UT Extension, PB 1580. http://weeds.utk.edu

Stubbendik, J., G.Y. Friisoe, and M.R. Bolick. 1995. Weeds of Nebraska and the Great Plains. Nebraska Department of Agriculture. Lincoln, NE.

Unger, A.M., E.P. Tanner, C.A. Harper, P.D. Keyser, F.T. van Manen, and J.J. Morgan. 2015. Northern bobwhite seasonal habitat selection on a reclaimed surface coal mine in Kentucky. Journal of the Southeastern Association of Fish and Wildlife Agencies 2:235-246.

Uva, R.H., J.C. Neal, and J. M. DiTomaso. 1997. Weeds of the Northeast. Comstock Publishing Associates. Ithaca, NY.

Waldrop, T.A. and S.L. Goodrick. 2012. Introduction to prescribed fires in Southern ecosystems. Science Update SRS-054. Asheville, NC: U.S. Dept. of Agriculture Forest Service, Southern Research Station.

Yarrow, G.K. and D.T. Yarrow. 2005. Managing Wildlife. Sweetwater Press, Birmingham, AL.

*Plant
Identification
Guide*

Identification and description of plants commonly found in food plots and early successional plant communities

This guide is intended to help you identify common weeds and other plants found in food plots, fallow fields, old-fields, and other early successional plant communities and to provide information on the wildlife value of those plants. Everyone wants to know how to get rid of weeds in their food plots, but few are able to identify the plants present, and most people do not realize the wildlife value of many of these plants. Likewise, most land managers recognize the importance of early successional plant communities for many wildlife species, but they cannot identify the plants present. Therefore, the quality of these plant communities is not as good as it could be because managers are unable to promote desirable plants and reduce coverage of undesirable plants. Identification is necessary in order to know the value of these plants as well as the appropriate management technique to either control them or promote them.

The information on plant identification given in the pages that follow is relatively general and is not intended to duplicate or replace more comprehensive information found in other plant identification guides. I suggest you obtain one or more plant or weed identification guides to provide you with more comprehensive and detailed information on plant identification. Several of the guides I recommend are provided in *References and recommended reading* on page 260. In particular, I use *Weed of the South, Weeds of the Northeast, Weeds of Nebraska and the Great Plains,* and *Forest Plants of the Southeast,* regularly. There are many other excellent plant identification guides, many with emphasis on particular regions of the country. Also, there are several websites that are excellent references for plant identification. Many of these are specific to a particular state. Search 'plant identification in [your state]' in your web browser and see what you find. Plants in this guide are arranged alphabetically by family, then by genus. Common names are listed prominently, with Latin names underneath. Multiple common names are provided for some plants. Multiple species are provided within some plant profiles.

ASTERACEAE
Southern or lanceleaf ragweed
Ambrosia bidentata

What is it? Is it good or bad? Do you need to spray it or not?

Plant identification is critical when managing food plots and early successional plant communities.

In the plot above, orchardgrass, purple deadnettle, and henbit are choking out planted clovers. In this situation, a selective herbicide application (imazethapyr and clethodim) would control the competing weeds and allow the clovers to produce more forage. In the corn plot below, boneset, common ragweed, and foxtail grasses are complementing the corn by providing natural forage, seed, and desirable cover for many wildlife species. The main difference in the "weeds" found in the two plots is their value for wildlife. Those above have little or no value for wildlife, whereas those below do. However, you cannot know this unless you can identify the plants.

Wild petunia
Ruellia caroliniensis
R. strepens

Description and ecology
Erect, perennial, warm-season forb that may grow to 3 feet tall. Stem is usually branched and sparsely hairy. Leaves are opposite. Flowers appear May – October and are typically blue with 5 petals (one petal is slightly larger than the other 4). One or two flowers occur on the flowering stalk, and individual flowers persist only for one day, but new flowers will continually appear.

Origin and distribution
Native; PA to MI to NE and south.

Considerations as a weed and for wildlife
Wild petunia may be found in perennial food plots or fallow plots. It does not usually present a competition problem with planted crops. In the rare event it is problematic, it can be controlled with a forb-selective herbicide. It is a moderate- to high-preference forage for white-tailed deer.

Pigweeds

Amaranthus albus (tumble pigweed),
A. blitoides (prostrate pigweed),
A. hybridus (smooth pigweed),
A. palmeri (Palmer amaranth),
A. powellii (Powell amaranth),
A. retroflexus (redroot pigweed),
A. spinosus (spiny amaranth),
A. tuberculatus (tall waterhemp),
A. viridis (slender amaranth or green pigweed)

Description and ecology

Erect (except for prostrate pigweed), annual, warm-season forbs. Leaves are alternate. Tumble pigweed is multi-branched and bushy in appearance when mature. Prostrate pigweed spreads over ground from central taproot. Smooth pigweed, Palmer and Powell amaranth, redroot pigweed, and waterhemp all are similar in appearance with relatively stout stems that produce terminal flower clusters. All have a shallow taproot that is pinkish or reddish. Spiny amaranth is freely branched with sharp spines along stem at base of leaves. Slender amaranth may be erect or spreading with terminal (at end of stem) and axillary (where leaf meets stem) flower clusters. See *Weeds of the Northeast* and *Weeds of the South* for complete description and comparison of pigweeds.

Origin and distribution

Native; throughout U.S. (*A. albus, A. blitoides, A. retroflexus*); eastern U.S., Great Plains, and portions of western U.S. (*A. hybridus*); southern half of U.S. and into Northeast (*A. palmeri*); eastern U.S. (*A. spinosus*); eastern U.S., except for southern Atlantic Coastal Plain (*A. tuberculatus*); southeastern U.S. (*A. viridis*)

Considerations as a weed and for wildlife

There are several species of pigweeds. All can become problematic in food plots. Some, such as Palmer amaranth, have become glyphosate resistant in many fields that have been sprayed with glyphosate repeatedly for many years. The seed of pigweeds are eaten by many bird species. In fact, pigweed seed is readily eaten by mourning dove and northern bobwhite. In fields managed for doves, I don't mind having some pigweeds if they are not too dense. Pigweeds can be controlled with several preemergence herbicides, such as atrazine and S-metolachlor, pendimethalin, and trifluralin, and with forb-selective postemergence herbicides, such as carfentrazone-ethyl, dicamba, bentazon, and 2,4-D. Most pigweeds are controlled with glyphosate in Roundup-Ready crops, but you should use other herbicides occasionally to help prevent glyphosate resistance. Acifluorfen, chlorimuron methyl, and pyroxasulfone are herbicide options to help control glyphosate-resistant pigweeds (see Appendix 2). Pigweeds are a relatively low-preference forage for white-tailed deer, but are occasionally grazed by deer.

Photo above: Palmer amaranth in soybeans. Photos on following page by L. Steckel except bottom right.

From left: Green pigweed, smooth pigweed, tumble pigweed.

From left: Palmer amaranth, smooth pigweed, redroot pigweed, green pigweed

From left: tall waterhemp, Palmer amaranth

Smooth pigweed

Spiny amaranth

Southern chervil
Southern wild chervil
Hairy-fruit chervil
Chaerophyllum tainturieri

Wild chervil
C. procumbens

Description and ecology
Cool-season annual forbs with alternate compound leaves that have a fern-like appearance. Southern chervil grows upright 1 – 3 feet tall with green, hairy stem that may be branched or unbranched. Wild chervil branches from base of plant with stems somewhat hairy, light green or purplish-green, and decumbent (lying on ground) but growing upward 6 – 18 inches tall. Leaves may turn pinkish-red in late spring/early summer after flowering. Small, white flowers with 5 petals occur at tip of stems. Fruits are green and cylindrical, up to ½-inch long, single-seeded, and typically point upward. Fruits of Southern chervil are broadest below the middle, whereas fruits of wild chervil are broadest at the middle.

Origin and distribution
Hairy-fruit chervil native; MD to NE south to AZ and FL
Wild chervil native; NY and Ontario south to NE, OK, and FL

Considerations as a weed and for wildlife
These chervils often appear in open areas with relatively recent soil disturbance. They may compete with establishing clovers, but are not often problematic. Southern chervil will grow tall enough to provide structure for game bird poults in late spring. The foliage is eaten occasionally by white-tailed deer, cottontail rabbits, and groundhogs. Chervil can be controlled with imazapic, 2,4-DB, or glyphosate in various food plot applications.

Poison hemlock
Conium maculatum

Description and ecology
Erect, biennial forb that may grow to 10 feet tall. Prefers moist areas but may be found on well-drained sites. Stem is stout, branching, smooth, splotched with purple, and is hollow except at branching nodes. Basal rosette is established first year. Leaves are 3 to 4 times compound, and strongly dissected. Flowers appear in July – August, are small and white and occur in flat-topped groups at end of branches. Reproduction is by seed only.

Origin and distribution
Nonnative (Eurasia); throughout U.S., except FL.

Considerations as a weed and for wildlife
Poison hemlock contains neurotoxins that are toxic to livestock, humans, and many wildlife species. As a noxious weed, poison hemlock should be eradicated wherever it occurs. It can be controlled preemergence with imazethapyr, S-metolachlor, and pendimethalin, and postemergence with forb-selective herbicides, such as dicamba, metsulfuron-methyl, 2,4-D, 2,4-DB, as well as glyphosate. Mowing soon after flowering may help reduce occurrence over time if the plant is not allowed to produce seed. However, disking will cause it to increase in density as cut roots will sprout.

From top: mature plant, flowering plant, and plant spreading in brassica food plot.

273

Queen Anne's lace
Wild carrot
Daucus carota

Description and ecology
Biennial warm-season forb. Produces rosette first year, then erect flowering stem second year. Rosette leaves resemble a fern. Flowering stem is hollow and has few leaves. Many small white flowers compose a flat-topped umbel ("flower head") at the top of the stem. Mature umbels curl, turn brown, and resemble a bird's nest.

Origin and distribution
Nonnative (Asia and Mediterranean region); throughout U.S. except southwest.

Considerations as a weed and for wildlife
Queen Anne's lace can be controlled with several preemergence herbicides, such as imazethapyr, with glyphosate in Roundup-Ready crops, and with forb-selective postemergence herbicides. In perennial forage plots, Queen Anne's lace should be sprayed with imazethapyr or 2,4-DB when it is a rosette. Once the flowering stem is produced, it can be set-back by mowing. Queen Anne's lace is a low- to moderately selected forage for white-tailed deer. It provides little benefit for other wildlife species.

Wild parsnip
Pastinaca sativa

Description and ecology
Biennial forb, usually branching, arising from rosette in spring; reaches 5 feet or more in height. Flowers late spring through summer. Leaves are alternate and compound with several toothed leaflets. Lower leaves may be large. Stems are relatively stout, smooth, and furrowed. Flowers are yellow and arranged in flat-topped umbel.

Origin and distribution
Nonnative (Eurasia); throughout U.S. except portions of Deep South.

Considerations as a weed and for wildlife
Wild parsnip is an aggressive, invasive, nonnative plant that displaces other valuable plants. Although it does provide structure for brooding wild turkey and a few songbirds, it should be eradicated when found. The foliage of wild parsnip is toxic and not eaten by mammals. It is irritating and can cause blisters on the skin of humans. Various bees and butterflies obtain nectar from the flowers. It can be controlled with glyphosate, triclopyr, 2,4-D, dicamba, imazapyr, imazethapyr, and metsulfuron-methyl.

Hedge parsley
Common field hedge parsley
Torilis arvensis

Description and ecology
Cool-season annual forb that grows about 1 – 2 ½ feet tall. Stems are relatively slender with short white hairs. Leaves are alternate, compound, and up to about 6 inches long (leaves lower on plant are largest) including relatively long stems and 3 leaflets that are pointed, coarsely toothed, and covered with fine white hairs. Flowers are small and white, each with 5 petals, and appear in relatively flat-topped clusters about 2 – 3 inches across. Each flower produces a single oblong seed that looks like a pinkish-green bur covered with small bristles that may attach to clothing or fur.

Origin and distribution
Nonnative (Eurasia); occurs throughout most of U.S. except southwest

Considerations as a weed and for wildlife
Hedge parsley may be found in old-fields, fallow fields, and most any area with recent soil disturbance. It is considered invasive and is spreading in distribution. It does not often outcompete planted cool-season forages, but can displace other species in early successional plant communities. It is occasionally eaten by white-tailed deer and probably rabbits and groundhogs. It can be controlled with imazapic, 2,4-DB, or glyphosate in various food plot applications.

Hemp dogbane
Indian hemp
Apocynum cannabinum

Description and ecology
Erect, perennial, warm-season forb with a stout stem that branches above mid-plant. Leaves are opposite and have a white midvein. Stem and leaves have a milky sap. Flowers are small, white, and appear May – August in round-topped clusters on top of plant. Seed pods are about 6 – 9 inches long and look like a bean, but they are not. Seeds are twisted inside the pods, each seed with a tuft of white sllky hairs (pappus) that helps them blow in the wind. Dogbane is not a legume and does not fix nitrogen.

Origin and distribution
Native; throughout U.S.

Considerations as a weed and for wildlife
Indian hemp can be problematic in perennial forage plots. It usually occurs as scattered plants, but it can be dense and outcompete planted forages. It should be sprayed when young with 2,4-DB or imazethapyr for best control. If it gets tall, it can be set-back by mowing or by applying glyphosate or a forb-selective herbicide via spot-spraying. Disking dogbane in fallow plots actually can cause the plant to spread as roots and rhizomes are broken and will sprout. As with other erect forbs, such as goldenrod and marestail, some dogbane can enhance the structure of a perennial clover plot for wild turkey poults. However, weed density should be monitored—don't allow them to get dense enough to shade-out clovers—and you should mow or spray before dogbane produces seed. Dogbane is not eaten by white-tailed deer. Dogbane is somewhat similar in appearance to common milkweed, but the main stem of milkweed is unbranched.

Common milkweed
Asclepias syriaca

Description and ecology
Erect, perennial, warm-season forb with a stout stem that is usually unbranched and growing from deep taproot and rhizome. Leaves are opposite and have a white midvein. Stem and leaves have a milky sap. Flowers are purplish-pink and appear late June – August. Fruits are large pods that contain many seeds on tufts of long silky white hairs that enable them to disperse in the wind.

Origin and distribution
Native; throughout eastern U.S. from Mid-South northward and west to OK, north to MT.

Considerations as a weed and for wildlife
There are many species of milkweeds. Common milkweed most often occurs in perennial forage plots after the second year. It can be widespread, but usually not so dense that it outcompetes planted forages. If needed, it should be sprayed when young with a forb-slective herbicide, such as dicamba or carfentrazone-ethyl, for best control. If it gets tall, it can be set back by mowing or by applying glyphosate via spot-spraying or with a wick applicator. Disking milkweed in fallow plots can cause the plant to spread as roots and rhizomes are broken and will sprout. As with other erect forbs, such as goldenrod and marestail, some milkweed can enhance the structure of a perennial clover plot for wild turkey poults. However, don't let it get dense enough to shade-out clovers. Be sure to mow or spray before milkweed produces seed pods. Common milkweed provides great structure for many wildlife species in old-field communities. Common milkweed is not a selected forage by white-tailed deer. Numerous butterflies use milkweeds for nectar, especially Monarchs, which rely on common milkweed as a host plant.

Milkweed Family **ASCLEPIADACEAE**

Honeyvine swallowwort
Honeyvine milkweed
Cynanchum laeve

Black swallowwort
C. nigrum

Pale swallowwort
European swallowwort
C. rossicum

Description and ecology
Trailing or vining perennial warm-season
vine. Stem is usually not branched. Arises from relatively deep taproot. Leaves are
opposite, dark green, and pointed at tip with a very noticeable midrib and veins.
Honeyvine swallowwort flowers are white; black swallowwort flowers are black or
purplish; pale swallowwort has pinkish-red flowers. Fruit is a green follicle similar in
size and shape to a Roma tomato.

Origin and distribution
Native; honeyvine swallowwort throughout most of eastern U.S. from PA and IA
southward; black swallowwort in northeastern U.S.; pale swallowwort nonnative
(Europe), throughout most of northeastern U.S. and portions of Midwest

Considerations as a weed and for wildlife
Swallowworts are most often found in cultivated warm-season plots. They are
not usually competitive, but can become fairly dense over time if not controlled.
Swallowworts can be controlled with tillage and preemergence applications of
imazethapyr. They can be controlled postemergence with forb-selective herbicides,
such as dicamba and 2,4-DB, and with glyphosate in Roundup-Ready crops. Leaves
of swallowworts may occasionally be eaten by deer or rabbits, but they are not an
important food source and do not provide cover for wildlife.

Photo: Honeyvine swallowwort

Common yarrow
Achillea millefolium

Description and ecology
Upright, warm-season, perennial forb. Stems may be single or in clusters from rhizomes. Leaves are alternate and finely dissected, fernlike (similar to Queen Anne's lace, but more dissected). Basal leaves may remain green through winter. Flowers are white composite flower heads in flat-topped or rounded clusters.

Origin and distribution
Occurs throughout the northern hemisphere. Plants in the U.S. generally are considered a complex of both native and introduced plants and their hybrids.

Considerations as a weed and for wildlife
Yarrow spreads by rhizomes and can become widespread and competitive in perennial forage plots if not sprayed. Yarrow can be controlled with imazethapyr in clover/chicory/alfalfa plots and 2,4-DB in clover/alfalfa plots if sprayed when young. Yarrow leaves may be eaten sparingly by ruffed grouse and rabbits. Yarrow is a low-preference forage for white-tailed deer.

Common ragweed
Ambrosia artemisiifolia

Description and ecology
Erect, annual, warm-season forb. May grow to 7 feet tall. Stout central stem produces many branches. Leaves are deeply lobed and opposite near base, but are alternate up the stem. Flowers are small and inconspicuous. Seed is an achene with several projections resembling a crown.

Origin and distribution
Native; throughout U.S.

Considerations as a weed and for wildlife
Common ragweed provides seed eaten by many songbirds (including indigo buntings, juncos, northern cardinal, and American goldfinch) and is an important food source for northern bobwhite and mourning dove. The cover provided by common ragweed is second to none for bobwhite and wild turkey poults and is highly desirable when managing fallow fields for these and other wildlife species. However, ragweed can be competitive in some warm-season plantings and in perennial forage plots. Common ragweed can complement some crops, such as corn, cowpeas, or sunflowers, grown for mourning dove and northern bobwhite, but it can smother other species, such as sesame or millets. Density and height of ragweed should be considered before you spray it. Also, presence of undesirable weeds can influence whether you spray or not. Common ragweed is controlled preemergence with imazethapyr and S-metolachlor, and postemergence with forb-selective postemergence herbicides, such as dicamba, and 2,4-D, or with glyphosate in Roundup-Ready crops. In perennial forage plots, ragweed should be sprayed when young with imazethapyr or 2,4-DB. If it gets tall, it can be set back by mowing or by applying glyphosate via spot-spraying or with a wick applicator. Relatively sparse ragweed in a perennial clover plot can make the plot more attractive to wild turkey poults. Mow ragweed in a perennial clover plot in late summer before it produces seed. Common ragweed is a moderate- to high-preference forage for white-tailed deer, and can provide good fawning cover in fallow fields.

From top: Common ragweed in annual clover plot; common ragweed grazed in perennial clover plot; common ragweed leaf; structure of mature common ragweed.

Southern ragweed
Lanceleaf ragweed
Ambrosia bidentata

Description and ecology
Erect, annual, branching, warm-season forb. Grows 3 – 4 feet tall. Leaves are alternate, except some lower leaves may be opposite, lanceolate in shape, hairy, and may have a lobe or tooth near leaf base. Flowers are small and in long spikes at end of branches. Seed is an achene with several projections resembling a crown. Normally associated with relatively dry, disturbed areas; tolerates low-nutrient soils.

Origin and distribution
Native; VA to southern OH and eastern KS, south to eastern TX and north GA.

Considerations as a weed and for wildlife
Southern ragweed is commonly found on sites with recent soil disturbance. It is more likely found in a warm-season grain plot planted via conventional cultivation than a problem in perennial forage plot. Southern ragweed complements grain crops because it provides excellent structure for upland game birds and several species of ground-feeding sparrows. I cannot find any reference indicating it in the crop of birds. However, the seed is virtually identical to that of common ragweed, except slightly larger. I imagine the same birds that eat common ragweed seed also eat Southern ragweed seed. Southern ragweed can be controlled preemergence with imazethapyr and S-metolachlor, and postemergence with glyphosate and dicamba. 2,4-D is relatively weak on Southern ragweed. It can be controlled or suppressed in perennial clover and chicory plots with imazethapyr and 2,4-DB if sprayed when young. Southern ragweed is not a selected forage by white-tailed deer.

Western ragweed
Ambrosia psilostachya

Description and ecology
Erect, branching, perennial warm-season forb. Usually grows 1 – 2 feet tall. Leaves are dissected and variable. Upper leaves are alternate, whereas lower leaves are opposite. Stem is relatively stiff and hairy.

Origin and distribution
Native; may be found throughout most of U.S., but most abundant in Great Plains.

Considerations as a weed and for wildlife
Western ragweed may be common in annual and perennial food plots. As with common ragweed, treatment must consider objectives. Western ragweed provides outstanding cover for game bird broods and ground-feeding songbirds. The seed is eaten by many species of songbirds (especially several species of sparrows, horned lark, eastern goldfinch, and towhees) and is an important food source for northern bobwhite, mourning dove, ground dove, and prairie-chickens. It is a moderately selected forage for white-tailed deer and pronghorn. Ground squirrels and other small mammals species eat the seed. Western ragweed can be controlled preemergence with imazethapyr and S-metolachlor, and postemergence with forb-selective postemergence herbicides, such as dicamba, and 2,4-D, or with glyphosate in Roundup-Ready crops. In perennial forage plots, western ragweed should be sprayed when young with imazethapyr or 2,4-DB.

Top photo by D. Elmore

Giant ragweed
Ambrosia trifida

Description and ecology
Erect, branching, annual, warm-season forb. May grow to 15 feet or more in height in fertile soils, but most often 4 – 5 feet tall. Leaves are opposite and have 3 large lobes. Flowers are small and inconspicuous. Seed is an achene, resembling those of common ragweed, but about three times as large.

Origin and distribution
Native; throughout U.S.

Considerations as a weed and for wildlife
Giant ragweed can be controlled with several preemergence herbicides, such as imazethapyr and S-metolachlor, with glyphosate in Roundup-Ready crops, and with forb-selective postemergence herbicides, such as dicamba, and 2,4-D. In perennial forage plots, ragweed should be sprayed when young. If it gets tall, it can be set-back by mowing or by applying glyphosate via spot-spraying or with a wick applicator. The seed of giant ragweed is not as common in bird diets as common ragweed, perhaps because of the larger seed. It also grows much taller than common ragweed. The structure provided by giant ragweed is good for bobwhites and wild turkey poults, but the value is not as great as common ragweed. Giant ragweed is a low- to moderate-preference forage for white-tailed deer.

Left: Giant ragweed
Directly below: Giant ragweed with common ragweed

Mayweed chamomile
Mayweed dogfennel
Anthemis cotula

Description and ecology
Annual, cool-season forb that is usually erect or decumbent (stem lying on ground with tip pointing upward). Leaves on stem are alternate and finely dissected (which helps distinguish mayweed chamomile from oxeye daisy). Leaves also have a strong unpleasant odor when crushed. Seedheads are single and at the end of stem. Ray flowers are white, 3-toothed, and surround disk flowers (yellow center).

Origin and distribution
Native; throughout U.S.

Considerations as a weed and for wildlife
Mayweed chamomile may be common in perennial cool-season plots. It provides little cover value and little food value for wildlife. It can be controlled by mowing before it seeds out, by spraying a forb-selective herbicide in a cool-season grain plot, spraying clopyralid in a brassica/cool-season grain plot, or spraying imazethapyr or 2,4DB in a perennial clover/alfalfa plot when the plant is young. In fallow areas, it can be killed with glyphosate or any forb-selective herbicide, but spraying is rarely necessary because it is not usually a problem. White-tailed deer and rabbits may seldomly graze the leaves (see middle photo).

Common burdock
Arctium minus

Description and ecology
Erect, biennial, warm-season forb. Plant exists as basal rosette during first year with large heart-shaped leaves with wavy margins. Undersurface of basal leaves is pubescent or woolly. Flowers in second year from July – October. Flower heads dry into a bur of individual bracts with hooked tips that attach to fur and clothing (very bothersome!).

Origin and distribution
Nonnative (Europe); occurs throughout U.S. from Mid-South northward.

Considerations as a weed and for wildlife
Common burdock can become a problem in food plots, especially perennial forage plots. It also can become numerous around the edge of warm-season plots. The basal rosette leaves are large and shade-out perennial forages. The rosettes may resemble broadleaf dock, but docks lack the pubescence on the undersurface of the leaves. Burdock has no value for wildlife. You should aggressively try to eradicate this plant. Spot-spraying glyphosate or a forb-selective herbicide, such as triclopyr, dicamba, clopyralid, or carfentrazone-ethyl, is effective. Because it is a biennial, mowing after it flowers, but before seed mature, will help reduce burdock.

Mugwort
Artemisia vulgaris

Description and ecology
Erect, perennial, warm-season forb. Mature stem is stout, greenish to purple, with longitudinal ridges. Plant may reach 6 feet in height by end of growing season. Reproduction is generally by rhizomes, rarely by seed. Thus, it is not apt to spread far from local area unless disturbed and moved. Up to 20 stems may arise from the same root system. Flowers from June – September. Leaves are alternate, green on top, and silvery on bottom because of dense, white pubescence. Leaves lower on the stem have deeply cut lobes, whereas those at the top of the stem are more linear in shape and not lobed. Crushed leaves have a strong, sage-like odor.

Origin and distribution
Nonnative (Europe); throughout eastern U.S.

Considerations as a weed and for wildlife
Mugwort may be found around the edges of warm-season grain plots or in perennial cool-season forage plots. It does not typically compete with warm-season grain plots, but can reduce production of perennial clovers, chicory, and alfalfa. It can provide good cover for northern bobwhite (in warm-season grain plots and fallow fields), wild turkey broods, and some ground-feeding songbirds, similar to dogfennel or horseweed. It is a low-preference forage for white-tailed deer. Mugwort tolerates mowing, and disking may cause the plant to spread as broken pieces of rhizomes will sprout. Mugwort is a tough perennial weed and can be difficult to control with selective herbicides. It can be controlled with glyphosate and some forb-selective herbicides, such as dicamba, in fallow areas. However, in those areas, it is likely providing good cover for various wildlife species. In perennial clover, chicory, alfalfa plots, 2,4-DB will provide some control if sprayed when young, or it can be spot-sprayed with glyphosate.

287

Old-field aster
Frost aster
Aster pilosus (or *Symphyotrichum pilosum*)

Bushy aster
A. dumosus

Description and ecology
Upright perennial warm-season forbs. Grow to about 4 feet tall in late summer, at which time stems are semi-woody at base with many alternate branches. Upper stems and leaves of *A. pilosus* often fuzzy, whereas *A. dumosus* is smooth. Flowers August – November. Basal rosette persists through winter. Commonly found in old-fields and roadsides. Spreads by wind-dispersed seed.

Origin and distribution
Native; eastern U.S.

Considerations as a weed and for wildlife
These asters are rarely a problem in annual plots where conventional tillage is used or where plots are planted via no-till and Roundup-Ready crops planted. However, they may become prevalent in perennial forage plots by the third or fourth year. Although a moderate- to high-preference forage for white-tailed deer, these asters can reduce yield of perennial clovers or chicory because of growth form and shade. An application of imazethapyr or 2,4-DB applied when the asters are young will reduce prevalence in perennial plots. In perennial clover plots managed specifically for wild turkeys, some old-field or bushy aster will improve the structure of the plot for turkeys. Ruffed grouse, wild turkeys, and rabbits eat the foliage, flowers, and seed of these asters. In fallow fields, they should be considered highly desireable as they provide outstanding cover for small game, deer fawns, and many species of songbirds.

Photos: Old-field aster

288

Photos in left column: Old-field aster
Photos in right column: Bushy aster

289

Sticktights
Bidens aristosa, B. frondosa

Spanishneedles
B. bipinnata

Description and ecology
Erect, annual, warm-season forbs. Grow
3 – 4 feet tall on some sites. Leaves are
opposite and compound (*B. bipinnata*
leaves are dissected 2 or 3 times). Flowers
July – October. Seeds are dark brown,
linear, and prickly, with needle-like spines
that stick to clothing or fur.

Origin and distribution
Native; *B. aristosa* and *frondosa* throughout most of U.S.;
B. bipinnata from MA to IL to KS and NM south.

Considerations as a weed and for wildlife
Sticktights are common in low-lying areas with poor
drainage. Thus, they are common in plots planted for
ducks. Dabbling ducks, such as mallard, wood duck, and
gadwall, will eat sticktight seed, but the seed provides
little energy. Sticktights and Spanishneedles also may
occur in upland fields and are most commonly found
in grain plots planted via conventional tillage. They
are controlled with several preemergence herbicides,
including imazapic, imazethapyr, pendimethalin, and
trifluralin, as well as forb-selective postemergence
herbicides, such as dicamba and 2,4-D. Sticktights are a
low- to moderate-preference forage for white-tailed deer.
Spanishneedles are a high-preference forage for white-
tailed deer. Rabbits may eat the foliage sparingly and
northern bobwhite may eat the seed.

Top two photos on left: *B. bipinnata*; Bottom photo on left: *B. frondosa*
Both photos below: *B. aristosa*

Cornflower
Bachelor's button
Centaurea cyanus

Description and ecology
Annual cool-season forb growing 1 – 3 feet tall. Leaves are narrow, pointed, and about 3 – 8 inches long. Leaves and stem have short hairs. Flowers occur at the tip of stems and typically are blue, but some are purplish or white. Various cultivars with different colors have been developed for gardeners and these escape into adjacent fields. Flowers appear spring through summer.

Origin and distribution
Nonnative (Europe); throughout U.S.

Considerations as a weed and for wildlife
Cornflower can become relatively dense and compete with perennial clovers, but this is uncommon. Although nonnative, most people enjoy the blooms in old-field communities. American goldfinch and other songbirds eat the seed. It is not a selected forage by white-tailed deer or eastern cottontail. Various bees, flies, and butterflies obtain nectar from the flowers. Cornflower can be controlled with glyphosate, forb-selective herbicides, as well as imazethapyr and a number of preemergence herbicides commonly used in grain crops.

Thistles
Carduus nutans (Musk or Nodding thistle)
Cirsium arvense (Canada thistle)
Cirsium carolinianum (Carolina thistle)
Cirsium discolor (Field thistle)
Cirsium horridulum (Yellow thistle)
Cirsium nuttalli (Nuttall's thistle)
Cirsium vulgare (Bull thistle)

Description and ecology
Warm-season forbs arising from basal rosette.
Musk thistle: biennial or winter annual; stem appears winged because leaf bases extend down stem; leaves deeply lobed; flowers pink to purple (rarely white); flower head not constricted like other thistles, often nodding and solitary; spine-tipped bracts at base of flower. **Canada thistle:** perennial that spreads underground by creeping roots; erect branching stem with grooves; leaf bases surround stem; flowers pink to purple (rarely white); flower heads clustered. **Carolina thistle**: biennial; relatively slender main stem multi-branched in upper portion; leaves with spiny margins but not usually lobed, flowers pink-purple on end of long stalks. **Field thistle:** biennial; robust stem; leaves white underneath; flowers pink to purple. **Yellow thistle:** biennial; stout stem with leaves nearly to top, hairy, occasionally branched; flowers surrounded with whorl of leaflike bracts; flowers may be yellow (most often along Coastal Plain) or reddish purple (rarely white). **Bull thistle:** biennial or winter annual; robust stem with spiny wings from bases of the leaves; leaves deeply lobed with stiff spines; leaves have coarse hairs on upper surface and softer whitish hairs below.

Origin and distribution
Musk thistle is nonnative (Eurasia) and occurs throughout U.S. except Deep South; Canada thistle is nonnative (Eurasia) and occurs from Mid-South northward; Carolina thistle is native and occurs in most of southeastern U.S.; field thistle is native to U.S. and occurs in eastern U.S.; yellow thistle is native to U.S. and occurs in southeastern U.S. and eastern Northeast; bull thistle is nonnative (Europe) and occurs throughout U.S.

Considerations as a weed and for wildlife
There are many species of thistles. Some can become widespread and dense. They are not usually a problem in fields cropped annually, but species such as musk, Canada, field, and bull thistle can become problematic in perennial forage plots. Most thistles are biennial with a rosette remaining after the first growing season. Rosettes are easily sprayed and killed with glyphosate or a forb-selective herbicide in annual food plots after the crop is dead. In perennial plots, thistles should be sprayed with 2,4-DB when the plant is young. Stems will elongate (bolt) during summer of the second year. It is best to spray thistles before they bolt. If thistles bolt, mowing before thistles flower may be a better option than spraying. Thistle seed is a preferred food source for American goldfinch. Hummingbirds and butterflies feed on nectar of thistle flowers.

Photo above: Musk thistle

Yellow thistle

Canada thistle

Field thistle

Bull thistle

Field thistle

Carolina thistle

Nuttall's thistle

293

Spotted knapweed
Centaurea maculosa

Description and ecology
Upright, biennial or short-lived perennial forb. A basal rosette is formed first year, then 1 to many branched wiry stems arise the following year. Stems are often spreading, rough, and may have small hairs. Leaves are alternate, deeply divided into narrow segments, have no spines, with a rough surface. Leaf margins usually entire. Upper leaves near flowers are smaller, narrower, and may not be lobed. Flowers are pink to purple and are at the ends of main and axillary branches.

Origin and distribution
Nonnative (Eurasia); throughout northern half of U.S., from VA to OK, west to northern half of CA

Considerations as a weed and for wildlife
Spotted knapweed is terribly invasive and outcompetes desirable native plants. Spotted knapweed can become prolific and problematic in food plots, outcompeting planted forages. It is not a selected forage by wildlife, though many pollinators visit the flowers. The structure presented by spotted knapweed is good for several ground-feeding birds and rabbits until the plant becomes overly dense. Mourning dove, goldfinches, and various species of sparrows may eat the seed. Regardless, spotted knapweed should be controlled where it occurs because of its invasive nature. It is a prohibited noxious weed in many states. It can be controlled with glyphosate and several forb-selective herbicides, such as dicamba, triclopyr, clopyralid, and 2,4-D. Preemergence herbicides, such as S-metolachlor, are effective in preventing spotted knapweed from establishing in various warm-season food plots.

Bottom photo by J. Neal

294

Chicory
Cichorium intybus

Description and ecology
Upright perennial forb. Young plant forms rosette, similar to a dandelion. Young leaves are oblong or lance-shaped with smooth margins, but become wavy and toothed as they mature. Varieties of chicory developed for forage may have smooth or wavy margins, depending on variety. Plant produces a deep taproot. Stems are branched, hollow, and green. Flowers are blue (rarely white).

Origin and distribution
Nonnative (Mediterranean region); throughout U.S.

Considerations as a weed and for wildlife
Chicory often occurs in disturbed areas, including fallow warm-season plots. It is a highly preferred forage for white-tailed deer and cottontails. In fact, various varieties of chicory have been developed and are commonly planted in perennial forage plots. When occurring naturally, it does not compete strongly with planted forages. Although it is an "escaped" nonnative plant, it is not highly invasive and does not compete heavily with native forages, even when found along roadsides. I have never seen a need to spray chicory in any food plot setting. Goldfinches and chickadees will eat the seed; wild turkey, black bear, and groundhog readily eat the foliage.

All photos show wild, naturally occurring chicory, except for middle right photo, which shows variable leaf shape of different varieties of chicory developed for forage.

Horseweed
Marestail
Conyza canadensis

Description and ecology
Erect, annual, warm-season forb. Seedlings arise from a basal rosette. Leaves, usually toothed, are alternate, somewhat hairy, and crowded along a tall stem. Elongated panicle consists of many small flower heads, July through October.

Origin and distribution
Native; throughout U.S.

Considerations as a weed and for wildlife
Horseweed can be problematic in food plots. It can be very dense and grows tall, often outcompeting planted crops. Horseweed has become tolerant of glyphosate in some fields. Horseweed is best controlled with a forb-selective herbicide, such as dicamba, triclopyr, clopyralid, or 2,4-D. In perennial forage plots, applications of imazethapyr or 2,4-DB when horseweed is young will provide some control. If it gets tall, it can be set-back by mowing before seed mature or by applying the appropriate herbicide via spot-spraying or with a wick applicator. Horseweed is a low- to moderate-preference forage for white-tailed deer. In fallow fields, horseweed can provide good cover for eastern cottontail, deer fawns, northern bobwhite and wild turkey broods, and various ground-feeding songbirds. Horseweed typically responds (is released) after spraying imazapic to control undesirable plants, such as tall fescue.

Top photo: Fallow field dominated by horseweed and fleabane providing outstanding cover for wild turkey and northern bobwhite broods.

Bottom photo: Horseweed has completely overtaken this perennial clover plot. However, even though it is outcompeting the clovers, it is providing good umbrella cover for turkey broods. This is why I don't mind some horseweed in clover plots managed for turkeys, but not so much that it shades-out the clover.

296

Plains coreopsis
Golden tickseed
Calliopsis
Coreopsis tinctoria

Description and ecology
Annual forb that grows 1 – 3 feet tall. Stems are green and smooth. Leaves are compound and up to 6 inches long and 4 inches wide; however, the leaflets are what you notice, and they are linear and slender, about 2 inches long and no more than ¼-inch wide. Flowerheads appear at ends of stems April through June. Flowers have a dark center (disk florets) and surrounding petals are reddish-brown near center and golden-yellow at tips, appearing as a yellowish flower with reddish-brown center. Plains coreopsis tolerates wide range of soil conditions.

Origin and distribution
Native in western U.S., but has escaped from gardens and is found throughout U.S.

Considerations as a weed and for wildlife
Plains coreopsis often occurs in old-field communities and may spread into food plots (see picture at right). It can be controlled with glyphosate and forb-selective herbicides, such as 2,4-D, dicamba, clopyralid, thifensulfuron-methyl + tribenuron-methyl if it becomes too dense. The plant provides good structure for game bird broods and ground-feeding sparrows. It is not a selected plant by white-tailed deer or eastern cottontail. Several species of insects, including bees, wasps, flies, butterflies, beetles, and moth caterpillars obtain nectar from the flowers or feed on the foliage. Thus, plains coreopsis provides not only cover for brooding wild turkey and northern bobwhite, but insects for food as well.

Eclipta
False daisy
Eclipta prostrata (or *E. alba*)

Description and ecology
Annual, warm-season forb that may be upright, spreading, or prostrate. Stems are freely branched, may be green, greenish-brown, or purple, often grow to 4 feet long, and root at the nodes. Leaves are opposite, linear, and pointed. Leaves attach directly to stem with no petiole (leaf stem). Leaves have a fairly noticeable midvein and widely spaced toothed margins. Flower heads are small, round, green initially, then white, and occur at the ends of stems, usually as 2 solitary heads in August – October. Eclipta is often found in moist areas.

Origin and distribution
Most sources agree it is nonnative (Asia); throughout eastern U.S. and portions of the southwest U.S.

Considerations as a weed and for wildlife
Eclipta may be found in cultivated annual plots or in perennial plots, especially those in relatively moist bottomland sites. Eclipta can be controlled postemergence with glyphosate, bentazon, and/or acifluorfen. Eclipta is a low- to moderate-preference forage for white-tailed deer.

American burnweed
Fireweed
Erechtites hieraciifolius

Description and ecology
Erect, annual, warm-season forb that often grows to 8 feet tall. Stem is fleshy, grooved, has a white pith, and usually branched in upper portions of large plants. Leaves are alternate, toothed, 2 – 8 inches long, and numerous along stem. Flowers are white, appear July – October, occur in clusters at the ends of branches, and the plumed seed are easily disseminated by wind. American burnweed is a classic pioneering species, often occurring in dense colonies the growing season following a fire.

Origin and distribution
Native; throughout eastern U.S.

Considerations as a weed and for wildlife
American burnweed may occur in fallow plots, and often along the edges of perennial plots. It is rarely a problem in food plots, but is often seen along edges, especially where fire has occurred. If it occurs in an area where it is unwanted, it can be controlled with a forb-selective herbicide or with glyphosate. Burnweed can provide cover for northern bobwhite and wild turkey poults. American goldfinch may eat the seed. It is a low-preference forage for white-tailed deer, but deer sometimes graze leaves of young plants.

Annual fleabane
Erigeron annuus

Philadelphia (or common) fleabane
E. philadelphicus

Rough (or daisy) fleabane
E. strigosus

Description and ecology
Erect, annual (annual and rough fleabane) or biennial (Philadelphia) cool-season forbs. Grow to 2-3 feet tall. Stems arise from a basal rosette. Leaves along stem are alternate and relatively sparse, especially rough fleabane. Philadelphia fleabane seedhead buds typically nodding. Flower April – October. Disk flowers (center) are yellow; ray flowers (surrounding disk flowers) are usually white, but may be pinkish or bluish (rough fleabane).

Origin and distribution
Native; occur throughout U.S. except southwest U.S.

Considerations as a weed and for wildlife
The fleabanes do not usually outcompete planted crops, but they can become widespread. For many species, this is desireable because fleabanes are moderate- to high-preference forage plants for white-tailed deer and can provide good structure for northern bobwhite, wild turkey broods, and rabbits when the plants are relatively dense, especially in a corn field. Fleabanes can be controlled with forb-selective herbicides or preemergence with imazethapyr. They can be reduced in perennial clover plots, if desired, with 2,4-DB or mowing just before they flower and produce seed.

Top right photos: Rough fleabane
Below: Philadelphia fleabane (left); Philadelphia fleabane grazed by deer (right)

Dogfennel
Eupatorium capillifolium

Yankeeweed
E. compositifolium

Description and ecology
Erect, annual or short-lived perennial, warm-season forb (yankeeweed is perennial) with stout stem reaching 9 feet tall on some sites (yankeeweed to about 6 feet). Leaves are alternate and finely dissected into linear segments. They have a distinctive odor when crushed. Flowers September – November.

Origin and distribution
Native; dogfennel throughout eastern U.S., yankeeweed from NC to AR and south.

Considerations as a weed and for wildlife
Dogfennel and yankeeweed can be problematic in warm-season plots and in perennial forage plots, usually after 2 – 3 years. They can be controlled with forb-selective herbicides, such as dicamba and triclopyr, with glyphosate in Roundup-Ready crops, and with some preemergence herbicides, such as imazethapyr and S-metolachlor, in warm-season plots. In perennial forage plots, they should be sprayed when young with imazethapyr or 2,4-DB for some control. If they get tall, they can be set-back by mowing or by applying glyphosate via spot-spraying or with a wick applicator. Dogfennel and yankeeweed are poor deer forages, but can provide excellent cover for deer as well as northern bobwhite and wild turkey poults and ground-feeding songbirds. Deer may selectively bed in mature corn fields where dogfennel is relatively dense. Relatively sparse dogfennel or yankeeweed in a perennial clover plot can make the plot more attractive to wild turkey poults. Mow both in a perennial clover plot in late summer before they produce seed.

Top and middle left photo: Dogfennel
Middle right and bottom photo: Yankeeweed

301

Blue mistflower
Eupatorium coelestinum
(or *Conoclinium coelestinum*)

Description and ecology
Erect, warm-season, perennial forb growing
1 – 3 feet tall. Stem is relatively stout, has
short hairs, and is branched in upper portion.
Leaves are opposite with short stems, deltoid
in shape, with toothed margins, and pointed
tips. Flowers appear July – November and
are light purple and numerous at the ends of
stems.

Origin and distribution
Native; throughout eastern U.S.

Considerations as a weed and for wildlife
Blue mistflower is a common forb found in
fallow plots. It can provide good cover for
northern bobwhite and wild turkey broods,
as well as ground-feeding songbirds. The
flowers attract bees and butterflies. It is
a low- to moderate-preference forage for
white-tailed deer. It is rarely a problem in
perennial forage plots, but if necessary it can
be sprayed when young with imazethapyr
or 2,4-DB for some control, or spot-sprayed
later with glyphosate.

Joe-pye weed
Eupatorium fistulosum

Description and ecology
Erect, perennial, warm-season forb with stout unbranched stem that may reach 10 feet tall on some sites. Leaves are whorled (ringed around the stem). Flowers are pinkish-purple and appear July – October.

Origin and distribution
Native; throughout eastern U.S.

Considerations as a weed and for wildlife
Joe-pye weed is rarely a problem, but sometimes occurs in perennial forage plots 2 – 3 years after planting. It should be sprayed when young with imazethapyr or 2,4-DB in perennial clover/alfalfa plots. If it gets tall, it can be set back by mowing or by applying glyphosate via spot-spraying or with a wick applicator. When occurring in clover plots managed for wild turkeys, joe-pye weed can provide attractive structure for broods through summer. It is a low-preference forage for white-tailed deer. Many species of butterflies are attracted to joe-pye weed.

Boneset
Thoroughwort

Eupatorium serotinum (Late flowering thoroughwort or Late boneset)
E. perfoliatum (Common boneset)
E. altissimum (Tall thoroughwort)
E. torreyanum (Hyssopleaf thoroughwort)
E. tomentosum (Roundleaf thoroughwort)
E. sessilifolium (Upland boneset)

Description and ecology

There are many species of Eupatoriums. These are erect, perennial, warm-season forbs that may be locally widespread. They range in height from 3 – 6 feet. Stems are relatively stout and usually branched. Leaves are opposite. Leaves of *E. serotinum* have a distinct stem. Leaf bases of *E. perfoliatum* are united around the stem. Leaves of *E. altissimum* are palmately 3-nerved with virtually no stem and narrow at the base. Leaves of *E. hyssopifolium* are usually in whorls of 4 and extremely linear and narrow. Leaves of *E. rotundifolium* are broadly rounded and sometimes heart-shaped at base and narrowed at tip. Leaves of *E. sessilifolium* have no stem and are rounded at the base. Flowers appear late summer through fall (according to species), are white, and occur in clusters at the end of branches.

Origin and distribution

Native; throughout most of eastern U.S.

Considerations as a weed and for wildlife

The bonesets and thoroughworts are perennial forbs that may occur in cultivated warm-season plots, but more often in perennial plots after 2 or 3 years or in fallow fields. They can provide outstanding cover for northern bobwhite and wild turkey, several species of sparrows, indigo bunting, eastern cottontail, and deer fawns. These species are relatively low-preference forages for white-tailed deer. They can be controlled in perennial clover/alfalfa plots with imazethapyr and 2,4-DB if sprayed when young. It is actually beneficial to have some boneset or thoroughwort in plots managed for wild turkeys because of the structure they provide. They can be set-back by mowing in late summer, but like other perennials, mowing does not get rid of them.

Top photo: Late flowering thoroughwort or late boneset; middle photo: roundleaf thoroughwort; bottom photo: hyssopleaf thoroughwort.

304

Common boneset

Late flowering thoroughwort

Slender goldentop
Euthamia caroliniana

Grass-leaved goldenrod
Euthamia graminifolia

Description and ecology
Perennial warm-season forb that may be multi-branched with bushy appearance, 2 – 4 feet tall. Stem relatively stout and smooth. Leaves alternate, linear and narrow, smooth and pointed. Leaves usually single-veined, less often with 3 veins. *E. graminifolia* is similar, usually with wider leaves and 3 veins, less often 5 veins. Yellow flowers are clustered at the ends of the branches; bloom in late summer/early fall. Tolerant of wide range of soil conditions, clay to sand, moist to dry.

Origin and distribution
Native; occurs along Coastal Plain of eastern U.S. and into KY.

Considerations as a weed and for wildlife
Slender goldentop can be weedy in perennial food plots. However, it provides outstanding cover for northern bobwhite, wild turkey broods, white-tailed deer fawns, eastern cottontail, and ground-feeding songbirds, especially in old-field communities. Seed are eaten sparingly by various sparrows. It is a relatively low preference forage of white-tailed deer. Slender goldentop can be controlled with spot-applications of glyphosate, triclopyr, 2,4-D, and dicamba.

Hairy galinsoga
Fringed quickweed
Galinsoga quadriradiata

Description and ecology
Erect, annual, warm-season forb that may grow about 2 feet tall. Stem has numerous branches and very hairy. Leaves are opposite, hairy on top, coarsely toothed, with a sharp tip. Flowers have a yellow disk (center) with 5 white, 3-toothed ray flowers surrounding the disk flowers.

Origin and distribution
Nonnative (Central and South America); throughout most of U.S., except southwest U.S.

Considerations as a weed and for wildlife
Hairy galinsoga is found often in warm-season annual plots that are managed with conventional cultivation. It does not usually present a competition problem, but can be widespread. It can be controlled preemergence with S-metolachlor or imazethapyr, with forb-selective herbicides, or with glyphosate in Roundup-Ready crops. However, control may not be required because it is a moderate- to high-preference forage for white-tailed deer and also is eaten by eastern cottontails and groundhogs.

Rabbittobacco
Sweet everlasting
Gnaphalium obtusifolium (or *Pseudognapalium obtusifolium*)

Purple cudweed
Gnaphalium purpureum (or *Gamochaeta purpurea*)

Description and ecology
Erect annual or biennial forb, arising from a basal rosette, growing to about 4 feet tall on some sites. Stem and leaves appear chalky white from dense, fine hair underneath leaves and on stem. Leaves are alternate and gray-green on top. Flowers are white and appear at end of branches in August – October.

Origin and distribution
Native; throughout eastern U.S.

Considerations as a weed and for wildlife
Rabbittobacco responds to soil disturbance. It rarely competes with the planted crop, but can be frequent or widespread in cultivated warm-season annual plots. Rabbittobacco provides good cover for rabbits and ground-feeding birds, especially during fall and winter as it maintains its structure through winter and the following spring. It is a low-preference forage for white-tailed deer. Rabbittobacco can be controlled with forb-selective herbicides, such as 2,4-D, dicamba, and triclopyr.

Top photo: rabbittobacco; bottom left and middle: rabbittobacco; bottom right: purple cudweed

Common sneezeweed
Helenium autumnale

Bitter sneezeweed
Helenium amarum

Description and ecology
Common sneezeweed is an erect, perennial, warm-season forb. Stem is unbranched until near top of plant. Leaves are alternate, lanceolate in shape, 1 – 4 inches long, with a pointed tip. Flowers are yellow and in clusters at tips of branches. Disk flowers in center are domed and yellow-green. Ray flowers are yellow, 3-lobed at tip, and surround the disk flowers. Bitter sneezeweed is an erect, annual, warm-season forb. The stem has many branches, the leaves are alternate and fine (very narrow) or thread-like. Flowers are numerous and yellow; both disk (center) and ray flowers are yellow. Ray flowers are 3-lobed at the tips. Ray flowers fall off and leave the disk flowers looking like a yellowish ball before turning brown.

Origin and distribution
Native; common sneezeweed occurs throughout U.S., except portions of southwest U.S.; bitter sneezeweed occurs throughout eastern U.S.

Considerations as a weed and for wildlife
Sneezeweed may occur in warm-season annual grain plots or in perennial cool-season forage plots. It does not usually present a problem in corn or soybean fields, but will compete with cool-season perennial forages. On sites where growth is tall enough, sneezeweed provides good cover for northern bobwhite, wild turkey broods, and ground-feeding songbirds in grain plots, but it does not have any food value for wildlife. Sneezeweed contains a glycocide that may cause digestive and other physiological problems in livestock and likely in some wildlife species as well. Sneezeweed can be controlled with glyphosate or a variety of forb-selective herbicides, such as dicamba, triclopyr, and metsulfuron methyl.

Top photo: Common sneezeweed, photo by J. Hamlington
Middle and bottom photos: Bitter sneezeweed

Camphorweed
False goldenaster
Heterotheca subaxillaris

Description and ecology
Annual, warm-season forb that may grow to 7 feet tall. Plant is quite variable, leading some botanists to separate into different species. Primary stem typically with branches (often spreading widely), often reddish-brown, with yellow flowers at tips in late summer/early fall. Stem and branches are fuzzy. Leaves are alternate, often clasping stem, and produce camphor aroma when crushed. Leaves typically do not persist until flowering, or may be crinkled and brown. Camphorweed can be controlled with spot-applications of glyphosate, triclopyr, 2,4-D, and dicamba.

Origin and distribution
Native; NY to CA and south

Considerations as a weed and for wildlife
Camphorweed can spread and outcompete other plants. However, the structure is quite good for brooding game birds, ground-feeding songbirds, eastern cottontail, and white-tailed deer fawns. Camphorweed is not a selected forage of white-tailed deer or eastern cottontail. Camphorweed can be controlled with preemergence application of imazethapyr in established perennial clover plots, or postemergence with glyphsosate, triclopyr, 2,4-D, or dicamba if necessary in old-field communities. Alternatively, the site could be disked soon after germination to reduce density of camphorweed if too thick.

Smooth cat's ear
Hypochaeris glabra

Spotted cat's ear
Rough cat's ear
Cat's ear dandelion
H. radicata

Description and ecology
H. glabra is an annual forb arising from a basal rosette with leaves that are green to purple with serrated margins. Bracts on flowers resemble scales. *H. radicata* is a perennial cool-season forb arising from root crown. Basal leaves are 2 – 6 inches long with firm, coarse hairs. Stems have few if any branches or leaves. Flowers are yellow, appear May – September, and look much like dandelion flowers.

Origin and distribution
Nonnative (Eurasia); throughout most of U.S.

Considerations as a weed and for wildlife
Spotted cat's ear is spreading across much of U.S., sometimes popping up in perennial cool-season forage plots. It has no value for wildlife. It can be controlled with imazethapyr or 2,4-DB in perennial forage plots, with thifensulfuron-methyl+tribenuron-methyl in cool-season grain plots, or with glyphosate or a forb-selective herbicide in fallow fields.

Top photos: Spotted cat's ear, showing flowerhead about to open and seedhead; Middle photos left and right; flowering heads and rough leaf texture; Bottom photos: Smooth cat's ear, showing flower about to open and smooth leaf texture

Sumpweed
Annual marshelder
Annual Iva
Iva annua

Description and ecology
Erect, annual, warm-season forb. Stem is pubescent and may grow to 4 feet tall. Leaves are opposite, and margins are usually toothed, but not lobed (unlike the ragweeds). Seedheads are at end of branches and appear similar to those of *Ambrosia* (ragweeds). However, the leaf structure is completely different (see bottom photo for comparison with common ragweed).

Origin and distribution
Native; throughout Great Plains and eastern U.S., except portions of New England

Considerations as a weed and for wildlife
Sumpweed can be common in cultivated grain crops and may appear in cool-season plots, especially those established via disking or some other soil disturbance. Sumpweed provides excellent cover for northern bobwhite and wild turkey broods, as well as ground-feeding songbirds. However, neither the seed or foliage is an important food source for wildlife. Sumpweed can be controlled with imazethapyr pre- or postemergence, with forb-selective herbicides, or by mowing or disking before it produces seed. However, is not usually competitive with taller grain crops, such as corn or grain sorghum.

Bottom photo: sumpweed (left), common ragweed (right)

Potato dwarfdandelion
Krigia dandelion

Carolina falsedandelion
Pyrrhopappus carolinianus

Description and ecology
Potato dwarfdandelion is a perennial cool-season forb that flowers in spring. Flowerhead is orangish-yellow, up to 2 inches in diameter. Flowering stalk usually 10 – 18 inches tall. Basal leaves are smooth but may be slightly toothed.

Carolina falsedandelion is an annual cool-season forb. Basal leaves are alternate, narrow, and 3 – 10 inches long. Falsedandelion may have stem leaves, whereas dandelion only has basal leaves. Flowerhead is bright sulpher yellow and usually 12 – 18 inches tall.

Origin and distribution
Potato falsedandelion native; southern NJ west to IA, south to TX. Carolina falsedandelion native; MD to KS, south to TX and FL.

Considerations as a weed and for wildlife
Dwarfdandelion has an edible tuber, but I have found no reports of wildlife eating the tuber. Bees, wasps, butterflies, and moths get nectar from dwarfdandelion and falsedandelion, but there is no other reported use by wildlife. Both of these plants usually appear sparingly. I have never seen them problematic, but they could be controlled if needed with glyphosate or broadleaf-selected herbicides, such as 2,4-D, dicamba, and triclopyr.

Top photos: Potato dwarfdandelion
Bottom photos: Carolina falsedandelion; also see photo of Carolina falsedandelion in flower on page 320

313

Prickly lettuce
Lactuca serriola

Wildlettuce
L. canadensis

Description and ecology
Annual or biennial cool-season forbs. Flowering stem, usually 3 – 5 feet tall, arises from basal rosette. Young leaves in rosette are oblong, toothed, and may be wavy or slightly lobed. Stem is hollow and exudes milky juice if cut. Prickly lettuce leaves are alternate with **fine prickles on margins and stiff prickles on a prominent midrib of lower leaf surface**. Prickles on midrib of lower leaf surface help distinguish prickly lettuce from sowthistles before flowering. Mature leaf lobes often shaped like a half moon. Leaf bases clasp stem with ear-like lobes. Flowers are yellow and in an open panicle, usually appearing July through September. **Note:** There are several native wildlettuce species, but prickly lettuce is found most often following disturbance in agricultural settings. Wildlettuce is not spiny (or with soft bristles).

Origin and distribution
Prickly lettuce is nonnative (Europe); throughout U.S. Wildlettuce is native and occurs throughout U.S.

Considerations as a weed and for wildlife
Prickly lettuce is not usually problematic, but can get dense and competitive on some sites. Prickly lettuce is a high-preference forage for white-tailed deer, and also may be eaten by eastern cottontail and groundhog. Pronghorn eat leaves of wildlettuce and American goldfinch may eat the seed. Wildlettuces can be controlled preemergence with S-metolachlor, and postemergence with forb-selective herbicides, such as 2,4-DB and dicamba, and with glyphosate in Roundup-Ready crops.

Top photo: Prickly lettuce;
Second from top: Wildlettuce;
Third from top: Prickly lettuce;
Bottom photo: wildlettuce.

Oxeye daisy
Leucanthemum vulgare
(or *Chrysanthemum leucanthemum*)

False aster
Boltonia asteroides

Description and ecology
Oxeye daisy is an erect, perennial forb. Flowering stems arise from rosette in June-July. Leaves along stem are alternate and are variably blunt-toothed or with rounded lobes. Flowers are single and terminal on stem; seedhead consists of 20 – 30 white ray flowers surrounding yellow center (disk flowers). Similar to mayweed chamomile (*Anthemis cotula*) and corn chamomile (*Anthemis arvensis*), but the chamomiles have finely dissected leaves and mayweed chamomile has a strong offensive odor when the leaves are crushed. False aster is a slender, multi-branched perennial forb that appears similar to oxeye daisy, except leaves are linear in shape and not lobed. The white ray flowers fall off leaving disk flowers that look like yellow balls.

Origin and distribution
Oxeye daisy is nonnative (Europe); throughout U.S. except portions of southwest U.S. False aster is native; throughout eastern U.S.

Considerations as a weed and for wildlife
Both oxeye daisy and false aster are often present in grain crops, especially those that have been cultivated and in the first year of fallow growth. They also may be common and widespread in perennial forage plots. They may provide low cover for ground-feeding birds if not too sparse. White-tailed deer, rabbits, and groundhogs may eat the foliage, but it is a low-preference forage. They can be controlled with forb-selective herbicides, such as triclopyr, clopyralid, and dicamba. Imazethapyr and 2,4-DB also are effective if sprayed when the plants are young in rosette stage. Occasional mowing does not get rid of oxeye daisy.

Top and middle photos: Oxeye daisy; Bottom photo: False aster

315

Pineapple-weed
Matricaria discoidea (or *M. matricarioides,* or *Chamomilla suaveolens*)

Description and ecology
Erect annual forb about 10-15 inches tall with bushy appearance. Stem arises from basal rosette and is multi-branched. Rosettes may remain green through winter. Leaves along stem are alternate, 1-2 inches long, and finely pinnately divided, similar to mayweed chamomile when young. However, leaves of pineapple-weed have a sweet odor similar to pineapple when crushed, whereas mayweed chamomile has an offensive odor when crushed. Pineapple-weed flowers appear April through September and are usually numerous at the ends of branches. They look like a greenish-yellow cone in a pale shallow cup.

Origin and distribution
Native to Pacific states, but naturalized throughout most of U.S. except extreme southeast U.S. and portions of southwest U.S.

Considerations as a weed and for wildlife
Pineapple-weed may be found in fallow fields and perennial plots that are beginning to thin. The plant can provide some cover for ground-feeding songbirds and rabbits. Juncos and goldfinches eat the seed. Pineapple-weed can be difficult to control as it is resistant to several herbicides. It can be controlled with glyphosate in fallow areas or in Roundup-Ready crops, and hexazinone will provide some control in alfalfa and rangeland. Mowing occasionally will not control pineapple-weed in perennial forage plots.

Cressleaf groundsel
Ragwort
Yellowtop
Packera glabella

Description and ecology
Erect, annual, cool-season forb that grows 2 –
3 feet tall. Stem is smooth, succulent, hollow,
often with reddish-purple coloration near base.
Stem arises from a basal rosette. Basal leaves
4 – 8 inches long and toothed. Stem leaves
are lobed and are smaller on upper portion of
stem. Leaves along stem are alternate. Flowers
are yellow, occur in clusters, mostly at the top
of the plant. Cressleaf groundsel often grows
in fallow fields as well as relatively moist,
somewhat shaded areas.

Origin and distribution
Native; throughout the South

Considerations as a weed and for
wildlife
Cressleaf groundsel may be found in food
plots, particularly after soil disturbance in
relatively moist, partially shaded
areas. It does not usually present
competition problems, but may be
widespread. It can be controlled
with imazethapyr, a forb-selective
herbicide, or with glyphosate. It
also can be controlled by mowing
soon after flowering. Cressleaf
groundsel may provide some low
cover for various wildlife species,
but is not often eaten. It contains
toxins that cause liver disease in
livestock.

All photos *P. glabella*, except bottom
right photo, which is a similar species,
P. anonyma, Small's Ragwort.

317

Black-eyed Susan
Rudbeckia hirta

Brown-eyed Susan
R. triloba

Description and ecology
Biennial or short-lived perennial forbs. Form rosettes first year, flower second year. Black-eyed Susan 1 – 2 feet tall, whereas brown-eyed Susan may grow to 5 feet tall. *R. hirta* may branch near base, whereas *R. triloba* branches along the stem and often has bushy appearance at maturity. Stems of both have conspicuous white hairs. Leaves alternate, rough texture from small stiff hairs. *R. hirta* leaves up to 7 inches long and 2 inches wide; basal leaves have long hairy stems, upper leaves with short stems or may clasp main stem. *R. triloba* leaves 4 inches long, 2 inches wide; lower leaves may be lobed. *R. hirta* produce one flower at tip of each stem; *R. triloba* may have 1 – 2 flowers per stem. Center of flower (disk or cone) of both is dark brown with 8 – 20 yellow ray petals surrounding *R. hirta* and 6 – 12 surrounding *R. triloba*. *R. hirta* flowers about 2 – 3 inches wide; *R. triloba* 1 ½ - 2 inches wide.

Origin and distribution
Both native; *R. hirta* throughout U.S. except portions of southwest U.S.; *R. triloba* throughout Great Plains and eastern U.S.

Considerations as a weed and for wildlife
Neither of these species are likely to be competitive in food plots, but they are commonly found in old-field and other early successional communities. Structure for game bird broods is good and both attract insects that birds eat. Both plants attract numerous species of pollinators. White-tailed deer, eastern cottontail, and groundhog eat the foliage of both. Prior to flowering, both are moderately selected by white-tailed deer.

Top and middle photos: Black-eyed Susan;
Bottom photo: Brown-eyed Susan

Common groundsel
Common ragwort
Senecio vulgaris

Description and ecology
Erect, annual, cool-season forb growing 12-18 inches tall, occurring often on disturbed sites. Plant grows from a basal rosette with multiple branches and alternate leaves occurring along branches. Leaves are deeply lobed or dissected and irregularly toothed along margins. Flowering occurs spring through summer in loose clusters at ends of branches. Three or more generations may be produced in one growing season. Flowers are yellow, but open into seedheads when mature that appear somewhat like dandelion seedheads.

Origin and distribution
Nonnative (Eurasia); throughout the U.S. except portions of Southwest

Considerations as a weed and for wildlife
Common groundsel may occur in annual or perennial plots, as well as fallow areas. It may be widespread and compete with planted forages, especially perennial forages, such as clovers and alfalfa. Common groundsel contains alkaloids that may cause liver damage in cattle. However, other ruminants, such as goats and sheep, may consume the plant with no apparent ill effects. Its effect on deer has not been reported. It is a low-preference deer forage, if deer eat it at all. I have never seen evidence of white-tailed deer grazing the plant. Goats, hogs, and rabbits may eat the foliage, and various sparrows and finches may eat the seed. Common groundsel may be resistant to triazine herbicides, such as atrazine and simozine. It can be controlled with dicamba, 2,4-DB, hexazinone, and glyphosate if sprayed when young. Cultivation or mowing just before groundsel flowers will provide control. However, if groundsel flowers, they may continue to mature and develop seed even after mowing or spraying.

319

Goldenrod
Solidago spp.

Description and ecology
Erect, perennial, warm-season forbs growing to 6 feet tall on some sites. Spread by seed and rhizomes. Leaves alternate along a tall main stem. Flowers are yellow-golden and appear in late summer – fall. There are at least 125 species of goldenrods in the U.S.

Origin and distribution
Native; throughout U.S.

Considerations as a weed and for wildlife
Goldenrods can be common in perennial forage plots where they may shade-out clovers, alfalfa, and chicory. When managing clover plots specifically for wild turkeys, some goldenrod (not too much) can make the plot more attractive because of the cover provided for broods. Goldenrods can be controlled in perennial plots if sprayed when young with imazethapyr or 2,4-D. If they get too dense, goldenrods can be set-back by mowing or by applying glyphosate or a forb-selective herbicide, such as dicamba, via spot-spraying. Selectivity of goldenrods as forage for white-tailed deer varies by species; some are highly selected, whereas others are low to moderately selected. Ruffed grouse, spruce grouse, sharp-tailed grouse, prairie-chickens, and rabbits will eat the foliage. Small mammals (meadow mice) eat the foliage and seedheads. American goldfinch, dark-eyed junco, and various species of sparrows eat the seed. Goldenrod can provide great cover for rabbits, deer fawns, wild turkey, and northern bobwhite broods, and many species of songbirds in fallow fields.

It is common to find goldenrods with galls. Those in the stems are most often caused by the goldenrod gall fly (round galls) or the goldenrod gall moth (elongated galls). Inside the galls are developing larvae. Goldenrod gall fly larvae emerge in spring, whereas goldenrod gall moths emerge in the fall. Carolina chickadees and downy woodpeckers actively seek galls to eat the insect larvae. Twisted clusters of stunted goldenrod leaves at the top of the stem are caused by goldenrod gall midges.

Sowthistles

Sonchus arvensis (Perennial)
S. asper (Spiny)
S. oleraceus (Annual or Common)

Description and ecology

All three are similar in appearance with flowering stems arising from basal rosettes. Leaf margins of spiny sowthistle are very prickly, whereas perennial sowthistle leaf margins are considered prickly, and those of annual sowthistle weakly or sparsely prickly. Leaf bases clasp the stem. Leaves, stems, and roots of all three exude milky sap when cut. There are usually several flowers in terminal clusters (very different from prickly lettuce and wildlettuce). Flowerheads are urn-shaped before the yellow petals emerge. Sowthistles are typically much shorter than prickly lettuce and wildlettuce, often only several inches tall, but may reach 4 feet.

Origin and distribution

Nonnative (Europe); annual and spiny sowthistle throughout U.S.; perennial sowthistle from Mid-South northward.

Considerations as a weed and for wildlife

Annual and perennial sowthistles are common and may occur widespread across fields. They can be controlled preemergence with atrazine and S-metolachlor in some warm-season plots, or postemergence with forb-selective herbicides, such as dicamba, thifensulfuron-methyl + tribenuron-methyl, and 2,4-D in cool-season annual grain plots. Sowthistles are low-preference forages for white-tailed deer. Rabbits and groundhog may eat the foliage sparingly. American goldfinch may eat the seed.

Photos: Spiny sowthistle

Dandelion
Taraxacum officinale

Description and ecology
Perennial, cool-season forb with a basal rosette and taproot. Leaf margins are wavy, often with deep lobes, and sparsely toothed. Flowers are yellow and appear May – June on a leafless stalk. Seedheads are round, composed of achenes with feathery bristles that blow in the wind.

Origin and distribution
Dandelion nonnative (Europe); throughout U.S.

Considerations as a weed and for wildlife
Dandelions can be problematic in perennial forage plots. Imazethapyr, clopyralid, and 2,4-DB are effective if sprayed when dandelions are young. Dandelions are eaten sparingly by eastern cottontail and are a low-preference forage for white-tailed deer. Wild turkeys and ruffed grouse eat the foliage and the seedheads.

Top three photos: Dandelion
Bottom left: Wild turkeys were selectively eating dandelion flower heads in this clover food plot, which has been invaded by cool-season grasses, buttercup, and daisy fleabane.
Bottom right: Carolina falsedandelion for comparison

Western salsify
Goatsbeard
Tragopogon dubius

Description and ecology
Upright, biennial forb arising from rosette. Flowering stem arises in second year and is smooth, round, and somewhat fleshy. Leaves are alternate, linear, pointed, and grasslike, often with a greenish-silver hue. Unlike grass leaves, western salsify leaves may exude a milky juice when broken. Flowers are yellow and appear in June – July. Green bracts subtending the flowers are pointed and always longer than the flowers. The seedhead looks like a dandelion seedhead, but larger, sometimes up to 3-4 inches in diameter.

Origin and distribution
Nonnative (Eurasia and northern Africa); throughout most of U.S., except southeastern Coastal Plain; most common in western U.S.

Considerations as a weed and for wildlife
Western salsify may occur in annual or perennial plots. It is not usually aggressive, but may be widespread. It can be controlled with glyphosate, metsulfuron-methyl, 2,4-D, 2,4-DB, and dicamba. Western salsify is eaten by dusky grouse and small mammals. It is reportedly bitter and a low-preference forage of rabbits and mule deer.

Crownbeard or wingstem
Verbesina alternifolia (Wingstem)
V. enceloides (Crownbeard or Golden crownbeard)
V. occidentalis (Yellow crownbeard)
V. virginica (White crownbeard)

Description and ecology
Erect, perennial, warm-season forbs, except *V. enceloides*, which is annual. All 4 species appear somewhat similar and may grow to 7 – 9 feet tall on some sites. Wingstem (*V. alternifolia*) has stem with characteristic "wings" extending along main stem between leaves. All 4 species have alternate leaves except yellow crownbeard (*V. occidentalis*), which has opposite leaves and also has winged petioles extending down stem, but less prominent than wingstem (*V. alternifolia*). Crownbeard (*V. enceloides*) leaves are opposite on lower portion of plant, but alternate on upper portion of plant. Stems and lower leaves of crownbeard are covered with fine white hairs. White crownbeard (*V. virginica*) has winged stems, but leaves are much wider than *V. alternifolia* or *occidentalis* and with a prominent white midvein. Flowers of white crownbeard are white, whereas both disk (center) and ray (surrounding disk) flowers of the other 3 species are yellow. Flowers of all species occur at end of branches.

Origin and distribution
All 4 species are native; *V. alternifolia* occurs throughout Great Plains and eastern U.S., except upper New England; *V. occidentalis* occurs from PA to MO and south to TX and FL; *V. virginica* occurs from ML to IA and south to TX and FL; *V. enceloides* occurs primarily in Great Plains, southwest U.S., and southeastern Coastal Plain.

Considerations as a weed and for wildlife
The *Verbesinas* can become problematic in perennial forage plots. However, they can provide attractive cover for wild turkeys in a clover/chicory plot as long as they are not so dense to outcompete planted forages. They may be considered beneficial in fallow plots as cover for wild turkey, northern bobwhite, white-tailed deer, and several species of songbirds. They are low-preference forage for white-tailed deer, rabbits, and groundhog. They can be controlled with imazethapyr and 2,4-DB if sprayed when young, or with glyphosate or various forb-selective herbicides, such as triclopyr, dicamba, or metsulfuron methyl.

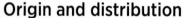

Top left photo is wingstem; Middle photo is yellow crownbeard; Bottom photo is white crownbeard

Tall ironweed
Vernonia altisima

Western ironweed
V. baldwinii

Description and ecology
Erect, perennial, warm-season forbs with stout stems. Tall ironweed can reach 10 feet tall on some sites, whereas western ironweed is generally 2-5 feet tall. Leaves of tall ironweed are alternate with toothed margins and a white midrib. Western ironweed leaves are similar, but wider and more dense along stem. Flowers are purple and appear in September – October (tall ironweed) and July – September (western ironweed).

Origin and distribution
Native; tall ironweed occurs southern NY to north FL west to IA and east TX; western ironweed occurs central U.S. from MN south to NM and LA.

Considerations as a weed and for wildlife
Tall and western ironweed can become widespread in perennial forage plots. They can be controlled by mowing when flowering or by spraying with imazethapyr or 2,4-DB when they are young. Tall and western ironweed are not selected forages by white-tailed and mule deer, but they provide attractive structure (cover) for

wild turkey poults feeding in perennial forage (clover, chicory, alfalfa) food plots. If plots are managed for wild turkeys, don't worry if you have some ironweed, just mow plot in late summer before ironweed produces seed. If ironweed begins to get dense and compete with planted forages, spot-spray as needed when young. Both tall and western ironweed provide outstanding cover for deer fawns, wild turkey, eastern cottontail, and northern bobwhite in fallow fields.

Top and bottom left photos: Tall ironweed; Bottom right photo: Western ironweed, photo by: Geir Friisoe

Common cocklebur
Xanthium strumarium

Description and ecology
Erect, annual, warm-season forb. Grows to 6 feet tall on some sites. Leaves are alternate. Flowers July – September. Fruit enclosed in a prickly bur that sticks to clothing and fur.

Origin and distribution
Native; occurs throughout U.S.

Considerations as a weed and for wildlife
Cocklebur is common in warm-season plots, especially those planted via conventional tillage and on bottomland sites. Cocklebur is very competitive with planted forages and can spread rapidly across a field as its seed are transported easily by equipment and by attaching to animals. It can be controlled preemergence with imazethapyr, imazamox, S-metolachlor, and atrazine. It can be controlled postemergence with dicamba, bentazon, 2,4-D, 2-4DB, imazethapyr, and clopyralid. Cocklebur can provide good cover for various species, including wild turkey, northern bobwhite, deer fawns, and ground-feeding songbirds. However, because it displaces more desireable plants, control may be needed. It is a low-preference forage for white-tailed deer and other mammals.

Bottom photo: Cocklebur outcompeting planted corn

Spring scorpiongrass
Spring forget-me-not
White forget-me-not
Myosotis verna (or *M. virginica*)

Description and ecology
Cool-season annual or biennial forb (although called scorpiongrass, it is not a grass) that grows about 6 – 18 inches tall. Stem is light green, fuzzy hairy, usually unbranched, but may be branched, and usually leaning. Leaves are alternate, toothless, oblong, hairy, up to 2 inches long and 1/3 inch wide with no stem, and with obvious linear central vein. Flowers are white, each with 5 lobes, and occur along stem and in clusters at ends of stems. The calyx (base of flower) densely covered with hairs. The fruit is a capsule with short stiff hairs that can cling to clothing and fur. Each capsule contains 4 seeds. Scorpiongrass may appear superficially similar to beaked cornsalad when flowering, but check the leaves. Beaked cornsalad has opposite leaves that are toothed near base.

Origin and distribution
Native; throughout most of U.S. except portions of Rocky Mtn states and FL.

Considerations as a weed and for wildlife
Spring scorpiongrass is often found in disturbed areas and rarely competes with planted cool-season forages. It can be controlled with imazapyr, 2,4-DB, or glyphosate in various food plot applications. Seed may be eaten by various sparrows. It is not a selected forage of white-tailed deer.

Wild mustard
Brassica kaber (or *Sinapis arvensis*)

Wild radish
Raphanus raphanistrum

Description and ecology
Erect, annual, cool-season forbs that may occur in winter or summer. Plants arise from basal rosette. Lower leaves are irregularly lobed and larger than upper leaves. Flowers are yellow and conspicuous and may occur May – August. Flowering stem is stout and may be 3 feet tall on some sites. Wild mustard and wild radish are very similar in appearance. Wild radish leaves have more and deeper lobes than wild mustard. Also, the seeds of wild radish are segmented in the pod and the segments break apart. Seedpods of wild mustard open lengthwise.

Origin and distribution
Nonnative (Europe); occurs throughout U.S.

Considerations as a weed and for wildlife
Wild mustard and wild radish can be very competitive in food plots, especially in perennial cool-season plots prepared via conventional tillage. Wild mustard can be controlled in clover and alfalfa plots with imazethapyr, 2,4-DB, and bentazon if sprayed when young. It can be controlled preemergence with imazethapyr, pendimethalin, and atrazine in various warm-season plots, and with forb-selective herbicides, such as thifensulfuron-methyl + tribenuron-methyl, dicamba, mesosulfuron-methyl, and bentazon in cool-season annual grain plots. Wild mustard is a low-preference forage for white-tailed deer and eastern cottontail.

Top photo: Wild mustard, photo by: J. DiTomaso
Bottom photos: Wild radish

Shepherd's-purse
Capsella bursa-pastoris

Description and ecology
Annual, cool-season forb with flowering stem arising from basal rosette. Flowers spring through early summer. Fruits are triangular to heart-shaped in appearance. Similar to Virginia pepperweed, but Virginia pepperweed fruits are round and main stem has multiple branches.

Origin and distribution
Nonnative (Europe); occurs throughout U.S.

Considerations as a weed and for wildlife
Shepherd's-purse can be common in cool-season forage plots, especially those established via conventional cultivation. Along with other cool-season annual forbs, it can compete with clover seedlings trying to establish. Imazethapyr and 2,4-DB work well in clover plots and other forb-selective herbicides, such as thifensulfuron-methyl + tribenuron-methyl, and dicamba, are effective in cool-season grain plots (such as wheat and oats). Shepherd's-purse seed is eaten by goldfinches and horned lark.

Hairy bittercress
Cardamine hirsuta

Description and ecology
Annual, cool-season forb arising from a basal rosette. Basal leaves may or may not be hairy, so don't look for fuzzy leaves. Leaves have 1 – 3 pairs of leaflets and a terminal leaflet that is larger than the leaflets along the side. Flowering usually occurs mid- to late spring. Seeds enclosed in an elongated capsule and "explode" out of the capsule when hit (you can hear and see the tiny seeds flying through the air 5 – 10 feet as you walk through an area with hairy bittercress).

Origin and distribution
Nonnative (Eurasia); eastern U.S.

Considerations as a weed and for wildlife
Hairy bittercress is not usually overly competitive with planted cool-season forages, but it can be widespread and sometimes dense. It usually grows in association with chickweeds, henbit, purple deadnettle, and geranium. It should be sprayed when young or before it germinates with other cool-season forb weeds with imazethapyr in perennial clover/alfalfa/chicory plots or with a forb-selective herbicide in cool-season annual grain plots. It provides no value for wildlife.

Bottom photo: Hairy bittercress and Carolina geranium

Pinnate tansymustard
Descurainia pinnata

Flixweed
D. sophia

Description and ecology
Upright, annual cool-season forbs growing to 2 – 3 feet tall on some sites. They usually begin growth in the fall from a rosette. Stems are often multi-branched, thick, fleshy, and pubescent. Leaves are alternate, green, and compound, divided 2 or 3 times into narrow segments. Pinnate tansymustard flowerheads are elongated racemes with yellow flowers at the tip of the stem. Flixweed flowerheads also are at the end of the stem and yellow, but usually ascend at right angles. Leaves of flixweed often are more grayish-blue than green and have an aromatic odor when crushed. Tansymustard seed pods are about ½-inch long and the seed occur in 2 rows within the pod, whereas flixweed seed pods are 1 – 1 ½-inch long and the seed occur in 1 row within the pod.

Origin and distribution
Pinnate tansymustard is native and occurs throughout U.S., but primarily in Great Plains and southwest U.S. Flixweed is nonnative (Eurasia) and found throughout U.S. except Deep South.

Considerations as a weed and for wildlife
These plants may occur on a wide variety of sites, from dry to moist, from sterile to fertile. They may occur in annual plots or established perennial plots. They do not usually outcompete planted forages, but may be widespread. They may be controlled with glyphosate or forb-selective herbicides. Tansymustard seed is eaten by Gambel and scaled quail. The foliage is eaten sparingly by pronghorn, elk, mule deer, and Canada goose.

Photos: Pinnate tanseymustard, photos by: J. DiTomaso

Virginia pepperweed
Lepidium virginicum

Field pepperweed
L. campestre

Description and ecology

Erect, annual, cool-season forbs that may occur in winter or summer. Virginia pepperweed arises from basal rosette with multiple-branched stem. Basal leaves do not persist on mature plants. Field pepperweed arises from basal rosette with a central main stem with larger leaves than Virginia pepperweed. Stem leaves of field pepperweed clasp the stem, whereas leaves of Virginia pepperweed do not clasp the stem. **All leaves of field pepperweed are densely covered with short hairs, whereas Virginia pepperweed leaves are smooth.** Both flower May – June. Virginia pepperweed fruit (seed pods) are round, flat, slightly notched at the tip, and contains small, tan seeds. Field pepperweed seed pods are egg-shaped with wing-like structures around the edge. Rosettes persist through winter and have a peppery taste, as do the seed pods.

Origin and distribution

Native; occur throughout U.S.

Considerations as a weed and for wildlife

Virginia pepperweed and field pepperweed are commonly found in fields planted via conventional techniques (plowing, disking); however, they rarely present a problem. Pepperweed can be controlled preemergence with imazethapyr or postemergence with forb-selective herbicides, such as dicamba, 2,4-D, 2,4-DB, and imazethapyr, and with glyphosate in Roundup-Ready crops. Pepperweeds rarely are eaten by white-tailed deer and the seed is of little use to any bird species.

Photos: Virginia pepperweed

Marsh yellowcress
Bog yellowcress
Rorippa palustris

Description and ecology
Annual (sometimes biennial and rarely perennial) mustard-like forb most often found on recently disturbed moist sites, but tolerates a range of soil conditions. It is usually upright, but occasionally reclining. Stem is usually branched, light green or reddish, smooth, and furrowed. Leaves are alternate and variable in shape; up to 7 inches long and 2 inches wide, coarsely toothed, with a few to several narrow lobes and deeper lobes near base. Leaf tips may be pointed or rounded. Depending on variety of the species, leaves may or may not have hairs on underside. Clusters of small yellow flowers occur at end of stems in summer.

Origin and distribution
Native; throughout U.S.

Considerations as a weed and for wildlife
Marsh yellowcress can become competitive in food plots on moist sites, usually along river bottoms that occasionally flood. Other than providing nectar to various bees and flies, marsh yellowcress has no reported wildlife value. Marsh yellowcress can be controlled with glyphosate, 2,4-D, 2,4-DB, imazethapyr, dicamba, bentazon, and pendimethalin.

Field pennycress
Thlaspi arvense

Description and ecology
Erect, annual, cool-season forb, typically 8 – 30 inches tall, that arises from a basal rosette. Leaves on stem are alternate, smooth, and clasp the stem. Similar in appearance to field pepperweed, but field pepperweed leaves are covered with fuzzy hairs. Leaves lower on the plant typically have strong wavy margins. Flowers are white and occur on a raceme at the end of the stem. Each seed is enclosed in a flattened pod that is oval to heart-shaped and narrowly notched with a winged edge all the way around. Field pennycress has a foul odor when crushed.

Origin and distribution
Nonnative (Eurasia); throughout U.S.

Considerations as a weed and for wildlife
Field pennycress may occur in annual or perennial plots. It rarely presents a competition problem, but may be widespread. It can be controlled preemergence with imazethapyr, and postemergence with a forb-selective herbicide. It has no reported value for wildlife.

Top two photos by S. Askew

Venus' lookingglass
Clasping bellwort
Clasping bellflower
Triodanis perfoliata

Description and ecology
Erect, annual, warm-season forb. Stem has 5 sides. Stem leaves may be opposite near ground, but upper leaves are alternate. Flower is purple with 5 lobes. Male and female flowers are on different plants.

Origin and distribution
Native; throughout the U.S., except for portions of southwestern U.S.

Considerations as a weed and for wildlife
Venus' lookingglass may be common in warm-season grain plots that are cultivated. It can become widespread in the field, but does not pose a competition problem with planted grain crops. It can provide low cover for northern bobwhite and ground-feeding songbirds, if relatively dense. It is a relatively low-preference forage for white-tailed deer. It can be controlled with various forb-selective herbicides, including dicamba, 2,4-D, and carfentrazone-ethyl.

Mouseear chickweed
Cerastium vulgare or *C. fontanum* ssp. *vulgare*

Common chickweed
Stellaria media

Description and ecology
Mouseear chickweed is a perennial cool-season forb. It grows prostrate and spreads as the stems root at the nodes. Common chickweed is an annual cool-season forb. It may grow erect or prostrate and usually forms dense patches. The two species are somewhat similar in appearance except the leaves of mouseear chickweed are distinctly hairy (fuzzy). Common chickweed flowers from early spring through early summer and during fall and early winter. Mouseear chickweed flowers April – October.

Origin and distribution
Both species are nonnative (Eurasia) and occur throughout U.S.

Considerations as a weed and for wildlife
Chickweeds can be very competitive in cool-season plots. If not controlled, chickweeds can become very dense and outcompete planted forages. They should be sprayed when young (in fall or winter). Imazethapyr works well in clover plots, and other forb-selective herbicides, such as thifensulfuron-methyl+tribenuron-methyl, and dicamba, are effective in cool-season grain plots, such as wheat and oats. Chickweeds can be controlled preemergence with imazethapyr, pendimethalin, pronamide, and trifluralin. 2,4-DB does not control chickweeds. Chickweeds usually grow in association with henbit and purple deadnettle. Chickweeds are not selected forages of white-tailed deer or eastern cottontail. Several songbird species eat the seed and leaves sparingly.

From top: Common chickweed, common chickweed in clover, mouseear chickweed

Deptford pink
Dianthus armeria

Description and ecology
Erect, annual, warm-season forb. Stem is sparsely branched and usually hairy along bottom portion. Leaves are alternate with fine hair, and clasp the stem. Flowers occur at the end of branches, appear May – August, are pink, and have 5 petals.

Origin and distribution
Nonnative (England); throughout most of U.S., except Deep South and southwestern U.S.

Considerations as a weed and for wildlife
Deptford pink is often found in warm-season food plots that were established with conventional cultivation. It can be controlled with a forb-selective herbicide, but is fairly nondescript and does not present a competition problem with any planting. It has no reported value to wildlife.

Common lambsquarters
Chenopodium album

Description and ecology
Annual warm-season forb. Stems are erect and branching. Leaves alternate, irregularly toothed, and usually with a gray powdery appearance. Flowers June through September.

Origin and distribution
Some sources say native; some say nonnative (Eurasia); throughout U.S.

Considerations as a weed and for wildlife
Lambsquarters is a common warm-season forb, especially in plots planted via conventional tillage. Lambsquarters sometimes can be relatively dense and present competition for planted forages. Lambsquarters can be controlled preemergence with pendimethalin, trifluralin, and S-metolachlor+atrazine, and it can be controlled postemergence with dicamba, 2,4-D, carfentrazone-ethyl, and with glyphsate in Roundup-Ready crops. The seed of lambsquarters is eaten by many bird species, including northern bobwhite and mourning dove, and the foliage is a low- to moderate-preference forage for white-tailed deer. Lambsquarters provides outstanding cover for bobwhite and wild turkey broods. It complements corn for structure and added seed for birds; thus, unless there are other undesirable forbs growing in association with lambsquarters or it is so dense that it is outcompeting planted crops, I do not spray it.

Left: photo showing lambsquarters grazed by white-tailed deer.

Kochia
Kochia scoparia

Description and ecology
Erect, fast-growing, annual, warm-season forb. Stem is stout, usually green with red tinge, with many ascending branches. Leaves are alternate and linear in shape. Flowers are small, green to red, and inconspicuous. Mature plant is often relatively round in shape; may break off at ground at maturity. Kochia is somewhat similar to mugwort and common lambsquarters, but leaves of kochia are linear and not toothed.

Origin and distribution
Nonnative (Eurasia); throughout U.S., except southeastern Coastal Plain

Considerations as a weed and for wildlife
Kochia may occur in warm-season grain crops or in perennial cool-season forage plots. Kochia can provide cover for wild turkey poults, eastern cottontail, and various sparrows, but it can quickly reduce forage available in a forage food plot. It is a low-preference forage for white-tailed deer. Kochia can be controlled preemergence with pendimethalin and imazethapyr, and postemergence with dicamba, carfentrazone-ethyl, and 2,4-D, and with glyphosate in Roundup-Ready crops or by spot-spraying. Imazethapyr and 2,4-DB may help control kochia in perennial forage plots if sprayed when young.

Asiatic dayflower
Commelina communis

Slender dayflower
Commelina erecta

Spreading dayflower
Climbing dayflower
Commelina diffusa

Description and ecology
Annual, warm-season forbs. Asiatic dayflower may
be upright or spreading. Slender dayflower is upright.
Spreading dayflower typically decumbent (sprawling on
ground) with creeping stems and tips pointing upward.
Leaves are alternate, smooth, pointed at the end, and
attached directly to the main stem. Asiatic dayflower leaves
up to 5 inches long and 2 inches wide; slender dayflower
leaves up to 6 inches long and 1 inch wide; has more slender
leaves; spreading dayflower leaves up to 2 inches long and
1 inch wide. Stems are round and smooth. These plants
flower mid-summer through early fall and each flower
persists only one morning, thus the common name. Asiatic
and slender dayflower have 2 blue petals and one white
petal, whereas all 3 petals on spreading dayflower are blue.
Asiatic and spreading dayflower root at the nodes, but
slender dayflower does not. Asiatic dayflower more likely
found on relatively moist sites, but may be found in uplands,
especially where there is soil disturbance. Slender dayflower
more likely on drier, relatively sandy sites.

Origin and distribution
Asiatic dayflower nonnative (Asia); slender and spreading
dayflower native. There are several species of dayflower.
These 3 are relatively common in the eastern U.S.

Considerations as a weed and for wildlife
Asiatic and spreading dayflower are most likely found
occurring in food plots established with cultivation, fallow
fields, or other sites with recent soil disturbance. They may
compete with planted forages. Mourning dove, northern
bobwhite, and various sparrows eat the seed of all 3 species.
White-tailed deer and eastern cottontail may eat the foliage
of all species, but Asiatic dayflower is highly selected
by deer. Dayflower can be controlled with glyphosate,
bentazon, imazaquin, imazethapyr, simazine, and dicamba.

Top photo: Asiatic dayflower
grazed by white-tailed deer
Middle and bottom photos:
Slender dayflower

Dodder
Dodder flower
Cuscuta spp.

Description and ecology
Parasitic annual vine that lacks chlorophyll or leaves. Plant arises from seed, and a short-lived root system develops. The plant quickly twines around other vegetation. The vine is yellow-to-red. Appendages called haustoria develop along the vine where the vine presses against host plants. The haustoria penetrate the host plant and absorb nutrients. Numerous, small, white flowers are produced along the vine.

Origin and distribution
Native; there are many species of *Cuscuta* that are found throughout the U.S.

Considerations as a weed and for wildlife
Dodder does not kill the host plant, but weakens it, which will reduce growth and yield of the host plant. Dodder has little benefit for wildlife. Various amphibians and small mammals may be found under relatively dense mats of dodder, and white-tailed deer have been recorded eating it, but it is not an important plant for wildlife and should be controlled when it occurs in food plots as it will reduce plant vigor and yield. Dodder can be controlled preemergence with imazethapyr, pendimethalin, pronamide, and trifluralin, and postemergence with imazethapyr or imazamox.

Morningglories
Bindweeds

Ipomoea coccinea (Red morningglory)
I. hederacea (Ivyleaf morningglory)
I. lacunosa (Pitted morningglory)
I. pandurata (Bigroot morningglory)
I. purpurea (Tall morningglory)
Jacquemontia tamnifolia
(Smallflower morningglory)
Calystegia sylvatica (Hedge bindweed)
Convolvulus arvensis (Field bindweed)

Description and ecology

There are at least 10 species of morningglory and 2 species of bindweeds found in upland areas in various portions of the U.S. Morningglories and bindweeds are climbing or trailing warm-season vines. The morningglories are annual, except bigroot morningglory, which is perennial. Bindweeds are perennial. Flowers are funnel-shaped and may be white, purple, pink, or red, depending upon species.

Origin and distribution

Morningglories are native to tropical or subtropical America. Some sources show hedge bindweed is native to U.S. as well as Eurasia; some sources say it is not native to U.S. Field bindweed is nonnative (Eurasia). Morningglories and bindweeds occur over most of the U.S.

Considerations as a weed and for wildlife

Morningglories and bindweeds can be very problematic in warm-season food plots. They will vine over planted crops, reduce available sunlight, compete for nutrients, and reduce crop yield. Preemergence herbicide applications are recommended to help control morningglory and bindweed coverage in warm-season food plots. Imazethapyr and S-metolchlor+atrazine perform fairly well on annual morningglories, but are less effective on bindweeds. Morningglories can be controlled postemergence with dicamba, carfentrazone ethyl, 2,4-D, 2,4-DB, imazethapyr, and glyphosate in Roundup-Ready crops. If you spray postemergence, morningglories and bindweeds should be sprayed when young. Morningglories are low- to moderate-preference deer forage plants. Bindweeds are moderately selected. Mourning dove, northern bobwhite, scaled quail, and various songbirds eat morningglory seed.

From top: Tall morningglory, ivyleaf morningglory, and hedge bindweed
See next page for additional photos

Red morningglory

Bigroot morningglory

Pitted morningglory

Smallflower morningglory

Top two photos: Bur cucumber
Bottom two photos: Wild cucumber

Bur cucumber
Sicyos angulatus

Wild cucumber
Echinocystis lobata

Description and ecology
Annual vines that grow up to 25 feet long. Both will climb over adjacent vegetation or sprawl along the ground. Stem of bur cucumber is hairy, whereas stem of wild cucumber is smooth. Leaves are alternate and somewhat star-shaped, resembling leaves of cultivated cucumber. Leaves of wild cucumber have 5 – 7 lobes that are rather pointed. Bur cucumber leaves have 3 – 5 lobes that are less pointed. Bur cucumber leaves are strongly indented at the base where connected to leaf stem. Wild cucumber flowers are pale yellowish-white, whereas those of bur cucumber are greenish-white. Flowering occurs in late summer. The fruit of wild cucumber is a pulpy green seed pod about 2 inches long covered with spines that contains 4 black seeds. The fruit of bur cucumber is yellowish, also covered with spines, but occurring in clusters of 3 – 10, each about ¾ inch long, containing only one seed that is brown and flattened. The fruit of bur and wild cucumber is not edible. Bur and wild cucumber occur most often in moist bottomlands, especially in forested openings.

Origin and distribution
Native; bur cucumber throughout eastern U.S.; wild cucumber from VA to UT and northward.

Considerations as a weed and for wildlife
Both bur and wild cucumber can become problematic in food plots, especially those located along river bottoms and other relatively moist areas. They have no real benefit to vertebrate wildlife. Bur and wild cucumber can be controlled with glyphosate and triclopyr, especially if spot-spraying in or adjacent to forested areas where soil-active herbicides may injure desirable trees at the edge of food plots. Dicamba can be used in grass crops, such as corn and small grains. Thifensulfuron + tribenuron also is effective in wheat and oats. Imazethapyr can be used in perennial clover and chicory and in soybeans and cowpeas. Imazamox can be used in alfalfa.

344

Yellow nutsedge
Cyperus esculentus

Description and ecology
Erect, perennial, warm-season sedge. Stem is triangular. Leaves have prominent midvein with long tip. Flowers are yellow-brown and occur in clusters on ends of stalked spikes.

Origin and distribution
Sources conflict; some say native, others say nonnative (Europe or Africa); throughout U.S.

Considerations as a weed and for wildlife
Yellow nutsedge, as well as a few other sedge species, can be problematic in both warm-season annual plots and perennial cool-season forage plots. Yellow nutsedge can become widespread, dense, and outcompete some planted crops. It can be difficult to get rid of, especially in perennial forage plots. Preemergence applications of imazethapyr (such as Pursuit) and imazamox (such as Raptor) are effective in controlling yellow nutsedge when planting soybeans or cowpeas (imazethapyr) or Clearfield™ corn or sunflowers (imazamox). Postemergence applications of imazethapyr in perennial clover/chicory/alfalfa plots will suppress yellow nutsedge, but will not get rid

of it. Bentazon (Basagran) can be used postemergence in various legume plots. S-metolachlor (Dual Magnum) applied preplant incorporated is effective in corn plots and halosulfuron methyl (Permit or Sandea) applied postemergence in corn or grain sorghum is effective in controlling yellow nutsedge. Glyphosate in Roundup-Ready crops will suppress yellow nutsedge, but expect regrowth. Planting wheat and annual clovers, such as crimson and arrowleaf, allow herbicide applications of your choice during summer and is a good technique for eradicating yellow nutsedge. Yellow nutsedge tubers are eaten by wild turkeys, dabbling ducks, and wild pigs, but the tubers are small and not as attractive as those of chufa, which is a variety of yellow nutsedge that produces larger tubers. Yellow nutsedge is not eaten by white-tailed deer.

Top photo: Yellow nutsedge
Bottom photo: Several species of sedges can be problematic in perennial forage plots, such as this flat-spiked sedge (*Carex amphibola*) in a clover plot

Common teasel
Dipsacus fullonum

Description and ecology
Erect, biennial, warm-season forb. Stem is relatively stout, angled, and covered with prickles. Stem arises from a rosette; leaves on stem are opposite. Seedhead is round to cylindrical and covered with relatively long, stiff awns. Linear bracts coming from bottom of seedhead often curve upward around seedhead.

Origin and distribution
Nonnative (Europe); throughout most of U.S., except portions of southeastern U.S.

Considerations as a weed and for wildlife
May be widespread in fallow warm-season grain crop fields, and may become problematic in cool-season perennial forage plots. When relatively dense, common teasel can provide cover for rabbits, northern bobwhite, and several species of songbirds. However, it has no food value and should not be considered a valuable plant for wildlife. You should try and eradicate this invasive plant on your property. Common teasel can be controlled postemergence with glyphosate and a variety of forb-selective herbicides, including triclopyr, dicamba, metsulfuron methyl, clopyralid, and 2,4-D.

346

Hophornbeam copperleaf
Acalypha ostryifolia

Description and ecology
Erect, annual, warm-season forb. Grows upright about 1 – 3 feet tall. Leaves are alternate with serrate (fine toothed) margins and somewhat heart-shaped at base. Flowers July – October.

Origin and distribution
Native; eastern U.S.; PA to IA southward to TX

Considerations as a weed and for wildlife
Hophornbeam copperleaf is most often found In warm-season food plots prepared via conventional tillage. It is not usually dense enough to compete with planted crops, but it can be on some sites. When relatively dense, provides excellent cover for game bird poults as well as rabbits. Mourning dove and various sparrows readily eat the seed. It is not a selected forage by white-tailed deer, rabbits, or groundhog. Hophornbeam copperleaf can be controlled preemergence with imazethapyr, postemergence forb-selective herbicides, including dicamba, carfentrazone-ethyl, 2,4-D, and glyphosate in Roundup-Ready crops.

3-seeded mercury
Copperleaf
Acalypha virginica

Description and ecology

Erect, annual, warm-season forb. May grow to 2 feet tall on some sites, but usually about 12 inches. Upright, but may appear spreading. Leaves are alternate. Flowers June – November. Relatively common on recently disturbed sites.

Origin and distribution

Native; eastern U.S.

Considerations as a weed and for wildlife

3-seeded mercury can be common in food plots prepared via conventional tillage. It is not usually widespread or dense enough to compete with planted seed, but it can be on some sites. It can complement grain plots, especially those grown for northern bobwhite or mourning dove. Various sparrows eat the seed of 3-seeded mercury and it is a low to moderately preferred forage for white-tailed deer. When competing with forage food plots, such as perennial clovers and chicory, 3-seeded mercury should be sprayed when it is young with imazethapyr or 2,4-DB.

Hyssop spurge
Chamaesyce hyssopifolia

Spotted spurge
C. maculata

Nodding spurge
C. nutans

Description and ecology
Prostrate, spreading, or upright, annual, warm-season forbs. Spotted spurge grows close or flat to the ground, spreading with many branches. Leaves of spotted spurge often have a purple mark in the center on top. Leaves of all 3 species are opposite and contain a milky sap. All 3 species grow from a taproot. Stems of nodding and hyssop spurge are upright or generally curve upward. Nodding spurge and hyssop spurge are very similar. Nodding spurge has fine hairs on the upper surface of the young leaves, whereas leaves of hyssop spurge are smooth.

Origin and distribution
Native; hyssop spurge throughout southeastern U.S.; spotted spurge and nodding spurge throughout eastern U.S. and portions of western U.S.

Considerations as a weed and for wildlife
Spurges are commonly found in warm-season plots planted via conventional tillage. They may become widespread, but rarely compete with planted crops. They can be controlled preemergence with imazethapyr, imazamox, and S-metolachlor, and postemergence with forb-selective herbicides, especially those containing metsulfuron, such as Escort and Cimarron Plus. Spurges are not selected by white-tailed deer, rabbits, or groundhog. However, northern bobwhite, mourning dove, prairie-chicken, and several species of songbirds may eat the seed.

Top two photos: spotted spurge;
Bottom photo: nodding spurge in arrowleaf clover

349

Woolly croton
Doveweed
Croton capitatus, C. monanthogynus

Tropic croton
Doveweed
Croton glandulosus

Description and ecology
Upright, annual, warm-season forbs. Woolly croton stem is covered with fuzzy hairs that may be orange-brown. Leaves are chalky-green, alternate, and fuzzy on top and bottom. Tropic croton stem is green, somewhat rough with sparse hairs, and branches from upper leaf whorl. Leaves are alternate and slightly toothed along the edges. Flowers are small and terminal (at the end of the stem) with leaves just beneath.

Origin and distribution
Native; eastern U.S.

Considerations as a weed and for wildlife
Woolly and tropic croton can be found in all types of food plots. They are rarely dense enough to compete with planted crops, but they can be quite widespread. Woolly and tropic croton can be controlled preemergence with imazethapyr and S-metolachlor in warm-season plots, and controlled postemergence with forb-selective postemergence herbicides and glyphosate in Roundy-Ready crops. I do not selectively spray woolly or tropic croton in fields managed for mourning doves because the seed of both are highly selected by doves. In perennial forage plots, they should be sprayed when young with imazethapyr or 2,4-DB. If relatively sparse, I don't worry about spraying them in perennial forage plots managed for deer. They are not usually competitive. Deer may occasionally eat woolly croton, but tropic croton is a highly selected deer forage. Several species of doves, northern bobwhite, prairie-chickens, and various songbirds eat the seed of both species.

Top photo: Woolly croton; Bottom photos: Tropic croton

False flowering spurge
Euphorbia pubentissima

Description and ecology
Erect, perennial, warm-season forb with a stout rootcrown. Stem is very slender, greenish to purple (at base with age) and usually branched near top. Stem contains milky sap that is poisonous to livestock. Leaves are alternate lower on the stem and whorled (in circular arrangement around the stem) near the top. Leaves are linear in shape, smooth, and have a whitish midrib; often droop downward. Flowers are small and terminal on the stems, with 5 white sepals (that look like petals) and yellow in the center. Flowers appear spring through summer.

Origin and distribution
Native; eastern U.S.

Considerations as a weed and for wildlife
False flowering spurge may be found in perennial food plots, especially those several years old with little management, or in fallow fields. I have never seen it dense enough to compete with planted forages. It can be controlled with forb-selective herbicides or with glyphosate, but there is rarely a need to spray this plant. The seed are eaten by mourning dove, ground dove, greater prairie-chicken, northern bobwhite, wild turkey, horned lark, and various buntings and sparrows. The foliage is eaten sparingly by white-tailed deer and pronghorn.

351

Partridge pea
Chamaecrista fasciculata (or *Cassia fasciculata*)
Chamaecrista nictitans (or *Cassia nictitans*)

Description and ecology
Erect, annual, warm-season forb. Stem is multi-branched and may become somewhat woody near base by the end of the growing season. Plant usually grows 1 – 2 feet tall, but some cultivars may grow 6 feet tall or more. Stem of *C. fasciculata* is usually hairy, whereas the stem of *C. nictitans* is smooth. Leaves are pinnately compound. Leaves of *C. fasciculata* are longer than *C. nictitans* (not more than 2 inches). Flowers are yellow and appear May – September. Flowers of *C. fasciculata* are larger than *C. nictitans* (¼ – ½-inch wide). Seedpods mature in late fall and contain several seed that are flattened and black.

Origin and distribution
Native; throughout eastern U.S.

Considerations as a weed and for wildlife
Partridge pea is relatively common and widespread, especially in the South. If not present, it is often planted, particulary along firebreaks as a source of food and cover for northern bobwhite. It is stimulated by burning and disking. It is a moderate-preference forage for white-tailed deer. It provides good cover for northern bobwhite and wild turkey broods as well as ground-feeding songbirds. Where various "improved" cultivars of partridge pea have been planted, it is common for it to get dense and grow 5-6 feet tall. It can be thinned or controlled by disking after the plants are about 6 inches tall or by applying triclopyr, metsulfuron methyl, dicamba, or glyphosate. Alternatively, it can be released from competition with pre- or postemergence applications of imazethapyr, imazapic, and grass-selective herbicides.

Showy crotalaria
Rattlebox
Crotalaria spectabilis

Description and ecology
Erect, warm-season, annual forb. Stem is stout, green to purplish, and may grow to 7 feet tall. Leaves are alternate, wider at tip than base, and pubescent underneath. Flowers are yellow and usually clustered near top of flowering stalk. Bean pods are cylindrical, appear fat, and hang down from stalk.

Origin and distribution
Nonnative (India); southeastern U.S.

Considerations as a weed and for wildlife
Rattlebox can be common in warm-season plots planted via conventional cultivation. It can be controlled preemergence with S-metolachlor+atrazine, and controlled postemergence with forb-selective herbicides, and with glyphosate in Roundup-Ready crops. Rattlebox may provide some cover, but it has no food value for wildlife. Plant contains alkaloids that may be poisonous to various animals.

Beggar's-lice
Ticktrefoil
Tickclover
Beggarweed
Desmodium glabellum, D. obtusum,
D. laevigatum, D. paniculatum

Description and ecology
There are many species of *Desmodium*.
The species listed here are erect, perennial,
warm-season forbs. Stems of *glabellum* and
paniculatum stems may or may not be hairy,
whereas stems of *obtusum* are sparsely to
densely hairy and *laevigatum* are smooth. Leaves
of each are alternate with 3 leaflets that are about 1 ½-3 ½
inches long with smooth margins. Leaves of *paniculatum* are
distinctly linear in shape and 3–6 times longer than wide.
Flowers are pinkish-purple and appear in late summer. *D.*
glabellum and *obtusum* flowers are on terminal panicles
that may be 1–2 feet long. *D. laevigatum* flowers are on long,
branching stalks, whereas *paniculatum* has a narrow panicle.
Seed appear in fall and are enclosed in a flat pod with hook-
ended hairs that enable it to stick to fur and clothing (hence,
beggar's-lice). *D. glabellum* seedpods contain 2–5 seeds;
obtusum 1–3 seeds; *paniculatum* 2–6; *laevigatum* 3–5.

Origin and distribution
Native; various species of *Desmodium* occur throughout the
eastern U.S. and into Great Plains, but especially along the
southeastern Coastal Plain. The species listed here are found
throughout the eastern U.S.

Considerations as a weed and for wildlife
The *Desmodiums* mostly occur in woodlands and old-fields
where prescribed fire is used, but these occur commonly
in fallow fields. *Desmodiums* are a moderately to highly
selected forage for white-tailed deer and eastern cottontail. The seed is an important
food source for northern bobwhite, and also is eaten by mourning dove, wild turkey,
and ruffed grouse. *Desmodiums* fix nitrogen, which improves soil for other plants.
Desmodiums do not normally present problems in food plots, but may be seen in old
perennial plots. In the rare situation where they need to be controlled, spot-spraying
with a forb-selective herbicide, such as dicamba or clopyralid, would be sufficient.
Alternatively, they can be released from competition with a grass-selective herbicide,
such as clethodim, or a broad-spectrum selective herbicide, such as imazapic or
imazethapyr.

From top: *D. glabellum* (top two photos), *D. paniculatum* (bottom photo)

Common (or annual) lespedeza
Kummerowia striata

Korean (or Kobe) lespedeza
K. stipulacea

Description and ecology
Low-growing and spreading annual, warm-season forbs. Stems are wiry and become semi-woody. Stems are branched. *K. striata* differs from *K. stipulacea* in that stems of *K. stipulacea* are branched at the base of the plant and are shorter (4–8 inches) than *K. striata* (4–16 inches). Leaves are alternate and compound with 3 leaflets, which have parallel and striate veins. *K. stipulacea* leaves differ from *K. striata* in that *K. stipulacea* leaflet tips are notched and have fine hairs along the margins. Flowers appear July – September and are pinkish-purple with white. A single seed is produced in a pod that matures in later summer/fall. Pod tip of *K. striata* is pointed, whereas that of *K. stipulacea* is rounded.

Origin and distribution
Nonnative (eastern Asia); throughout most of eastern U.S., from PA west to IA and southward

Considerations as a weed and for wildlife
Both of these annual lespedezas once were widely planted, especially throughout the South, as a food source for northern bobwhite. They have since naturalized and are widespread. Northern bobwhite relish the seed. Wild turkey, mourning dove, and ruffed grouse may eat the seed sparingly. The foliage is a low- to moderate-preference forage for white-tailed deer and eastern cottontail. Both of these lespedezas can become relatively dense and compete with other plants, especially in areas where they were planted in the past. They can be controlled with forb-selective herbicides, such as dicamba, metsulfuron methyl, triclopyr, and clopyralid. Alternatively, they can be released with imazapic, imazethapyr, imazamox, and grass-selective herbicides.

Photo: Common lespedeza

Everlasting peavine
Lathyrus latifolius

Description and ecology
Vining perennial forb. Stem is green, smooth, winged, and somewhat succulent, about 4 – 8 feet long, vining and climbing vegetation or structures, such as fence posts. Leaves are compound with 2 linear leaflets 2 – 4 inches long per leaf. Flowers are pinkish-purple in clusters of 5 – 15 at the end of branches. Seed pods are 3 – 4 inches long and about 1/4-inch wide with 10 – 25 seeds per pod.

Origin and distribution
Nonnative (Europe); throughout U.S., except for FL and portions of southwest U.S.

Considerations as a weed and for wildlife
Everlasting peavine is relatively common around field borders and along grown-up fencerows. It may occur in fallow plots, especially those established via conventional cultivation. It is not normally dense and does not normally compete with planted crops. Although it is relatively succulent, it is a low-preference forage for white-tailed deer. It can be controlled via spot-spraying with forb-selective herbicides.

Middle photo by: S. Askew

356

Sericea lespedeza
Lespedeza cuneata

Description and ecology
Erect, perennial, warm-season forb. Stem becomes semi-woody upon maturity and arises from taproot. Leaves are alternate with 3 leaflets. Flowers are white, but may have pinkish or purplish center.

Origin and distribution
Nonnative (Asia); throughout most of U.S.

Considerations as a weed and for wildlife
Sericea lespedeza can become widespread and problematic in perennial forage plots. It is very difficult to control because there is no herbicide that will control sericea lespedeza postemergence without killing perennial forages (such as perennial clovers, alfalfa, chicory, and perennial peanut). It is an extremely low-preference forage of white-tailed deer (if you see them eat it, nutrition is limiting on the site). Birds will eat sericea seed, but they get no nutritional value from it because the seed is so hard it passes through the digestive system without breaking down. This is problematic because acids in the digestive system break-down the seed coat and increase germination after the seed is passed from the animal. If sericea lespedeza becomes problematic in your forage plots, it is time to rotate the plot to another crop. Kill the entire plot with glyphosate, then plant an annual grass, such as corn, wheat, or millets. If you kill sericea with glyphoste, it will brown out, but may sucker from the roots. If you are managing early succession, instead of a food plot, triclopyr+fluroxypyr is recommended. Sericea will return from the seedbank. However, fluroxypyr is soil active for 1 - 2 months, so any sericea that germinates soon after spraying will be killed. As additional sericea appears, spot-spray with triclopyr+fluroxypyr. Imazapic also may provide preemergence control, but will not kill sericea postemergence. It is best to kill sericea lespedeza as soon as you find it in a field or it will only add to the seedbank and persist for many years. You should aggressively try to eradicate this plant. Spot-spraying with triclopyr + fluroxypyr is effective, but you have to be determined and persistent.

357

Birdsfoot trefoil
Lotus corniculatus

Description and ecology
Prostrate to ascending perennial, cool-season forb.
Stem is round at base and squarish at top. Leaves
are alternate and compound with 3 terminal
leaflets. Leaflets are narrow at base and wider at
tip, which is somewhat pointed. Flowers appear
in summer, are yellow, and occur at the end of
slender stalks. Seedpods spread from end of
branch and resemble a bird's foot, hence name.

Origin and distribution
Nonnative (Europe); throughout U.S., but most
common in northern states

Considerations as a weed and for wildlife
Birdsfoot trefoil has been planted widely over the
years to supplement grasses for livestock forage
and for soil stabilization and reclamation. It often
occurs in fields where food plots are planted and
is sometimes included in plantings for wildlife.
Birdsfoot trefoil is eaten to some extent by eastern
cottontail and groundhog, and is a low-preference
forage for white-tailed deer. If fields contain
extensive coverage of birdsfoot trefoil, and more
desirable forages are wanted for white-tailed
deer, spray the field with glyphosate, and plant
the desired mixture. I would not include birdsfoot
trefoil in any mixture planted for white-tailed
deer. There is no need to. Many more desirable
species are available. Birdsfoot trefoil also can be
controlled with forb-selective herbicides, such as
dicamba, metsulfuron methyl, and clopyralid.

Sweetclover
Yellow sweetclover
White sweetclover
Melilotus officinalis

Description and ecology
Erect, annual, or biennial forb. Stem is green, branching, and grows to 7 feet tall on some sites. Leaves are alternate and compound with 3 leaflets that are finely toothed along the margins. Flowers are yellow or white and occur on a raceme at the end of branches with 35 – 75 flowers per raceme.

Origin and distribution
Nonnative (Europe); throughout U.S.

Considerations as a weed and for wildlife
Sweetclover has been planted widely for livestock forage and nitrogen fixation. Sweetclover is not a true clover (Trifoliums are true clovers). Sweetclover is quite invasive, spreading into areas where it is not wanted and difficult to control. Sweetclover may be eaten by white-tailed deer and eastern cottontail, but it is a relatively low-preference forage. Where relatively dense, it provides decent cover for rabbits, game bird poults, and ground-feeding songbirds. Despite these potential benefits, there is no need to plant it. There are much better forage options available. Sweetclover should be controlled where it exists, or it will spread into unintended areas. Sweetclover is best controlled with forb-selective herbicides, such as dicamba, clopyralid, and thifensulfuron-methyl+tribenuron-methyl, or glyphosate.

Wild senna
American wild senna
Northern wild senna
Senna hebecarpa (or *Cassia hebecarpa*)

Southern wild senna
S. marilandica

Description and ecology
Erect, perennial, warm-season forbs that may grow to 7 feet tall on some sites. The two species are virtually identical, distinguished by flower structures. Stem is stout and light green. Leaves are compound with 10 – 20 oblong-shaped leaflets with smooth margins and a pointed tip. Flowers are yellow and occur on top of the stem in mid- to late summer. Seedpods are about 4 inches long, contain 10 – 18 seeds (segments), and occur along stem where leaves meet stem.

Origin and distribution
Native; throughout most of eastern states

Considerations as a weed and for wildlife
Wild senna is often found in fallow fields or in food plots that have been established with conventional tillage, but also might occur in perennial plots that are 3 or more years old. Wild senna can provide good cover for wild turkey and northern bobwhite broods as well as ground-feeding songbirds, especially in fallow plots. Wild senna contains powerful laxatives (anthraquinones); therefore, the foliage and seed are rarely eaten by wildlife. Wild senna can be controlled preemergence with S-metolachlor+atrazine, and postemergence with dicamba, clopyralid, and 2,4-D, and with glyphosate in Roundup-Ready crops. If wild senna occurs in perennial forage plots, it can be spot-sprayed with glyphosate. Mowing will not get rid of wild senna.

Sicklepod
Coffeeweed
Coffeebean
Senna obtusifolia (or *Cassia obtusifolia*)

Description and ecology
Upright, annual, warm-season forb. Stem is green and branched. Leaves are compound (usually 6 leaflets) and alternate. Flowers are yellow and usually on stem arising at leaf axil (where leaf meets main stem). Seeds are produced in long, slender, curvilinear pods. Plant has foul odor when crushed.

Origin and distribution
Native to American tropics; eastern U.S., from PA to IA southward

Considerations as a weed and for wildlife
Sicklepod is common in warm-season plots planted via conventional tillage. Sicklepod can be controlled preemergence with S-metolachlor+atrazine, and postemergence with dicamba, clopyralid, and 2,4-D, and with glyphosate in Roundup-Ready crops. If sicklepod occurs in perennial forage plots, it can be spot-sprayed with glyphosate, or mowed before it produces seed. Although sicklepod may provide canopy cover for ground-feeding birds, it has no food value for wildlife, and because it is so invasive and outcompetes other plants, you should aggressively try to eradicate this plant.

Hemp sesbania
Sesbania herbacea

Bagpod sesbania
Glottidum vesicarium

Description and ecology
Erect, annual, warm-season forb, often growing to 12 feet tall. Stem is stout, green, round, and branched (bagpod sesbania stems are brown at base). Leaves are compound and alternate and usually about 12 inches long (bagpod sesbania leaves may be longer). Leaflets are opposite and about an inch long. Flowers are yellow, often with purple spots (bagpod sesbania flowers orangish tinged with maroon). Hemp sesbania bean pods are usually 6 – 10 inches long, slender, and curved with about 30 – 40 seeds, which are mottled brown and black. Bagpod sesbania pods hang down and are relatively short and wide with only 1 or 2 seeds per pod.

Origin and distribution
Native to Coastal Plain of southeastern U.S.

Considerations as a weed and for wildlife
Sesbania is common in sandy soils along the southeastern Coastal Plain, especially moist sites that are periodically flooded. Sesbania may provide cover for wood ducks and other dabbling ducks in flooded fields, as well as cover for northern bobwhite and wild turkey broods in upland fields. Sesbania seed may be eaten by dabbling ducks as well as mourning dove, northern bobwhite, and wild turkey. Sesbania can be problematic because it is prolific and spreads very easily. It can quickly take-over a field and out-compete other species (such as soybeans or other crops). It should be managed with care, or it will invade into areas where it may be undesirable. Sesbania contains toxins (saponins) that may cause physiological problems in livestock. The foliage is not selected by wildlife. It can be controlled with glyphosate and various forb-selective herbicides, such as 2,4-D, triclopyr, acifluorfen, and dicamba. However, as with all herbicides, follow label instructions for application carefully, especially around wetlands and drainages.

Photos: Bagpod sesbania

Rabbitfoot clover
Trifolium arvense

Description and ecology
Erect annual forb that grows about 6 – 12 inches in height. Stems are branched, covered with fine, soft, hairs and may be green to reddish. Leaves are alternate and compound with 3 leaflets. Flowers are grayish-pink and occur in clusters at the end of stems. Rabbitfoot clover usually occurs on relatively poor, dry sites.

Origin and distribution
Nonnative (Europe); throughout most of U.S.

Considerations as a weed and for wildlife
Rabbitfoot clover may occur in fallow warm-season grain plots and in perennial plots that have not been managed. It is a low-preference forage for white-tailed deer, eastern cottontail, and groundhog. Soil amendment and replanting is likely needed if rabbitfoot clover is prevalent in a field. It can be controlled with glyphosate, dicamba, clopyralid, and 2,4-D.

Hop clover
Trifolium aureum, T. campestre, or *T. dubium*

Black medic
Medicago lupulina

Description and ecology
Low-growing, annual or biennial forbs. Stems are much branched and become relatively tough and wiry upon maturity. Leaves are alternate and compound with 3 leaflets that are wider at tip than base. Flowers are small (less than ½-inch wide), yellow, and occur May – September. Hop clover and black medic are very similar but flower at different times of year. Hop clover flowers in the spring through early summer, then declines in the heat. Black medic is more heat tolerant, flowering in summer. After flowering, the flower petals of black medic fall off and reveal a cluster of small, recurved, black seedpods, whereas the flowers of hop clover turn brown and remain attached. Reproduction of both is by seed.

Origin and distribution
Nonnative (Eurasia); throughout most of U.S.

Considerations as a weed and for wildlife
Hop clover and black medic are common on soils with relatively low pH and low nutrient availability. They often pop up in perennial clover plots that have not been managed over the years. They are not overly competitive, but can become widespread. They can be controlled by disking before they flower or with forb-selective herbicides, such as dicamba , 2,4-D, and thifensulfuron-methyl+tribenuron-methyl. They are eaten sparingly by rabbits and groundhog. They are a low-preference forage for white-tailed deer.

From top: Hop clover, perennial clover plot with invading hopclover
Bottom photos: Black medic

Narrowleaf vetch
Common vetch
Vicia sativa

Hairy vetch
Vicia villosa

Description and ecology

Spreading and vining annual forbs. Stems are slender, green, and usually smooth. Leaves are alternate and compound; leaflets are slender and narrow. Narrowleaf vetch has 6 – 12 pairs of leaflets, whereas hairy vetch has 8 – 20 pairs of leaflets. Narrowleaf vetch flowers are pale pinkish-purple, paired, and occur between the leaf and the stem. Hairy vetch flowers are pinkish-purple and occur in a raceme at the end of branches with 10 – 40 flowers.

Origin and distribution

Nonnative (Europe); hairy vetch occurs throughout eastern U.S.; narrowleaf vetch occurs throughout U.S.

Considerations as a weed and for wildlife

There are many species of native vetches. These two are nonnative and have become widespread and common by sowing in hayfields throughout much of the U.S. However, they are considered invasive and compete with more desirable plants for wildlife. They do not provide cover and they are low-preference forages for white-tailed deer. Seeds of these vetches may cause skin and neurological problems in some ruminant animals. Hairy and narrowleaf vetch can be controlled with forb-selective herbicides, such as dicamba, clopyralid, and thifensulfuron-methyl+tribenuron-methyl. If you are considering including a nonnative vetch in a planting mixture for wildlife, don't! Read the other sections of this book. There are many more desirable plants that you can use.

Top photos: Narrowleaf vetch
Bottom photo: Hairy vetch

Redstem filaree
Common storksbill
Erodium cicutarium

Description and ecology
Prostrate or ascending annual cool-season forb growing 4 – 20 inches tall. Plant arises from a basal rosette, which usually establishes in fall. Basal leaves are approximately 3 – 8 inches long, hairy, and compound with deeply cut lobes (almost fernlike in appearance). Flowering stems arise in spring and are reddish and fleshy with fine hairs. Flowers appear April – June on the ends of stalks that are about 4 inches long. Flowers are about ½-inch in diameter with 5 pinkish-purple petals. Each flower produces a fruit with a beak-like projection that appears similar to a stork's bill or crane's bill.

Origin and distribution
Nonnative (Europe and north Africa); throughout U.S.

Considerations as a weed and for wildlife
Redstem filaree can be widespread in perennial forage plots. It may be grazed by mule deer, elk, and pronghorn in western rangeland where forage availability is relatively low. It is a low-preference forage for white-tailed deer. It can be controlled postemergence with 2,4-DB if sprayed when young. It can be controlled preemergence or postemergence with imazethapyr or imazamox if sprayed when young. It can be controlled with glyphosate and forb-selective herbicides, such as dicamba or triclopyr.

Carolina geranium
Cranesbill
Geranium carolinianum

Cutleaf geranium
G. dissectum

Smallflower geranium
G. molle or *G. pusillum*

Description and ecology
Erect, annual or biennial, cool-season forbs that may grow to a height of 1 – 3 feet by late spring. Plants arise from a branched rosette. Stems are multi-brached. Leaves are alternate near base, opposite above, hairy (cutleaf is hairy), and deeply lobed or toothed. Smallflower leaves are not deeply lobed like Carolina and cutleaf. Carolina geranium flowers are whitish pale pink to purple, have 5 petals, and appear April – August. Cutleaf geranium flowers are dark-pinkish-purple and also have 5 petals, but the petals are paired, which makes it appear as if it has 10 petals. Sepals (leaflike structures subtending the flower petals) of smallflower are not awn-tipped like those of Carolina and cutleaf. The fruit has an elongated beak, which has led to some people to call these plants cranesbill.

Origin and distribution
Carolina geranium is native and occurs throughout U.S.; Cutleaf is nonnative (Europe) and occurs in portions of eastern U.S.; Smallflower is nonnative (Europe) and occurs throughout most of U.S.

Considerations as a weed and for wildlife
These geraniums are commonly found in cool-season forage plots, but do not normally grow dense enough to outcompete planted forages. In my opinion, they are of the few naturally occurring cool-season plants that actually complement planted forages. Deer and rabbits eat Carolina geranium about as readily as they do clovers. Therefore, I do not specifically try to get rid of geranium. However, they often are in association with other cool-season weeds, such as henbit and chickweeds, so they often get sprayed with them. When occurring in warm-season plots, I also "let 'em go." The forage is eaten by deer and rabbits, and the seed is eaten by a number of birds, including northern bobwhite and mourning dove.

Top photo: Carolina geranium; Middle left photo: Carolina geranium; Middle right photo: Cutleaf geranium; Bottom photo: Smallflower geranium

Spotted St. Johnswort
Hypericum punctatum

St. Andrew's Cross
H. hypericoides

Description and ecology
There are several species of *Hypericum*, some herbaceous and some more woody, such as St. Andrew's cross. Spotted St. Johnswort is a perennial forb that may grow to about 2.5 feet tall. Leaves are opposite and usually without stems (if there are stems, they are very short). There are small dots on the underside of the leaves. The stem is green or red and smooth. Flowers appear in mid-summer and are yellow with 5 petals and are only about ½-inch across.

Origin and distribution
Native; occurs throughout eastern U.S.

Considerations as a weed and for wildlife
Spotted St. Johnswort may occur in perennial plots. It is not usually dense, but can outcompete planted forages. There are many pollinators that use spotted St. Johnswort, but it is not a selected forage for deer or other mammals. The leaves of spotted St. Johnswort contains hypericin, which is a toxin to some animals. Spotted St. Johnswort can be controlled in perennial clover/alfalfa plots with 2,4-DB or imazethapyr if sprayed when young.

Top two photos: Spotted St. Johnswort; third photo: St. Andrew's cross, bottom photo: pineweed

St. Andrew's cross (*H. hypericoides*) and pineweed (*H. gentianoides*) are two other *Hypericum* species that are commonly found in old-fields and fallow plant communities throughout the eastern U.S. St. Andrew's cross is a small shrub growing 3 – 4 feet tall, primarily along the Coastal Plain, whereas pineweed is an annual forb.

Common (or soft) rush
Watergrass
Juncus effusus

Slender rush
Path rush
J. tenuis

Description and ecology
Rushes are clump-forming, grass-like perennial plants. Rush stems are wiry, relatively stiff, hollow, and round, whereas sedge stems are triangular, and most grass stems are round but not hollow. Flowering occurs late spring through summer, with flowers appearing in clusters near the ends of branches. Seeds are small, relatively round, and tan in color. Upon maturity, the tips of rush stems turn tan or brown. Tips of common rush stems are sharp and often injure eyes of large animals foraging on other plants amongst the rushes. Slender rush is 4 - 12 inches tall and may be found in moist or dry fields; common rush is 2 - 4 feet tall and most often found in moist areas. Seed remain viable in the seedbank for many years.

Origin and distribution
Native; throughout U.S.

Considerations as a weed and for wildlife
Rushes can be problematic in perennial forage plots, especially those in bottomland fields that are relatively moist. Rushes often become dense and outcompete planted forages. Rushes are difficult to control with herbicides in perennial plots. Spot-spraying with glyphosate or 2,4-D can be effective. Use of imazethapyr to control other weeds often will suppress growth and spread of rushes. Mowing is not effective at all.

Photos: Common rush

If rushes are too dense for spot-spraying, it usually is best to spray the entire plot with glyphosate or cultivate and replant. Annual plantings should be considered where rushes are problematic. Rushes are eaten by marsh and swamp rabbit, beaver, porcupine, and common muskrat. Various songbirds, such as marsh wren, swamp sparrow, and indigo bunting may nest in rushes. Upland chorus frog, spring peeper, and American toad, and Fowler's toad use rushes for cover in bottomland fields.

369

Ground ivy
Glechoma hederacea

Description and ecology
Creeping, perennial, cool-season forb. Leaves are opposite and kidney-shaped or rounded with broad rounded teeth. Stems are square and may grow to 3 feet long. Stems root at the nodes. Flowers are purple, occur in clusters of 2 – 7, and appear April – June. Can be confused with speedwell, henbit, and common mallow. However, speedwell has round stems, henbit does not creep along the ground or root at the nodes, and mallow has alternate leaves and round stems. Also, being in the Mint Family, ground ivy also smells of mint when crushed.

Origin and distribution
Nonnative (Eurasia); throughout U.S., except southwestern U.S.

Considerations as a weed and for wildlife
Ground ivy can be very competitive in cool-season plots. It should be sprayed when young with 2,4-DB or imazethapyr in perennial clover/alfalfa plots or with glyphosate or a forb-selective herbicide, such as dicamba, metsulfuron-methyl, or triclopyr, in fallow plots. If not controlled, it can become very dense and outcompete planted forages. Mowing will not control ground ivy. It usually grows in association with chickweeds, speedwell, henbit, and purple deadnettle. Ground ivy provides no value to wildlife except as cover for small animals, such as voles.

370

Henbit
Lamium amplexicaule

Description and ecology
Erect, annual, cool-season forb. Grows to 12 – 18 inches tall. Stems are square. Leaves are opposite with rounded-toothed margins, dark green above and pale below. Flowers are pinkish purple and occur most commonly in late winter through spring, but may appear in fall/winter. May superficially appear similar to purple deadnettle except flowers are erect and arising from top leaves.

Orlgin and distribution
Nonnative (Eurasia); occurs throughout U.S.

Considerations as a weed and for wildlife
Henbit can be very competitive in cool-season plots. Established perennial clover/alfalfa plots can be sprayed with imazethapyr (such as Pursuit) in late summer/early fall before henbit and other cool-season annual weeds germinate. Pronamide (Kerb) can be sprayed preemergence when planting clovers and alfalfa and winter peas, or can be sprayed over the top of established clovers and alfalfa in early fall to control henbit when it germinates. If the weeds have already germinated, they should be sprayed when young with imazethapyr in perennial clover/alfalfa plots or with glyphosate or a forb-selective herbicide, such as dicamba, metsulfuron-methyl, or triclopyr in fallow plots. 2,4-D or 2,4-DB does not control henbit. If not controlled, henbit can become very dense and outcompete planted forages. Mowing will not control henbit. It usually grows in association with purple deadnettle and chickweeds. Butterflies and bees get nectar from the flowers in spring, but henbit is not a selected forage of white-tailed deer, eastern cottontail, groundhog, or wild turkey.

Top and middle photo: deadnettle
Bottom photo: deadnettle (left in photo)
and henbit (center in photo)

Purple deadnettle
Lamium purpureum

Description and ecology
Erect, annual, cool-season forb. Grows to about 18 inches tall on some sites. Stems are square. Leaves are opposite with rounded-toothed margins, are broader at base than tip, and upper leaves have a purplish hue. Flowers are pinkish purple and occur most commonly in late winter through spring, but may appear in fall/winter. Superficially similar to henbit except flowers are almost hidden under a "hood" of leaves.

Origin and distribution
Nonnative (Eurasia); occurs throughout U.S.

Considerations as a weed and for wildlife
Purple deadnettle can be very competitive in cool-season plots. Established perennial clover/alfalfa plots can be sprayed with imazethapyr (such as Pursuit) in late summer/early fall before purple deadnettle and other cool-season annual weeds germinate. Pronamide (Kerb) can be sprayed preemergence when planting clovers and alfalfa and winter peas, or imazethapyr can be sprayed over the top of established clovers and alfalfa in early fall to control purple deadnettle when it germinates. If the weeds have already germinated, they should be sprayed when young with imazethapyr in perennial clover/alfalfa plots or with glyphosate or a forb-selective herbicide, such as dicamba, metsulfuron-methyl, or triclopyr in fallow plots. 2,4-D or 2,4-DB does not control purple deadnettle. If not controlled, deadnettle can become very dense and outcompete planted forages. Mowing will not control purple deadnettle. It usually grows in association with henbit and chickweeds. Butterflies and bees get nectar from the flowers in spring, but purple deadnettle is not a selected forage of white-tailed deer, eastern cottontail, groundhog, or wild turkey.

Wild mint
Field mint
Mentha arvensis

Description and ecology
Erect, perennial, warm-season forb that may grow 2 – 3 feet tall. Stem is square, may or may not be hairy, and may or may not be branching. Leaves are opposite, smooth on top, possibly sparsely hairy below, and pointed. Flowers appear June – October, are purplish-white, and occur in clusters between leaf and stem and at end of stems. Wild mint tends to be found on relatively moist sites.

Origin and distribution
Native; throughout U.S. except the Deep South

Considerations as a weed and for wildlife
A fairly common mint that may be found in perennial food plots. Wild mint provides nectar for various bees and butterflies, but little else for wildlife. It is considered a low-preference forage for white-tailed deer, eastern cottontail, and groundhog. It may compete with planted forages. Wild mint may be controlled with forb-selective herbicides, such as dicamba, if sprayed when young, or with glyphosate. Mowing does not control wild mint.

Perilla mint
Beefsteak plant
Perilla frutescens

Description and ecology
Erect, annual, warm-season forb. Grows to about 2 feet tall. Stem is square, purple, or green. Leaves are opposite, toothed along edges, rounded, but longer than wide (2 – 4 inches wide), deep green and often tinged in purple. Leaves are pungent when crushed. Flowers are at ends of branches, and are white or purplish.

Origin and distribution
Nonnative (eastern Asia, India); occurs throughout eastern U.S., except Florida and upper New England

Considerations as a weed and for wildlife
May occur in warm-season grain plots or in perennial cool-season forage plots. Perilla mint may provide some cover for bobwhite and wild turkey broods, but has no food value for wildlife. It contains toxins that cause physiological problems in livestock. Perilla mint can be controlled by mowing before it produces seed or with imazapic (preemergence) or 2,4-D, triclopyr, dicamba, or glyphosate postemergence.

Healall
Prunella vulgaris

Description and ecology
Erect or ascending perennial forb. The stem is square and branched. Leaves are opposite. Flowerhead is elongated and at the end of branches. A few to several flowers usually appear on flowering head at once. Flowers are pinkish-purple and appear June – September.

Origin and distribution
Nonnative (Europe); throughout U.S.

Considerations as a weed and for wildlife
Healall is sometimes found in perennial forage plots that have been established 2 or more years. It does not usually outcompete forages, but can be widespread, especially in plots that need renovation. It can be controlled in perennial clover/alfalfa plots if sprayed when young with imazethapyr. Healall has no real food or cover value for mammals or birds. Bees and butterflies obtain nectar from the flowers.

Lyreleaf sage
Salvia lyrata

Description and ecology
Erect, biennial, or perennial forb that grows to about 30 inches tall on some sites. Arises from basal rosette. Stem is square and has only 1 or 2 pairs of thin leaves, which are opposite. Most of leaves are in the basal rosette, 4 – 8 inches long. Leaves are usually greenish-blue to purple. Flowers are pinkish-purple, tubular, and occur in whorls around the stem at 90-degree angle.

Origin and distribution
Some sources say native, some say nonnative (Europe); throughout eastern U.S.

Considerations as a weed and for wildlife
Lyreleaf sage is most often found in long-established perennial food plots and fallow plots. It rarely competes with planted forages, but if so, you can control it with imazethapyr or imazamox, in perennial clover, chicory, or alfalfa. You also can spot-spray with glyphosate, 2,4-D, dicamba, or triclopyr if it becomes too thick in early successional areas. Hummingbirds and butterflies are attracted to the flowers.

Top photo by S. Askew

Lanceleaf sage
Salvia reflexa

Description and ecology
Erect, annual, warm-season forb growing to approximately 2 feet tall. Stem is square and usually bushy-branched in upper portion of plant. Leaves are opposite, about 3 inches long and relatively narrow. Leaf veins are noticeable. Flowers are relatively small, blue to white, and appear as though they have noticeable upper and lower "lips." Flowers occur in whorls around the stem in summer and fall.

Origin and distribution
Native; portions of Mid-Atlantic and Midwest, throughout Great Plains and Rocky Mountain states.

Considerations as a weed and for wildlife
Lanceleaf sage may be problematic in some annual crops. It can be controlled preemergence with imazethapyr, and postemergence with glyphosate, dicamba, imazethapyr, and 2,4-D if sprayed when young. If not too dense, lanceleaf sage can provide good groundcover for quail, wild turkey poults, and ground-feeding songbirds.

Photos by Peter Dziuk, Minnesota Wildflowers

Downy skullcap
Hoary skullcap
Scutellaria incana

Mad-dog skullcap
S. laterifolia

Description and ecology
Downy skullcap is perennial forb that grows 2 – 4 feet tall. Leaves opposite and toothed and up to 3 inches long. Stems 4-angled. Stems and leaves covered with fine hair. Upper portion of main stem often branched with several flowering stems (racemes). Flowers light to dark purple, tubular, with upper portion appearing as a hood and lower portion as a broad lip with white center. Blooms mid- to late summer. Usually found in old-fields, savannas, and woodlands. Mad-dog skullcap similar in appearance to downy skullcap but grows only to 2 ½ feet tall, does not have hairy leaves or stems, and occurs on moist or wet sites.

Origin and distribution
Both are native; throughout eastern U.S.

Considerations as a weed and for wildlife
Skullcaps are pollinated primarily by bumblebees. Not eaten by deer or other herbivores because of bitter taste. Rarely a problem in food plots. May complement other plants by providing suitable structure for wild turkey and northern bobwhite broods and ground-feeding songbirds. Can be controlled if needed with glyphosate, imazethapyr, or broadleaf-selective herbicides, such as 2,4-D, 2,4-DB, or dicamba.

American germander
Canada germander
Teucrium canadense

Description and ecology
Perennial, warm-season forb that grows 2 – 3 feet tall. Main stem usually is unbranched, stout, hollow, hairy, and square. Leaves are opposite, toothed, 2 – 5 inches long, pointed at end, with fairly prominent venation. Purplish-pink flowers are somewhat tubular with relatively large lower lip and are produced on a terminal spike mid-June through September. Most often found in relatively moist fertile sites in full sun.

Origin and distribution
Native; throughout U.S.

Considerations as a weed and for wildlife
American germander spreads by rhizomes as well as seed. It can spread and form relatively large colonies in old-field communities, competing with other plants. The plant and associated colonies provide good cover for fawns as well as upland game bird poults. Various songbirds, such as field sparrow and indigo bunting, will readily nest in the associated cover. The plant is not eaten by deer, rabbits, or groundhogs, but several species of bees and butterflies visit the flowers for nectar. It can be controlled by spot-spraying glyphosate, triclopyr, 2,4-D, metsulfuron-methyl, or dicamba.

Wild onion
Allium canadense

Wild garlic
A. vineale

Description and ecology

Upright, perennial, cool-season forbs. Roots are fibrous and attached to a bulb. Wild onion bulbs have a fibrous outer coat that is tan or brown. Leaves arise from bulb. Wild garlic leaves are smooth, round, and hollow. Wild onion leaves are flat in cross-section and not hollow. Flowering stems of both wild onion and wild garlic are solid. Reproduction is largely from bulblets produced at top of stems. Flowers also may be present. Bulblets of wild garlic develop green leaves that look like long "wild hairs" coming from the bulblet "head." False garlic (*Nothoscordum bivalve*) may be confused with wild onion or garlic before flowering, but false garlic flowers before wild onion or wild garlic. It has no value for wildlife, and does not compete with planted forages.

Origin and distribution

Wild onion is native and common across eastern U.S.; wild garlic is nonnative (Europe) and occurs across most of eastern U.S. and portions of western U.S.

Considerations as a weed and for wildlife

Wild onion and wild garlic are often prolific and widespread. Wild onion and garlic rarely compete with planted crops, but they can make cool-season forage plots less attractive and palatable when onion or garlic are dense. Where there are bad infestations of wild onion or garlic, a grass forage crop, such as wheat or oats, should be planted and the onion or garlic can be controlled with 2,4-D and dicamba. Thifensulfuron-methyl+tribenuron-methyl (Harmony Extra) controls wild garlic. In perennial white clover plots where onion/garlic infestation is bad, an application of 2,4-D at 1 quart/ac should provide some control without killing the clover. Wild onion and wild garlic have no food or cover value for wildlife.

Top photo: Wild garlic seedhead; Middle photo: Wild garlic in clover; Bottom left photo: hollow stem of wild garlic; Bottom right photo: False garlic (*Nothoscordum bivalve*)

Wild asparagus
Asparagus officinalis

Description and ecology
Perennial forb with stout central primary stem and multiple secondary stems that may be pointing upward, spreading, or drooping. Plants may reach 6 – 7 feet in height. Stems are green and smooth. Needle-like leaves (actually stemlets) are alternate and slender. Flower clusters develop where leaves meet the stem. Fruits are little round yellowish-green berries that droop down, turning red upon maturity with 2 – 4 seeds inside. Wild asparagus is naturalized and often found along roads and other open areas not in cultivation. Stems may be cut and eaten when they are about 6 inches tall in spring.

Origin and distribution
Nonnative (Europe); throughout U.S.

Considerations as a weed and for wildlife
Wild asparagus can spread into unwanted areas. It provides little value for wildlife. However, white-tailed deer do graze the leaves occasionally. It can be controlled by spot-spraying glyphosate.

Velvetleaf
Abutilon theophrasti

Description and ecology
Upright, annual, warm-season forb. Stem is relatively robust and upright and usually unbranched. Leaves are alternate and heart-shaped, covered with fine velvety hairs, and with long, pointed tips. Flowers are yellow with 5 petals. Fruit is an enclosed cup-shaped capsule with beaks around the outer edge containing 9 – 15 seeds. Flowers and capsules are produced on short stalks from the leaf axils (where leaf stem meets main stem).

Origin and distribution
Nonnative (Asia); throughout U.S.

Considerations as a weed and for wildlife
Velvetleaf sometimes occurs in warm-season food plots, especially those prepared via conventional tillage. It can become widespread and relatively dense and can compete with planted crops. Velvetleaf can be controlled preemergence with imazethapyr, and postemergence with forb-selective herbicides, such as dicamba and clopyralid, or with glyphosate when planting Roundup-Ready crops. Velvetleaf has no value for wildlife.

Venice mallow
Hibiscus trionum

Description and ecology
Annual, warm-season forb. Stem is usually erect, branched, and sparsely pubescent. Leaves are lobed and coarsely toothed. Flowers are white or yellowish with 5 petals and purplish center. Seeds are enclosed in a membranous capsule with purplish stripes.

Origin and distribution
Nonnative (Europe); throughout most of U.S., except portions of southeastern Coastal Plain and southwestern U.S.

Considerations as a weed and for wildlife
Venice mallow commonly occurs in cultivated areas and sometimes in perennial forage plots. It can be controlled with imazethapyr or 2,4-DB if sprayed when young, or with glyphosate or a forb-selective herbicide. It has no value for wildlife other than providing bees with nectar.

Common mallow
Malva neglecta

Description and ecology
Cool-season annual or biennal forb that may occur in winter and summer. Plant is usually spreading along ground, but may be erect. Leaves alternate, round to heart-shaped, and shallowly lobed with toothed margins. Stems are round. Ground ivy is similar in appearance, but ground ivy leaves are opposite and has square stems.

Origin and distribution
Nonnative (north Africa and Europe); throughout U.S. except Deep South

Considerations as a weed and for wildlife
Mallow can be very competitive in cool-season plots. It can be controlled with imazethapyr or 2,4-DB in clover/alfalfa plots if sprayed when young. If not controlled, it can become very dense and outcompete planted forages. It has no wildlife value.

All photos on left by Peter Dziuk, Minnesota Wildflowers

Arrowleaf sida
Sida rhombifolia

Prickly sida
Teaweed
S. spinosa

Description and ecology
Erect, multi-branched, warm-season annual forbs that grow to 2 – 3 feet tall. Prickly sida stem has short spine-like projections at the base of each leaf. Leaves are alternate and usually finely toothed along the margins. Flowers appear June – September and have whitish petals with a yellow center. Flowering stem of prickly sida is shorter than leaf stem below, whereas the flowering stem of arrowleaf sida is longer than leaf stem below.

Origin and distribution
Native; eastern U.S.

Considerations as a weed and for wildlife
Prickly and arrowleaf sida commonly occur in warm-season plots, particularly those established with conventional tillage techniques. They can be controlled preemergence or postemergence with imazethapyr or postemergence with dicamba. Herbicides containing aminopyralid, such as ForeFront R&P HL/GrazonNext HL, or Chaparral, are effective. Plants should be sprayed when young, before exceeding 3 inches in height. Glyphosate does not usually provide complete control at the relatively low rates used with Roundup-Ready crops. However, these plants do not usually present competition issues and may make a food plot more attractive to some species. They can provide low cover for northern bobwhite and wild turkey broods, as well as ground-feeding songbirds. They are moderate-preference forage for white-tailed deer and may be eaten by eastern cottontail.

Top photo: Prickly sida grazed by white-tailed deer
Bottom two photos: Arrowleaf sida

Maryland meadowbeauty
Rhexia mariana

Virginia meadowbeauty
Rhexia virginica

Description and ecology
Erect, warm-season, perennial forbs that may get 2 feet tall. Stems are square or weakly square and somewhat finely hairy. Leaves are opposite and 3-nerved with very short leaf stems. *R. mariana* leaves have spots of red pigment that are barely noticeable. *R. mariana* flowers are pale pink to slightly purple, whereas *R. virginica* flowers are dark pink. Flowers have 4 petals and prominent yellow anthers. The fruit is a capsule that is urn-shaped.

Origin and distribution
Native; throughout most of eastern U.S.

Considerations as a weed and for wildlife
The meadowbeauties are relatively delicate wildflowers that are sometimes found in fallow fields and perennial plots that have not been managed. They pose no problem; they do not compete with planted forages. They provide nectar for bees and butterflies, but do not provide much other benefit for wildlife. White-tailed deer may eat the leaves and flowers sparingly. In the odd event it is necessary, they can be controlled with forb-selective herbicides.

Top two photos: Maryland meadowbeauty
Bottom photo: Virginia meadowbeauty

Carpetweed
Mollugo verticillata

Description and ecology
Spreading, prostrate, annual, warm-season forb; often forms mats. Stem is green, round, and smooth. Leaves are whorled (occur all the way around stem). Flowers are small, white, have 5 petals, and appear July – September. May be superficially similar to catchweed bedstraw (*Galium aparine*), but stem of catchweed bedstraw is square and covered with fine bristles.

Origin and distribution
Native to tropical America; naturalized northward throughout eastern U.S. and into western states

Considerations as a weed and for wildlife
Carpetweed can be widespread and dense. It doesn't usually grow dense enough to outcompete planted crops, but it can on occasion. Carpetweed is common in warm-season food plots managed with conventional tillage. It can be controlled with various preemergence herbicides, such as imazethapyr, trifluralin, and pendimethalin, and with selective postemergence herbicides, such as carfentrazone-ethyl, 2,4-D, and dicamba. Carpetweed has no real food or cover value for wildlife.

Common eveningprimrose
Oenothera biennis

Description and ecology
Erect, biennial forb that grows 3 to 7 feet tall on some sites. Stem is relatively stout and arises from basal rosette in second year. Leaves along stem are alternate and appear to spiral up the stem. They are largest at base of stem (up to 7 inches long) and gradually get smaller toward the top of the stem (1-2 inches long). Leaf margins are smooth. Flowers are yellow (pale reddish at maturity) and occur along a terminal spike, June through October. Seeds are enclosed in a capsule that opens upon maturity. The stem and seed capsules are woody and remain standing through winter.

Origin and distribution
Native; Great Plains east and along Pacific coast

Considerations as a weed and for wildlife
As a biennial, common eveningprimrose is most often seen in fallow fields or perennial plots. Common eveningprimrose rarely competes with planted forages except if it becomes widespread in a perennial forage plot. It can be controlled with forb-selective herbicides, such as dicamba, 2,4-D, and 2,4-DB if sprayed when in the rosette. It may be killed with glyphosate, but it may require a relatively strong rate. Mowing just after flowering and prior to seed maturing will provide control. Common eveningprimrose can provide good structure for ground-feeding songbirds, northern bobwhite chicks, and wild turkey poults. The seed is eaten by mourning dove, northern bobwhite, American goldfinch, and dark-eyed junco. It is a low- to moderate preference forage for white-tailed deer, mule deer, and pronghorn. Hummingbirds, bees, butterflies, and moths frequent the flowers for nectar.

Cutleaf eveningprimrose
Oenothera laciniata

Showy eveningprimrose
O. speciosa

Description and ecology
Cutleaf primrose is a weakly ascending or prostrate annual, cool-season forb. Stems are hairy and branch from a basal rosette. Stems may grow 3 feet in length. Leaves are alternate, oblong, 1–3 inches in length, and coarsely toothed or irregularly lobed with a white midvein. Flowers are single, yellow to red, and appear March – August. Seed occur in a cylindrical capsule that is about an inch long and often curved. Showy eveningprimrose is an erect or spreading short-lived perennial that may reach 2 feet in height. Stem is light green and circular in cross section. Leaves are alternate, 2–3 inches long, 3/4 inch wide, and margins may be toothed or smooth. Flowers consist of 4 broad petals that are white to pink with prominent veins and a yellow center. Flowers appear continuously from March – August.

Origin and distribution
Cutleaf eveningprimrose is native; New Hampshire east to South Dakota and south, including southwestern U.S. Showy eveningprimrose is native to Great Plains and southwestern U.S., but has naturalized throughout much of eastern U.S.

Considerations as a weed and for wildlife
Both cutleaf and showy eveningprimrose are found often in cultivated annual plots and other disturbed areas, and may be found in perennial plots. They can be widespread and compete with various planted forages. They can be controlled with forb-selective herbicides, such as dicamba, 2,4-D, and 2,4-DB if sprayed when in the rosette. They may be killed with glyphosate, but require a relatively strong rate. Cutleaf eveningprimrose seed are eaten sparingly by mourning dove and northern bobwhite. Bees and butterflies frequent the flowers of both species for nectar. They are relatively low-preference forages for white-tailed deer, cottontails, and groundhog.

Top photo: Cutleaf eveningprimrose
Bottom photo: Showy eveningprimrose (*O. speciosa*), photo by D. Elmore

Yellow woodsorrel
Oxalis stricta

Violet woodsorrel
Oxalis violacea

Description and ecology
Erect, annual or perennial, cool-season forbs that may reach 12 inches tall. Stem is relatively slight and hairy. Leaves are alternate and compound with 3 leaflets with a shamrock shape. Flowers are yellow or purple, with 5 petals, and appear April – September.

Origin and distribution
Native; throughout U.S.

Considerations as a weed and for wildlife
There are several species of woodsorrel. Yellow woodsorrel is widespread and common. It may occur in warm-season or cool-season plots, especially fallow warm-season plots and perennial cool-season plots that are at least 2 – 3 years old. It rarely competes with planted forages, but rather takes advantage of available open space when planted forages decline. It is a low- to moderate-preference forage for eastern cottontail, groundhog, and white-tailed deer. Mourning dove, ground dove, ruffed grouse, northern bobwhite, juncos, and various sparrows eat the seed. It may be controlled with 2,4-DB in perennial clover/alfalfa plots, but spraying is rarely needed.

Top two photos: Yellow woodsorrel
Bottom two photo: Violet woodsorrel

390

Maypop passionflower
Passiflora incarnata

Description and ecology
Perennial warm-season vine. Stems may be upright, spreading, or climbing. Leaves are alternate and usually 3-lobed with finely serrate (sharp-toothed) margins. Flowers are purple and white, large, and showy. Fruit is relatively large and oblong (2 – 3 inches), green, and edible.

Origin and distribution
Native; southeastern U.S.

Considerations as a weed and for wildlife
Passionflower is often found in warm-season plots planted via conventional tillage. It can be controlled postemergence with various forb-selective herbicides and with glyphosate in Roundup-Ready crops. Passionflower is resistant to postemergence applications of imazethapyr, which can make it difficult to control in perennial forage plots or leguminous warm-season plots. Some control may be realized with 2,4-DB in perennial forage plots if sprayed when young. Passionflower is not an important plant for wildlife, but bobwhite and some songbirds do eat the seed, and white-tailed deer and eastern cottontail will occasionally eat a few of the leaves.

Pokeweed
Phytolacca americana

Description and ecology
Erect, perennial, warm-season forb with many branches. May grow to 9 feet tall. Stem is smooth and red in late summer. Leaves are alternate. Flowers mid-summer through fall, producing a raceme of berries that turn from green to purple when mature. Older portions of plant, including fruit, are at least mildly toxic to some animals and humans.

Origin and distribution
Native; throughout South and into northeastern U.S.

Considerations as a weed and for wildlife

Pokeweed can be common in warm-season food plots, especially those prepared via conventional tillage, and in perennial forage plots 2 – 3 years after planting. Pokeweed can be controlled preemergence with imazethapyr, or postemergence with various forb-selective herbicides and with glyphosate in Roundup-Ready crops. It can be controlled in perennial clover/alfalfa plots with imazethapyr or 2,4-DB if sprayed when young. If it gets tall, it can be set back by mowing or by applying glyphosate via spot-spraying or with a wick applicator. Pokeweed is an important plant for many wildlife species. The fruit is eaten by many songbird species, including northern mockingbird and eastern bluebird. Mourning doves and northern bobwhite relish pokeweed seed. Mourning doves are strongly attracted to pokeweed seed. Unless there are noxious broadleaf weeds growing in association with pokeweed, I do not spray it when it occurs in dove fields. Pokeweed is a moderate- to high-preference forage for white-tailed deer. Young leaves are very nutritious with protein commonly exceeding 30%. I manage for pokeweed in early successional areas, but because of the amount of shade cast, I usually spray it in forage plots managed for deer. Relatively sparse pokeweed can make a perennial forage plot more attractive for wild turkey broods. Gray and red fox, opossum, raccoon, and white-footed mouse also eat the fruit.

Bracted plantain
Plantago aristata

Southern plantain
Paleseed plantain
P. virginica

Description and ecology
P. aristata is an upright, perennial, cool-season forb. Leaves occur in rosette. Leaves are lanceolate and narrower than buckhorn plantain and usually more upright and less spreading. Flowering stems arise from rosette. Flowers occur in a terminal cylindrical spike with a prominent bract under each flower. *P. virginica* is an annual cool-season forb. Leaves are densely hairy and elliptic, but narrower than broadleaf plantain. Flowering stems arise from rosette.

Origin and distribution
Both bracted and Southern plantain are native; they occur throughout eastern U.S. and isolated areas in western U.S.

Considerations as a weed and for wildlife
Bracted and Southern plantain may occur in various types of food plots, including cultivated warm-season plots as well as perennial forage plots. Southern plantain can compete with clovers in spring. Both of these plantains can be controlled in clover/alfalfa plots with 2,4-DB or imazethapyr if sprayed when young. Bracted and Southern plantain may be eaten sparingly by rabbits.

Top two photos: Bracted plantain
Bottom photos: Southern plantain

Narrowleaf plantain
Buckhorn plantain
Staghorn plantain
Plantago lanceolata

Description and ecology
Upright perennial forb. Leaves occur in rosette. Leaves are slender and linear or lanceolate. Numerous flowering stems arise from basal rosette. Flowers are very small and occur in a terminal dense spike.

Origin and distribution
Nonnative (Europe); throughout U.S.

Considerations as a weed and for wildlife
Narrowleaf plantain can become widespread, dense, and competitive in perennial forage plots. It can be controlled with imazethapyr or imazapic in clover/chicory/alfalfa plots or 2,4-DB in clover/alfalfa plots if sprayed when young with heavy rate. It can be controlled with dicamba in various grass plots, and with imazapic or triclopyr in fallow plots or other early successional areas. Seed of narrowleaf plantain may be eaten by several species of songbirds, including northern cardinal and several species of sparrows. Foliage is eaten by ruffed grouse, rabbits, and small rodents. However, it is a low-preference white-tailed deer forage.

Narrowleaf plantain (above) and broadleaf plantain (right)

Broadleaf plantain
Plantago major

Blackseed plantain
P. rugelii

Description and ecology
Upright, perennial, cool-season forbs. Leaves occur in rosette. Leaves are wide and smooth with a prominent midvein. Flowering stems arise from center of rosette with very small flowers along a terminal spike. Broadleaf and blackseed plantain are very similar. Leaf stem bases of blackseed plantain tend to be reddish purple and the leaf margins are usually wavy-toothed.

Origin and distribution
Broadleaf plantain is nonnative (Eurasia) and occurs throughout U.S.; blackseed plantain is native and occurs throughout eastern U.S.

Considerations as a weed and for wildlife
Broadleaf plantain can become widespread and competitive in perennial forage plots. These plantains can be controlled with 2,4-DB or imazethapyr if sprayed when young in perennial clover/alfalfa plots. They also can be controlled with forb-selective herbicides, such as triclopyr, in grass plots or fallow areas. Glyphosate and dicamba are relatively weak on plantains (relatively heavy rates required). Seed of broadleaf and blackseed plantain may be eaten by several species of songbirds. Foliage is eaten by ruffed grouse, rabbits, and small rodents. Broadleaf and blackseed plantain are low-preference white-tailed deer forages.

Top photo: Broadleaf plantain; Bottom photos: blackseed plantain

395

Creeping bentgrass
Redtop
Agrostis stolonifera or *A. palustris*

Description and ecology
Perennial cool-season grass that may be erect or decumbent (spreading along ground with tips pointed upward). It spreads from rhizomes (stem below soil surface) and from stolons (rooting stem that spreads on soil surface). Flowering stems may reach nearly 4 feet tall. Leaves are blue-green, linear, typically 4 – 7 inches long, and smooth on the upper side and rough on the underside. Flower head is a dense panicle, which is open when young, but closes upon maturity, and may be green to reddish-purple or straw-colored.

Origin and distribution
Nonnative (Europe, north Africa); throughout U.S. except FL and south TX.

Considerations as a weed and for wildlife
Creeping bentgrass (or redtop) is highly invasive, spreads rapidly, and can outcompete planted crops. It is often used with golf courses. It should be controlled with glyphosate or grass-selective herbicides. Disking or tilling breaks up the stolons and rhizomes and causes more sprouts.

Broomsedge
Broomsedge bluestem
Broomstraw
Andropogon virginicus

Description and ecology
Upright, clump-forming, perennial, warm-season grass. Stems reach 1-5 feet in height; typically taller on better sites. Leaves are smooth, linear (except few sparse hairs near leaf base), about 12 inches long, pointed, and pale green. Leaf blades are folded (or keeled) at the base. Flowering occurs September through November. Flowers are feathery and silky-white in appearance, and occur along the upper half of the stems at the axil (where the leaf meets the stem). Upon senescence, broomsedge turns a brilliant tan and remains upright through winter and the following spring. Broomsedge grows on low pH soils, but is more vigorous and taller in productive soils with soil pH 6.0-7.0. Broomsedge is outcompeted by nonnative perennial cool-season grasses, such as tall fescue and orchardgrass, because they begin growth approximately 6 weeks prior to broomsedge in late winter. Broomsedge may outcompete nonnative perennial cool-season grasses when soil pH begins to drop below 5.7, which stresses the nonnative cool-season grasses.

Origin and distribution
Native; throughout eastern U.S.

Considerations as a weed and for wildlife
Broomsedge commonly occurs in fallow fields and may be found in perennial forage plots that have begun to thin. Broomsedge is one of the most important grasses for wildlife in the eastern U.S. because of the structure (cover) it provides. Northern bobwhite, eastern meadowlark, grasshopper sparrows, and others construct nests at the base of broomsedge. The height of broomsedge, along with the open interstitial space between the grass bunches, provides good cover for northern bobwhite, wild turkey, and ground-feeding songbirds, especially when desirable forb cover also is present. White-tailed deer commonly bed and hide fawns in fields containing broomsedge. Broomsedge is not an important food source for wildlife, though the foliage is eaten by various small mammals, such as cotton rat. Broomsedge can be controlled with glyphosate and can be set-back with reduced control with grass-selective herbicides, such as clethodim. Heavy disking and continued late growing-season fire also may reduce broomsedge density in early successional plant communities.

Sweet vernalgrass
Vanilla grass
Anthoxanthum odoratum

Description and ecology
Upright, tufted, perennial, cool-season grass that grows best in full sun but will tolerate light shade. Leaves are flat with pointed tips and may be smooth or somewhat hairy. Flowering stem reaches 1-3 feet tall. Seedhead is a spike with awned spikelets. Flowers April through June. Plant has a sweet aroma when crushed, hence the alternate common name, vanilla grass. Reproduces by seed. Tolerates low pH soils.

Origin and distribution
Nonnative (Eurasia); throughout eastern U.S., portions of southwest U.S., and Pacific coastal states

Considerations as a weed and for wildlife
Sweet vernalgrass can become competitive in perennial forage plantings. It provides no documented wildlife value. It can be controlled with grass-selective herbicides, glyphosate, and likely with several preemergence broad-spectrum selective herbicides, but data are lacking. Sweet vernalgrass should be controlled because it can spread rather vigorously and compete with more desirable plants.

Bromegrasses
Bromus arvensis (Field brome, Japanese brome)
B. catharticus (Rescuegrass)
B. commutatus (Hairy chess)
B. hordeaceus (Soft brome)
B. secalinus (Common chess or cheat)
B. tectorum (Downy brome or cheatgrass)

Description and ecology
All are upright, annual, cool-season grasses. Leaves of field brome are hairy on both sides. Rescuegrass spikelets are laterally compressed (flattened) and leaves are pubescent (fine hairs) on upper side only. Hairy chess leaves pubescent on both sides. Soft brome often decumbent (stem and branches flat on ground with tips pointing upward). Common chess leaves smooth with sparse hairs near base. Leaves of Japanese brome pubescent on upper and under surface; young leaves often twisted; has longer awns (½ – ¾ inch) than other bromes. Leaves of downy brome seedlings are twisted.

Origin and distribution
None of the bromes listed here are native. All except rescuegrass (South America) are native to Europe. All occur throughout most of U.S. except rescuegrass, which is found throughout the southern half of the U.S.

Considerations as a weed and for wildlife
These annual bromegrasses can become dense and very competitive in food plots. Bromegrasses are not an important food source for wildlife. The seed of bromes are eaten sparingly by a few small rodents, and the foliage is eaten sparingly by elk and mule deer in the western U.S., where other forages may be limited. None of the bromes are selected forage for white-tailed deer. Annual bromegrasses can be controlled postemergence with grass-selective herbicides and glyphosate. Imazapic can be sprayed preemergence or postemergence when plants are young. It may be necessary to grow forb-only food plots (such as clovers) for 2 – 3 years to completely get rid of annual bromegrasses. Be diligent! You should aggressively try to eradicate these grasses.

Top photo: field brome; Middle photo: downy brome; Bottom photo: common chess

Smooth brome
Bromus inermis

Description and ecology
Perennial cool-season grass. Grows to 3 – 4 feet in height. Produces sod-forming rhizomes.

Origin and distribution
Nonnative (Europe); throughout U.S. Widely planted throughout Midwest.

Considerations as a weed and for wildlife
Smooth brome can become competitive in perennial cool-season food plots. Smooth brome is most problematic in old-fields and along crop-field edges where it was planted previously. Smooth brome usually is very dense near ground level, which impedes travel by small wildlife and makes finding food (seed, insects) difficult. Dense sod and thatch prevent the seedbank from germinating, thus reducing forb coverage and species diversity. Not selected as forage by white-tailed deer, rabbits, or groundhogs. A few songbirds, such as red-winged blackbird and pheasants may nest in smooth brome. However, structure for brooding is exceptionally poor. Postemergence applications of glyphosate in fall are recommended. Spring applications also are effective, but spot-spraying recurring smooth brome will be necessary as spring applications may not be as effective at killing the root systems as fall applications. It can be controlled in clover/chicory plots with a grass-selective herbicide, such as clethodim. However, periodic retreatment will be necessary because grass-selective herbicides do not completely kill most perennial grasses. Spot-spraying with glyphosate also is effective. Smooth brome is **not** controlled with imazapic. In my opinion, smooth brome is the tall fescue of the Midwest. If you want to improve fields for deer and wild turkey, you need to eradicate smooth brome.

Southern sandbur
Cenchrus echinatus

Longspine sandbur
C. longispinus

Field sandbur
C. spinifex

Description and ecology
Tufted, annual, warm-season grasses. Field sandbur may be a short-lived perennial from stolons (stem that grows along the ground with occasional roots where it touches soil). Burs of field sandbur are ovoid with no more than 40 spines, whereas those of longspine sandbur are more round and have more than 40 spines. Burs of Southern sandbur have one row of united flattened spines.

Origin and distribution
Native. Southern sandbur occurs throughout Deep South and southwestern U.S. Longspine sandbur occurs throughout U.S. Field sandbur occurs throughout South, into Midwest, and Southwest.

Photos: Field sandbur

Considerations as a weed and for wildlife
Sandburs can be problematic in all types of food plots—cultivated, drilled, annual, or perennial. They can be controlled preemergence with pendimethalin, imazethapyr, imazapic, imazamox, and S-metolachlor, and postemergence with grass-selective herbicides, or with glyphosate in Roundup-Ready crops. Sandbur can become a real problem if not controlled. You should aggressively try to eradicate this weed. Sandbur has no wildlife value.

Bermudagrass
Cynodon dactylon

Description and ecology
Spreading, prostrate, warm-season, perennial grass that spreads via rhizomes (horizontal stems belowground) and stolons (aboveground stems growing horizontally along the ground, rooting wherever they contact mineral soil). Stems are wiry and tough. Leaves are narrow and bluish green. Flowering stems are upright. Seedhead is 3 – 7 spikes on the end of the stem, producing very small flowers and seed.

Origin and distribution
Nonnative (Eurasia); throughout U.S. except upper Great Plains.

Considerations as a weed and for wildlife
Bermudagrass is very invasive. It can be problematic in warm-season forage plots and in perennial forage plots. In warm-season plots, such as iron-clay cowpeas or soybeans, spray bermudagrass with a grass-selective herbicide, which will reduce coverage of bermudagrass and help control it over time, but grass-selective herbicides will not get rid of bermudagrass. It will come back. Grass-selective herbicides also can be used in perennial clover plots, alfalfa, and chicory. Glyphosate can be used in Roundup-Ready crops to control bermudagrass. For information on getting rid of bermudagrass prior to planting food plots, see pages 42-43. You should aggressively try to eradicate bermudagrass. It has no wildlife value.

Top photo: Bermudagrass (seedheads shown) growing with broadleaf signalgrass and common lespedeza.

Orchardgrass
Dactylis glomerata

Timothy
Phleum pratense

Description and ecology
Upright, perennial, cool-season bunchgrasses. Orchardgrass leaves are bluish green with prominent midrib. Leaf sheaths (lower part of leaf that surrounds stem) are flattened. Stem is smooth. Seedhead is a panicle with clustered flowers. Timothy leaves are hairless with rough margins and taper to a sharp point. Stem is pale and often appears swollen. Timothy has short rhizomes (creeping underground stems) and many have short stolons (stems from roots just above ground). Seedhead is a terminal spike-like panicle.

Origin and distribution
Nonnative (Eurasia); throughout U.S.

Considerations as a weed and for wildlife
Orchardgrass and timothy can become increasingly dense and competitive in perennial cool-season forage plots. They should be sprayed in the fall or spring with a grass-selective herbicide when occurring in perennial forage food plots. For best results, orchardgrass and timothy should be sprayed with glyphosate in the fall before perennial plots are planted. Orchardgrass and timothy may be eaten to some extent by rabbits, groundhogs, deer, and elk when young and tender if little other food is available. They are not selected forages by white-tailed deer (see Tables 8.2 and 8.3 on page 141) and should not be planted in a food plot. Orchardgrass and timothy provide structure similar to tall fescue and should not be considered "wildlife-friendly" grasses.

Top photo: Orchardgrass; Middle photo: Timothy; Bottom photo: Dense structure and growth habit of orchardgrass, timothy, and tall fescue choke out clover quickly.

Crabgrass
Digitaria bicornis (Tropical crabgrass)
D. ciliaris (Southern crabgrass)
D. ischaemum (Smooth crabgrass)
D. sanguinalis (Large crabgrass)

Description and ecology
Annual warm-season grasses that are usually prostrate and spreading, but may be ascending. All except smooth crabgrass will root at lower leaf nodes when spreading. Large crabgrass has hairs on the leaf blade and sheath (lower part of leaf that surrounds stem). Smooth crabgrass has only few hairs near base of leaf. Southern crabgrass lacks hairs on the leaf blade. Seedheads consist of several spikes clustered at top of stems.

Origin and distribution
Nonnative (Eurasia and Europe); tropical crabgrass occurs along southeastern Coastal Plain; southern crabgrass occurs throughout the South and into the Northeast and Midwest; smooth crabgrass occurs throughout U.S. except most of TX; large crabgrass occurs throughout U.S.

Considerations as a weed and for wildlife
Crabgrasses are very common and can be very competitive for planted grains and forages. If not controlled, crabgrasses can lead to reduced crop yields or even planting failures. Crabgrasses are controlled relatively easily with several preemergence herbicides, such as pendimethalin, trifluralin, imazethapyr, and S-metolachlor, when planting warm-season plots, and with grass-selective postemergence herbicides. When spraying postemergence, spray crabgrass when young for best control. Mowing does not control crabgrasses. Several bird species, including mourning dove, northern bobwhite, wild turkey, and various sparrows (especially chipping, field, savannah, clay-colored, and Lincoln) and juncos, eat crabgrass seed. When present around shallow wetlands, dabbling ducks (especially green-winged teal) eat the seed. Rabbits and groundhogs eat the foliage sparingly. Crabgrass is not a selected forage for white-tailed deer.

All photos: large crabgrass

Barnyardgrass
Wild millet
Echinochloa crus-galli

Japanese millet
Duck millet
E. frumentacea

Junglerice
E. colona

Description and ecology
Upright, annual, warm-season grasses. *E. crus-galli* may grow to 6 feet tall, whereas *E. colona* is normally no more than 3 feet tall. Leaves of both species are smooth and there is no ligule. Stems are smooth and often bent. Seedhead is a tight panicle with several branches. Usually found in moist areas. *E. colona* does not have awns (hair-like structures coming from base of flower). *E. frumentacea* was considered a variety of *E. crus-galli*, but is now considered a separate species. *E. frumentacea* does not have awns, whereas *E. crus-galli* usually does, and the panicle of *E. crus-galli* usually breaks apart in segments, whereas the panicle of *E. frumentacea* does not.

Origin and distribution
Nonnative (Eurasia); throughout U.S.

Considerations as a weed and for wildlife
All three species may be found in warm-season food plots, especially those prepared via conventional tillage and those in moist areas. They do not usually present a competition problem. Nonetheless, they can be controlled with several preemergence herbicides, such as pendimethalin, trifluralin, imazapic, imazethapyr, imazamox, and S-metolachlor, and with postemergence grass-selective herbicides and glyphosate in Roundup-Ready crops. The seed of these grasses is an important food for many bird species. They should be considered a bonus in any warm-season food plot intended for birds, including mourning dove, northern bobwhite, dabbling ducks, and wild turkey. *Echinochloas* are not a selected forage for white-tailed deer.

Top photo: Japanese millet or duck millet
Bottom left: Junglerice; Bottom right: Barnyardgrass or wild millet

Goosegrass
Eleusine indica

Description and ecology
Annual, warm-season grass that produces a prostrate rosette that is whitish in center. Stems are flattened. Seedhead comprises 2 – 6 spikes clustered at top of stem.

Origin and distribution
Nonnative (Eurasia); throughout U.S.

Considerations as a weed and for wildlife
Goosegrass is a tough annual grass that can be very problematic, especially in food plots prepared via conventional tillage. It spreads quickly and can outcompete planted crops. Some preemergence herbicides, such as S-metolachlor, pendimethalin, and trifluralin, work well to control goosegrass. Grass-selective postemergence herbicides will help control goosegrass, but repeat applications usually are necessary. Glyphosate is effective with Roundup-Ready crops. Various birds, such as northern cardinal, ground dove, and savannah sparrow, may eat goosegrass seed sparingly, but goosegrass provides little food value and no cover value for wildlife.

Skunkgrass
Stinkgrass
Eragrostis cilianensis or *E. megastachya*

Description and ecology
Annual warm-season grass that may be erect in a dense clump or spreading with tips pointed upwards. May reach 2 feet tall. Stem is smooth and branched, bent below. Leaves grow to 6 inches long, smooth above and rough below, and produce foul odor when crushed. Ligule (at base of leaf) is a ring of stiff hairs. Seedhead is a dense panicle that is slightly pyramid-shaped. Stinkgrass reproduces by seed and has a shallow root system.

Origin and distribution
Nonnative (Europe); throughout U.S.

Considerations as a weed and for wildlife
Skunkgrass can be widespread on cultivated sites, but does not usually compete with planted forages. It can be controlled with grass-selective herbicides, cultivation, or mowing before seedheads mature. It has no value for wildlife.

All photos: Skunkgrass

Purple lovegrass
Eragrostis spectabilis

Description and ecology
Erect, clump-forming, perennial, warm-season grass. Typically grows 1 – 2 feet tall with dense, fine-branched panicles (seedheads) that turn purple in late summer. The stem is stout with dense hairs. Leaves are about 8 – 20 inches long with tapering tips and rough margins.

Origin and distribution
Native; throughout eastern U.S. and Great Plains

Considerations as a weed and for wildlife
Purple lovegrass may occur in fallow warm-season plots and perennial plots that are at least 2 – 3 years old. Like any perennial grass, it will compete with planted clovers, alfalfa, chicory, etc. The foliage is rarely eaten by wildlife, but it can provide some cover for small wildlife in fallow fields. It can be controlled with a grass-selective herbicide, but if purple lovegrass is widespread, the plot likely needs renovation. It is a desirable species for the cover it provides in early successional areas.

Tall fescue
Festuca arundinacea, Lolium arundinaceum, or
Schedonorus phoenix

Description and ecology
Upright, perennial, cool-season bunchgrass.
Leaves are relatively rough, especially when
rubbed from tip to base. Leaves have very few
hairs, if any, and have prominent parallel veins.
Leaf collar (base of leaf where it meets sheath)
on seedlings is whitish and divided in center by
midrib. Stem is smooth. Seedhead is a panicle.

Origin and distribution
Nonnative (Europe); throughout U.S.

Considerations as a weed and for wildlife
Tall fescue can become problematic in perennial
forage plots, especially those where tall fescue
was growing prior to planting and where it
was not sprayed properly before preparing
the seedbed. In perennial forage plots, spray
tall fescue in the fall or spring with a grass-
selective herbicide. Although effective, this will
have to be repeated over time. Grass-selective
herbicides are not as effective as glyphosate
at killing tall fescue. For best results, eradicate
tall fescue by spraying with glyphosate in the
fall before planting perennial forage plots.
See pages 40-41 for additional information
on eradicating tall fescue prior to planting.
By whichever scientific name plant ecologists
choose to use, make no mistake, tall fescue is no
friend of wildlife. Neither the seed nor foliage
of tall fescue is selected by wildlife. Eating the
foliage and especially the seed of tall fescue
can have toxic effects on livestock and wildlife
because of an endophyte fungus associated
with tall fescue. You should aggressively try to
eradicate this plant wherever it occurs on any
property managed for wildlife.

409

Velvetgrass
Holcus lanatus

Description and ecology
Upright, perennial, cool-season bunchgrass. Leaves are grayish green and covered with velvety pubescence (fine hairs). Seedhead is a grayish purple-tinged panicle that is soft, similar to the leaves. Seedhead turns tan at maturity.

Origin and distribution
Nonnative (Europe); throughout U.S. except Great Plains

Considerations as a weed and for wildlife
Velvetgrass can become dense in perennial forage plots if not controlled and outcompete planted forages. It should be sprayed with a grass-selective herbicide or glyphosate in fallow areas in fall or spring. Velvetgrass may be eaten sparingly by rabbits. It is not a selected forage by white-tailed deer. It may provide cover for small rodents, such as voles, when relatively dense, but otherwise provides no value for wildlife.

Little barley
Hordeum pusillum

Description and ecology
Erect, annual, cool-season grass. Flowering stems grow to about 18 inches tall. Leaves are about 3 – 5 inches long. Seedhead is a spikelike raceme 1 – 3 inches long with crowded awns (hairlike structures).

Origin and distribution
Native; throughout U.S., except west coast

Considerations as a weed and for wildlife
Little barley is a common "weedy" grass that tolerates relatively poor soils. It often occurs in warm-season plots in the first fallow season and may occur in perennial cool-season plots. It provides little value for wildlife. It is not a selected forage for white-tailed deer. It can be controlled preemergence with imazethapyr, pendimethalin, S-metolachlor, and postemergence with grass-selective herbicides.

Bottom photo by: J. Neal

Red sprangletop
Leptochloa panicea

Description and ecology
Erect, tufted, annual, warm-season grass. Stem is often branched and bent at the nodes. Grows 3 – 4 feet in height. Sheaths and seedhead usually appear red. Seedhead is a panicle with extremely slender racemes and small seed.

Origin and distribution
Native; southern U.S.

Considerations as a weed and for wildlife
Red sprangletop is most common in cultivated food plots, especially those in lowland areas that may be flooded for waterfowl. It can become relatively dense and compete with planted crops, especially rice, millets, and small legumes. It can provide some low cover for various songbirds, rabbits, and small rodents, but is not an important plant for wildlife. It can be controlled with glyphosate, grass-selective herbicides, and most preemergence herbicides, such as imazethapyr, pendimethalin, and S-metolachlor, used for various legume and grain crops.

Perennial ryegrass
Lolium perenne

Annual ryegrass
Italian ryegrass
L. multiflorum or *L. italicum*

Description and ecology
Upright cool-season grasses. The seedhead is a spike with several spikelets. Perennial ryegrass spikelets have 2 – 10 florets (small flowers comprising the spikelet) with no awns (stiff, bristle-like structures; if present on perennial ryegrass spikelets, they are very short). Annual ryegrass spikelets have 10 – 20 florets with long conspicuous awns. Ryegrass leaves have a characteristic sheen on the leaf when relatively young, appearing almost as if it is wet.

Origin and distribution
Nonnative (Europe); throughout U.S.

Considerations as a weed and for wildlife
Ryegrass, particularly annual ryegrass, is often planted in the Deep South as a white-tailed deer forage, which is unfortunate because there are several other forage options that are much preferred over ryegrass by wildlife (see *White-tailed Deer* chapter for discussion on ryegrass) and are not invasive. Ryegrass is aggressive and can outcompete beneficial forbs, such as clovers, in a cool-season forage plot. Ryegrass can be difficult to eradicate from a field unless you are persistent. However, if a field is sprayed for 2 – 3 years in a row with a grass-selective herbicide, such as clethodim, **in the fall** with possible follow-up in the spring, it can be eradicated. Ryegrass is a low-preference forage for white-tailed deer (see *Table 8.3* on page 141) and the seed is not an important food source for birds.

413

Japangrass
Japanese stiltgrass
Nepalese browntop
Microstegium vimineum

Description and ecology
Spreading or upright annual, warm-season grass. May grow to 2 feet tall. Leaves are alternate and usually 2 – 4 inches long, growing from a slender, wiry stem. Flowers in late summer with seed produced September – October. Seedhead is a terminal raceme on the flowering stalk. Japangrass is shade-tolerant and extremely invasive. It often completely shades-out other plants in forests, woodlands, and along the edges of fields. It thrives in moist areas, such as bottomland forests with dappled sunlight.

Origin and distribution
Nonnative (Asia); naturalized throughout eastern U.S. from New York, Illinois, and Arkansas southward.

Considerations as a weed and for wildlife
Japangrass can be a problem in perennial cool-season plots, but it is easily controlled pre- or postemergence with imazethapyr or postemergence with grass-selective herbicides. Glyphosate and imazapic (see *Dealing with japangrass* on pages 220-221) are effective options when japangrass occurs along woods roads and in other areas. Japangrass provides no food value for wildlife and should be treated aggressively and as soon as it appears.

Nimblewill
Muhlenbergia schreberi

Description and ecology
Spreading to upright perennial warm-season grass, usually up to 8 – 10 inches tall. Leaves are very narrow, 1 – 2 inches long. Stems (stolons) spread and root at nodes. Appears similar to bermudagrass. However, nimblewill seedhead is a single, slender, spike-like, inconspicuous terminal panicle that appears in late summer, whereas bermudagrass seedhead comprises 3 – 7 spikes. Nimblewill is often found in fields, orchards, and woodlands; it is moderately shade tolerant.

Origin and distribution
Native; throughout eastern U.S.

Considerations as a weed and for wildlife
Nimblewill can become widespread and competitive in cool-season forage plots, especially perennial clovers. Nimblewill can be controlled with grass-selective herbicides, but not eradicated. Spot-spraying with glyphosate is effective. Postemergence application of mesotrione is effective. Nimblewill is not controlled with imazapic. Nimblewill seed may be eaten sparingly by wild turkey, juncos, and sparrows. Nimblewill can provide decent cover in early successional areas for wintering sparrows and eastern cottontail.

Witchgrass
Panicum capillare

Description and ecology
Upright or decumbent (stem lying on ground with tips pointing upward) annual, warm-season grass. Leaves are distinctly pubescent (fine hairy) on both sides. Leaf sheaths also are hairy. Seedhead is an open, spreading panicle. After maturity, it often breaks off and "tumbles" along the ground with the wind. Similar in appearance to fall panicum, but fall panicum does not have hairy leaves or stems.

Origin and distribution
Native; throughout U.S.

Considerations as a weed and for wildlife
Witchgrass does not usually pose a competitive problem with planted crops, but it can change the structure of cover presented in a field. Witchgrass can be controlled with preemergence herbicides, such as pendimethalin, trifluralin, and S-metolachlor, and postemergence with grass-selective herbicides and with glyphosate in Roundup-Ready crops. However, witchgrass (along with other Panicums) is an important food source for many bird species, including mourning dove, ground dove, northern bobwhite, wild turkey, field sparrow, clay-colored sparrow, chipping sparrow, grasshopper sparrow, Lincoln sparrow, song sparrow, white-crowned sparrow, white-throated sparrow, and several others. It (as well as fall panicum) can add much value to a warm-season food plot, especially corn fields, both in terms of food and cover. Deer often selectively bed in sections of corn fields that contain dense witchgrass or fall panicum. Witchgrass is not a selected by forage by white-tailed deer.

Fall panicum
Panicum dichotomiflorum

Description and ecology
Upright or sometimes decumbent (bent over or lying on ground with tips pointing upward) annual, warm-season grass. Stems may have waxy appearance and are bent at leaf nodes. Leaves are usually smooth on both sides when mature, as are sheaths (fall panicum is sometimes called smooth witchgrass). Midvein is light green or white. Seedhead is an open panicle with numerous branches.

Origin and distribution
Native to eastern U.S.; occurs throughout U.S.

Considerations as a weed and for wildlife
Fall panicum is commonly found in grain crops. It usually is not a problem with crop production or yield, especially with taller crops such as corn, but it can be widespread and relatively dense. Fall panicum can be controlled with preemergence herbicides, such as pendimethalin, trifluralin, and S-metolachlor, and postermergence with grass-slective herbicides and with glyphosate in Roundup-Ready crops. However, where fall panicum is relatively dense and tall and with other weeds, the structure can be attractive for many species, including eastern cottontails, white-tailed deer, and several songbirds. In fact, deer may selectively bed in areas of corn fields with dense fall panicum or witchgrass. Presence of fall panicum or witchgrass also makes it easy to burn corn fields in late winter/early spring if desired, depending on objectives. Fall panicum is easily controlled in perennial clover plots if sprayed when young with a grass-selective herbicide. Fall panicum is an important food source for many bird species, including mourning dove, ground dove, northern bobwhite, wild turkey, field sparrow, clay-colored sparrow, chipping sparrow, grasshopper sparrow, Lincoln sparrow, song sparrow, white-crowned sparrow, white-throated sparrow, red-winged blackbird, northern cardinal, and several others. Fall panicum is not a selected forage by white-tailed deer.

Dallisgrass
Paspalum dilatatum

Description and ecology
Upright, sod-forming, perennial, warm-season grass. Basal leaves are erect and spreading and shorter than flowering stem. Flowering stem is normally about 18 inches tall with 3 – 7 alternate branches (racemes) containing seeds. Flowers mid-summer through early fall.

Origin and distribution
Nonnative (South America); naturalized throughout the southeastern U.S.

Considerations as a weed and for wildlife
Dallisgrass may become weedy in perennial food plots. The seed may be eaten sparingly by northern bobwhite, wild turkey, mourning dove, ground dove, and various songbirds. However, dallisgrass does not provide cover for birds and usually excludes more beneficial plants, especially where it has been planted. It is not a selected forage by white-tailed deer. It should be eradicated on properties managed for wildlife. Glyphosate and imazapic can be used to kill dallisgrass. Grass-selective herbicides will suppress or control it in perennial forage plots.

Bahiagrass
Paspalum notatum

Description and ecology
Upright, sod-forming, perennial, warm-season grass. Basal leaves are erect and spreading and much shorter than the flowering stem. Flowers May through November with mature seed mid-summer through winter. Flowers occur on spike-like racemes that are paired and usually appear as a V.

Origin and distribution
Nonnative (South America); naturalized and common from NC to TX along the Coastal Plain, north to TN and AR.

Considerations as a weed and for wildlife
Bahiagrass is common in fields throughout the Gulf Coastal Plain because it is planted for livestock. It may become weedy in perennial plots and spread quickly into fallow fields. It is not a selected forage by white-tailed deer. The seed may be eaten sporadically by wild turkey, northern bobwhite, and several species of sparrows, but there is no cover value. Some landowners are led to believe bahiagrass has value for wild turkeys because they occasionally see turkeys in fields of bahiagrass. Turkeys are seen in the fields by humans (just as they are by predators) because there is no cover! Bahiagrass should be eradicated on properties managed for white-tailed deer, wild turkey, northern bobwhite, and rabbits. Various forbs should be promoted instead of this sod-forming grass that spreads and limits other plants that are much more valuable to wildlife. Metsulfuron methyl and glyphosate may be used to eradicate bahiagrass. Grass-selective herbicides can be used to control bahiagrass in perennial forage plots.

419

Vaseygrass
Paspalum urvillei

Description and ecology
Upright, perennial, warm-season bunchgrass that grows to about 6 feet tall. Leaf margins are rough and midvein is whitish. Flowers spring through December. Seedhead is a raceme of 12 – 25 branches pointing upward at the end of the flowering stem.

Origin and distribution
Nonnative (South America); naturalized and common throughout the Coastal Plain from SC to TX.

Considerations as a weed and for wildlife
Vaseygrass is fairly invasive and found frequently in old-fields and ditches. It can become problematic in perennial forage plots. It can be suppressed or controlled with a grass-selective herbicide. Glyphosate can be used to eradicate vaseygrass in fallow areas. It is not eaten by deer and is not an important food for rabbits or birds. The structure provided by vaseygrass can provide cover for northern bobwhite and wild turkey broods and various ground-feeding songbirds as long as it is not too dense to limit mobility or germination and growth of forbs.

Reed canarygrass
Phalaris arundinacea

Description and ecology
Erect, perennial, cool-season grass that spreads from thick rhizomes; generally occurs in colonies. Flowering stem is smooth and reaches 4 – 5 feet tall. Leaves are bluish-green, 2 – 15 inches long, rough along the edges, and without hairs. Seedhead is dense panicle, 3 – 6 inches long, and distinctively pale-yellow. Common in wetlands, ditches, and other moist areas, but also may occur in adjacent uplands.

Origin and distribution
Native to the Midwest and west coast; throughout U.S., except Deep South. It is believed that introduced genotypes from Eurasia have hybridized with the native genotype and it is extremely aggressive in wetlands throughout most of U.S.

Considerations as a weed and for wildlife
Reed canarygrass is aggressive and competitive in moist areas where it generally occurs. It is less aggressive as it spreads into uplands, but still will outcompete many planted forages, especially clovers. Reed canarygrass provides cover for many wildlife species in wetland areas (such as various chorus frogs and song sparrow), but few upland species benefit from this grass. Reed canarygrass may have high levels of alkaloids and it is not an important forage for any wildlife species. It is not eaten by white-tailed deer. It can be killed with glyphosate and imazapyr, and grass-selective herbicides can be used to control its spread into upland areas.

Bluegrass
Poa annua (Annual bluegrass)
P. pratensis (Kentucky bluegrass)
P. trivialis (Roughstalk bluegrass)

Description and ecology
Bluegrasses are upright, clump-forming, cool-season grasses. Annual bluegrass clumps enlarge by tillering (shoots growing from base of plant). Kentucky bluegrass is perennial and spreads from rhizomes (underground stems radiating from base of plant). Roughstalk bluegrass is perennial and spreads from stolons (horizontal stem on top of ground); the stalk and leaves have a rough feel. Kentucky and roughstalk bluegrass may get about 3 feet tall, whereas annual bluestem is shorter, no more than 1 foot tall. Leaf tips of the bluegrasses are slightly curved upward, like the bow of a boat. Seedheads are relatively small, open panicles.

Origin and distribution
Nonnative (Europe); throughout U.S.

Considerations as a weed and for wildlife
Bluegrasses can become relatively dense and competitive in perennial forage plots if not controlled. Rabbits and groundhogs may eat bluegrass when the grasses are young, but not as readily as clovers and other forbs, such as Carolina geranium. Various sparrows eat the seed sparingly, and mule deer, white-tailed deer, and elk will eat bluegrass in early spring if little other forage is available. Bluegrass is not a selected forage for white-tailed deer and should not be planted in a perennial forage plot because other forages are much preferred. Like all perennial cool-season grasses, they provide no forage during winter above the Deep South because they are dormant. Bluegrasses can be controlled with glyphosate and grass-selective herbicides.

Top photo: Annual bluegrass; middle photo: Annual bluegrass in clover plot; bottom photo: Roughstalk bluegrass

Foxtailgrasses
Bristlegrasses
Setaria faberi (Giant foxtail)
S. parviflora (Knotroot foxtail)
S. pumila (or *S. glauca*; Yellow foxtail)
S. verticillata (Bristly foxtail)
S. viridis (Green foxtail)

Description and ecology
All of the foxtailgrasses are upright annual grasses, except knotroot, which is perennial. Giant foxtail is the largest of these species, often growing 4 – 6 feet tall. The other species grow to about 3 feet tall. Giant foxtail seedheads are distinctive in that they droop over. Bristly foxtail seedheads also curve over, but not as much as giant foxtail. Seedheads of the other species are erect. Knotroot foxtail has short, knotty rhizomes. Yellow and green foxtail do not have rhizomes. Yellow foxtail has hairs at the base of the leaves; green foxtail does not.

Origin and distribution
Giant foxtail is nonnative (southeastern Asia) and occurs throughout eastern U.S. and into Great Plains; knotroot foxtail is native and occurs throughout most of eastern U.S. and California; yellow foxtail is nonnative and occurs throughout U.S.; bristly foxtail is nonnative and occurs throughout U.S. except in Southeast; green foxtail is nonnative and occurs throughout U.S.

Considerations as a weed and for wildlife
Foxtailgrasses are common in food plots. They rarely outcompete planted crops, except some millets, but can be widespread and become dense. The seed of foxtailgrasses is an important food source for many bird species, including mourning dove, ground dove, northern bobwhite, wild turkey, indigo bunting, snow bunting, northern cardinal, dickcissel, blue grosbeak, juncos, horned lark, Lapland longspur, eastern and western

From top: Giant foxtail, yellow foxtail, yellow foxtail roots

423

meadowlark, pyrrhuloxia, and many sparrows (field, clay-colored, grasshopper, lark, Lincoln, savannah, song, vesper, white-crowned). In fact, foxtailgrasses are considered the most important "weedy grasses" for wildlife in the U.S.. When occurring in a grain plot for mourning doves, they should be considered beneficial. Foxtail millet (*S. italica*) is often planted for doves. Foxtailgrasses are easily controlled postemergence in perennial forage plots with a grass-selective herbicide, and preemergence with s-metolachlor, pendimethalin, trifluralin, and imazethapyr in various grain crops. Foxtailgrasses are not a selected forage by white-tailed deer.

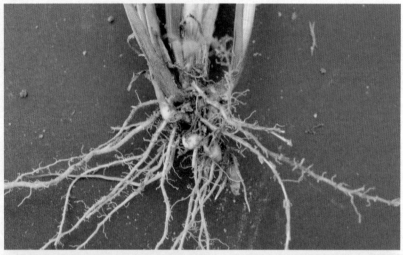

Photos: Knotroot foxtail

Johnsongrass
Sorghum halepense

Description and ecology
Upright, robust, perennial, warm-season grass. Flowering stem is large, and may reach 9+ feet tall. Produces large rhizomes that may produce additional stems. Leaves are long and have a prominent white midvein. Seedhead is a large open panicle, often purplish in color.

Origin and distribution
Nonnative (southern Eurasia); throughout U.S.

Considerations as a weed and for wildlife
Johnsongrass is commonly found in all types of food plots. Its density and height allow it to outcompete many types of food plot plantings and can reduce yields of grain crops drastically. Johnsongrass can be controlled preemergence with imazethapyr and pendimethalin, and postemergence with imazethapyr, a grass-selective herbicide, or with glyphosate in Roundup-Ready crops. Relatively sparse johnsongrass is not a problem, but don't let it spread. In fact, the structure of johnsongrass is not bad for bobwhites and wild turkey poults unless it becomes dense. Johnsongrass seed is eaten by many bird species. The problem is johnsongrass is quite invasive and if not controlled can become dense and widespread fairly quickly. Johnsongrass is not eaten by white-tailed deer.

Broadleaf signalgrass
Urochloa platyphylla

Description and ecology
Annual warm-season grass that is often decumbent (bent over with stem and leaves near or flat on ground with tips pointing upward). Stem is bent at nodes and may root at the lower nodes. Leaf sheaths are hairy. Leaves often appear as flags pointing straight off stem.

Origin and distribution
Native; southeastern U.S.

Considerations as a weed and for wildlife
Broadleaf signalgrass can become widespread and dense, outcompeting planted crops. It is common in conventionally prepared warm-season food plots. It can be controlled preemergence with S-metolachlor, pendimethalin, trifluralin, and imazethapyr, and postemergence with grass-selective herbicides and imazethapyr. It has no cover value for wildlife. Mourning dove eat the seed. It is not eaten by white-tailed deer or eastern cottontail.

Wild buckwheat
Climbing buckwheat
Black bindweed
Polygonum convolvulus (or *Fallopia convolvulus*)

Description and ecology
Annual vining forb that creeps along the ground or climbs other vegetation. Leaves alternate and arrowhead-shaped. Flowers are small, greenish-white, and appear at ends of stems or at axils of leaves and stems. Flowering begins in late June. A thin membrane (ocrea) surrounds stem at each node. Stems are reddish-green and smooth. Seed may remain viable in soil for 5 years. Wild buckwheat is similar in appearance to bindweed (*Convovulus* spp.) and morningglories (*Ipomoea* spp.). However, bindweeds and morningglories are longer vines, and their flowers are much larger and tubular-shaped.

Origin and distribution
Nonnative (Europe); throughout U.S.

Considerations as a weed and for wildlife
Wild buckwheat is a serious, problematic weed in many crops. It can be controlled with glyphosate at relatively strong rates if sprayed when young. Wild buckwheat is resistant to 2,4-D and 2,4-DB. Preplant or preemergence applications of trifluralin, pendimethalin, and S-metolachlor provide some, but incomplete control. Imazethapyr, imazaquin, and imazamox applied pre or postemergence when young provide better control. Thifensulfuron-methyl and tribenuron-methyl applied postemergence in wheat and oats provides control. Clopyralid can be used to control wild buckwheat in small grains and brassica plots. Seed of wild buckwheat may be eaten by a number of birds, including mourning dove, northern bobwhite, greater prairie chicken, ring-necked pheasant, horned lark, savanna sparrow, and lapland longspur.

Top two photos: Pennsylvania smartweed
Bottom photo: Pale smartweed

Pale smartweed
Polygonum lapathifolium (or *Persicaria lapathifolia*)

Pennsylvania smartweed
P. pensylvanicum (or *Persicaria pensylvanica*)

Lady's thumb
P. persicaria (or *Persicaria maculosa*)

Description and ecology
Erect or spreading annual forbs. Leaves alternate, pointed at tip. Pennsylvania smartweed usually with purple marks in center of leaf. Lady's thumb leaves usually, but not always, have what appears as a black smudge on upper portion of leaf. Flowers are pinkish or white and in dense spike-like clusters at the end of stems. Flowering racemes of lady's thumb relatively stout and pointing upward. Racemes of other species more slender and nodding. Stems may be green or reddish. Stems of all species within Polygonaceae are jointed and appear swollen at the nodes, which are covered with a clear membranous sheath called an ocrea. There are many species of smartweeds and knotweeds. Some occur in uplands, some in wetlands or moist areas, and some in both. Pennsylvania smartweed is most commonly found in low-lying areas, but it also may occur in upland fields. Pale smartweed, lady's thumb, and swamp smartweed (*Polygonum amphibium*) are perennial and more likely to occur in moist areas.

Origin and distribution
Pale, Pennsylvania, and swamp smartweed native to eastern U.S., but found throughout U.S. Lady's thumb nonnative (Europe).

Considerations as a weed and for wildlife
Smartweed and knotweed seeds are relished by dabbling ducks, however, smartweeds can smother crops planted for ducks. Several other birds eat the seed of Pennsylvania smartweed, including northern bobwhite, Wilson's snipe,

northern cardinal, several sparrows (fox, swamp, Vesper, savannah, white-crowned, white-throated, song, and seaside), bobolink, horned lark, and juncos. Smartweeds are not a selected forage for deer or rabbits. A host of bees, butterflies, and moths feed on the nectar of the flowers. Smartweeds can be difficult to control, especially in perennial forage plots. Imazethapyr and 2,4-DB may provide some control if sprayed when smartweeds are young. Glyphosate may provide control in Roundup-Ready crops, If smartweeds are sprayed when young. Smartweeds can be killed with forb-selective postemergence herbicides in grass crops or after annual cool-season forages have died in summer.

Both photos: Lady's thumb

Mile-a-minute vine
Asiatic tearthumb
Polygonum perfoliatum (or *Persicaria perfoliata*)

Description and ecology
Annual vine with light green, triangular-shaped leaves. Stem is reddish-green. Short, sharp, downward-curving spines occur along stem and main leaf veins. Fruit is cluster of blue berries. Germination begins in March and continues through April. Dormant seed may remain viable in the seedbank for at least 4 years. Forms a dense canopy over other vegetation.

Origin and distribution
Nonnative (Asia); New England to PA and OH, south to NC

Considerations as a weed and for wildlife
Mile-a-minute vine is not often a problem in food plots, but more often found around field edges, in old-field communities, recently logged, and other disturbed sites. It spreads quickly and should be controlled as soon as it is discovered. It can be controlled with glyphosate, imazapyr, triclopyr, sulfometuron-methyl, and hexazinone. Growing-season fire also may help control this annual vine. Fruits may be eaten, and thus spread, by various songbirds and small mammals. Foliage is not eaten by mammals. Mile-a-minute vine is extremely invasive and should be treated aggressively.

Red sorrel
Sheepsorrel
Dock
Sourgrass
Rumex acetosella

Description and ecology
Warm-season perennial forb. Leaves are primarily basal and arrowhead-shaped with two narrow and spreading basal leaves. Flowering stem is square, reddish, and may grow to 2 feet tall. Flowers May – September.

Origin and distribution
Nonnative (Eurasia); occurs throughout U.S.

Considerations as a weed and for wildlife
Rumex can become widespread and dense across fields if not controlled. However, it is often associated with poor-quality soils. It can be controlled with clopyralid in various food plot applications (see Appendix 2), with dicamba or thifensulfuron-methyl + tribenuron-methyl in various grass crops, or with glyphosate via spot-spraying in food plots or old-fields. Rumex seed is eaten by sparingly by wild turkey, ruffed grouse, northern bobwhite, and several songbirds. Rumex is not a selected forage by white-tailed deer.

Top two photos: *R. acetosella*
Bottom left photo: Curly dock for comparison
Bottom right: Broadleaf dock for comparison (see next page for plant descriptions)

Curly dock
Rumex crispus

Broadleaf dock
Rumex obtusifolius

Description and ecology
Perennial warm-season forbs that arise from a basal rosette. Leaves are shiny green with wavy margins and will become red-purplish through the season. Broadleaf dock is similar to curly dock except basal leaves of broadleaf dock have long smooth stems and leaf margins are only slightly wavy. Flowering stem is smooth and stout and may grow to 4 feet tall on some sites. Taproot is fleshy and yellow. Flowers June – August. Seeds are green and turn brown when mature.

Origin and distribution
Nonnative (Eurasia); occurs throughout U.S.

Considerations as a weed and for wildlife
Curly and broadleaf dock are some of the worst weeds you can have in food plots, especially perennial forage plots. They are invasive, outcompete planted forages, and do not provide food or cover for wildlife. They should be sprayed when very young (seedlings about 3 – 4 inches tall) in perennial forage plots with a strong rate of imazethapyr or 2,4-DB. They are most easily controlled in grass crops or in annual cool-season plots after the forages have died with glyphosate or with forb-selective herbicides, such as dicamba, clopyralid, or triclopyr. You should aggressively try to eradicate these plants.

Top: Curly dock
Middle: Broadleaf dock
Bottom: Curly dock in chicory

Common purslane
Portulaca
Little hogweed
Portulaca oleracea

Description and ecology
Prostrate, mat-forming, annual, warm-season forb. Stems are branched and spread along ground with stem ends ascending slightly. Stem is reddish-purple, smooth, and succulent. Somewhat similar to a spurge, but stem does not contain a milky latex like the spurges. Leaves are alternate, but may appear opposite. Flowers are small and yellow with 5 petals, appear July – September, and occur singly at base of leaves.

Origin and distribution
Nonnative (Europe); throughout U.S.

Considerations as a weed and for wildlife
Common purslane is widespread and frequently found in food plots, especially those established via conventional tillage. It is a prolific seed producer and several birds eat the seed, including mourning dove, ground dove, chipping sparrow, Vesper's sparrow, lapland longspur, and lark bunting. The foliage is eaten by cottontails and jackrabbits, and it is a moderate-preference forage of white-tailed deer. Common purslane can be competitive with some planted crops. It can be controlled preemergence with pendimethalin and postemergence with selective herbicides, such as dicamba and carfentrazone-ethyl.

Scarlet pimpernel
Poor-man's weatherglass
Red chickweed
Anagallis arvensis

Description and ecology
Low-growing, ascending, annual forb that is multi-branched and most commonly found in areas with relatively sandy soils. Stem is squarish and smooth. Leaves are somewhat egg-shaped, smooth, and opposite (or sometimes in whorls of 3) with small purplish dots on underside. Flowering occurs spring through late summer. Flowers are orangish with 5 petals and occur singly at the end of slender stems originating at leaf axil (where leaf meets stem). Flowers open when sun is shining, but close in the evening or when there is considerable cloud cover. Scarlet pimpernel appears much like common chickweed, hence the alternate common name, red chickweed. However, it should not be confused with chickweed because chickweed has round stems and pimpernel flowers are very different from chickweed flowers.

Origin and distribution
Nonnative (Europe); throughout U.S.

Considerations as a weed and for wildlife
Scarlet pimpernel can be widespread and may compete with low-growing forages, such as clovers. It is not eaten by wildlife, most likely because of saponin glycosides, which give it a bitter taste. It can be controlled with broadleaf-selective herbicides, imazethapyr, or with glyphosate. It tolerates close mowing, so selective herbicide applications may be necessary to remove it from perennial forage plantings.

Buttercup
Ranunculus abortivus (Smallflower buttercup)
R. arvensis (Corn buttercup)
R. bulbosus (Bulbous buttercup)
R. repens (Creeping buttercup)
R. sardous (Hairy buttercup)

Description and ecology
Buttercups are upright or spreading cool-season forbs. There are many species of buttercup. Most are annual, but bulbous buttercup is perennial and common across most of U.S. Reproduction is from seeds. Leaves are from a basal rosette and usually 3-parted. Flowers are yellow and solitary at the end of the flowering stem.

Origin and distribution
Nonnative (Europe); these 5 species occur at least in portions of eastern and western U.S.

Considerations as a weed and for wildlife
Buttercups are often found in cool-season forage plots. Wild turkey and white-tailed deer may eat the leaves in spring. They can become widespread, aggressive, and outcompete planted forages. Buttercups can be controlled with postemergence applications of imazethapyr in clover/chicory/alfalfa plots if sprayed when young, or with thifensulfuron-methyl (Harmony Extra), dicamba, or 2,4-D in wheat and oat plots. Imazethapyr and pendimethalin applied to established clover and alfalfa before buttercups germinate in fall and winter will help control spread of buttercups. You should aggressively try to eradicate buttercups.

Top two photos: Hairy buttercup
Second from bottom photo: Creeping buttercup
Bottom photo: Buttercup grazed by deer

435

Harvest-lice
Southern agrimony
Swamp agrimony
Agrimonia parviflora

Description and ecology
Perennial, warm-season forb. Leaves are alternate and compound. Leaflets are toothed with prominent venation. Pairs of small (secondary) leaflets are located between large (primary) leaflets along the leaf stem. The main stem is very hairy, and may grow to 5 feet tall. Clusters of yellow flowers occur at the end of the main stem in mid- to late summer. Fruits are about ¼-inch in diameter with numerous hooked prickles, hence common name harvest-lice. Immature fruits are green; mature fruits are brown. Generally found in relatively moist areas with full sunlight.

Origin and distribution
Native; throughout eastern U.S.

Considerations as a weed and for wildlife
Harvest-lice is rarely a problem in food plots, but may occur in perennial forage plots. It is relatively common in old-fields. It provides some structural value for game bird broods and ground-feeding songbirds. The foliage is commonly eaten by white-tailed deer, but it is not highly selected. If considered problematic, it can be controlled by spot-spraying glyphosate, triclopyr, 2,4-D, or dicamba.

Mock-strawberry
Indian mock-strawberry
Duchesnea indica

Description and ecology
Trailing, perennial, cool-season forb. Lacks a stem, but spreads by stolons (aboveground runners). Leaves are alternate and compound with 3 leaflets that are coarsely toothed. Flowers are yellow with 5 petals and appear April – June. Fruit is red and edible, but dry and tasteless.

Origin and distribution
Nonnative (Asia); throughout eastern U.S., as well as WA, OR, and CA.

Considerations as a weed and for wildlife
Mock-strawberry, like wild strawberry and common cinquefoil, often occurs in perennial forage plots at least a few years old. It is not terribly competitive, but seems to take advantage of space left when planted forages begin to thin, and if more aggressive weeds are not present. Taste matters! Whereas wild strawberry is eaten by many wildlife species, mock-strawberry is not. It has little if any wildlife value. Note: the species, indica, means from India, hence the common name Indian mock-strawberry. It is called a "mock" strawberry because it is not a true strawberry (*Fragaria*). The nutlets are on the surface of the fruit, whereas strawberry nutlets are inside the fruit. Mock strawberry can be controlled in clover and alfalfa plots postemergence with 2,4-DB if sprayed when young. It can be controlled with forb-selective herbicides, such as dicamba and thifensulfuron methyl, in various grass plots or in fallow areas.

Photo by: T. Jolly

Wild strawberry
Fragaria virginiana

Description and ecology
Spreading, perennial, cool-season forb. Stems are short and not apparent. Spreads by stolons (aboveground runners). Leaves are palmately compound with 3 coarsely toothed leaflets 1 – 4 inches long. Flowers are white with 5 petals and appear April – May. Flowering stem is separate from leaf stem. The strawberry is sweet and edible, but smaller than cultivated strawberries.

Origin and distribution
Native; throughout eastern U.S.

Considerations as a weed and for wildlife
Wild strawberry sometimes occurs in perennial plots that are 3 or more years old. They are not so much competitive with planted forages, such as clovers, but simply seem to occur where planted forages have thinned. The foliage and fruit of wild strawberry is relished by white-tailed deer, mule deer, elk, cottontails, groundhog, wild turkey, ruffed grouse, dusky grouse, sooty grouse, and greater prairie-chicken. The fruit is eaten by many birds, including gray catbird, American crow, swamp and white-throated sparrow, brown thrasher, American robin, eastern towhee, veery, and cedar waxwing, as well as opossum, fox squirrel, red squirrel, eastern chipmunk striped skunk, and likely many others. Wild strawberry is an important plant where it occurs for many wildlife species.

Sulfur cinquefoil
Potentilla recta

Description and ecology
Upright perennial forb that grows 1 – 2.5 feet tall. Leaves are alternate and palmately divided with 5 (sometimes 7) leaflets. Leaf margins are prolonged toothed. Usually 1, but up to 3 flowering stems per plant, with fine hairs. Flowers are sulfur-yellow and appear May through July.

Origin and distribution
Nonnative (Europe); naturalized throughout most of U.S.

Considerations as a weed and for wildlife
Sulfur cinquefoil may occur in perennial food plots and compete with planted clovers. It can be controlled when young with 2,4-DB. Sulfur cinquefoil is a low-preference forage for white-tailed deer and rabbits.

Common cinquefoil
Old-field cinquefoil
Potentilla simplex

Description and ecology
Perennial forb that grows low to the ground and spreads from wiry stolons. Leaves are alternate with 5 leaflets. Leaf margins are toothed. Appears in late winter/early spring. Flowers are solitary and have 5 yellow petals that are on the end of long slender stalks, spring through summer. There are many species of cinquefoils in the U.S. Mock-strawberry (*Duchesnia indica*) and wild strawberry (*Fragaria virginiana*) sometimes are confused with common cinquefoil, but they have 3 leaflets instead of 5 and wild strawberry has white flowers, whereas mock-strawberry and cinquefoil have yellow flowers.

Origin and distribution
Native; throughout eastern U.S.

Considerations as a weed and for wildlife
Cinquefoil can creep into perennial food plots and compete with planted clovers. It can be controlled when young with 2,4-DB. Common cinquefoil is eaten by white-tailed deer, rabbits, and groundhogs, but it is low-preference. Ruffed grouse in the southern and central Appalachians commonly eat the leaves in winter.

Top photo by: J. Neal

Poorjoe
Rough buttonweed
Diodia teres

Virginia buttonweed
D. virginiana

Description and ecology
Poorjoe is an annual, warm-season, spreading or upright forb especially common on sandy soils in the South. Virginia buttonweed is an annual (or perennial in warmer climates) warm-season, usually spreading (sometimes ascending) forb more often found on moist sites. Leaves of both are opposite, simple (without lobes), sessile (without a stem), and linear- to lance-shaped with a prominent midvein. Poorjoe leaves are considerably narrower (usually ¼-½ inch) than leaves of VA buttonweed (½-1 inch). Stems may be single or branched, and nearly round. Poorjoe stems are quite hairly; VA buttonweed less hairy. Flowers have 4 petals. Poorjoe flowers usually are pinkish-purple, whereas those of VA buttonweed are white.

Origin and distribution
Native; poorjoe is found throughout most of eastern U.S., typically on dry sites; Virginia buttonweed is found throughout the U.S., except the north-central U.S. from Wisconsin west to Montana and Wyoming.

Considerations as a weed and for wildlife
Poorjoe and buttonweed can become problematic in forage plots where they may spread over and outcompete planted forages. They can be controlled with a variety of broadleaf-selective herbicides as well as imazethapyr. Both poorjoe and Virginia buttonweed are low-preference deer forages. The seed of each are eaten sparingly by mourning dove, bobwhite, wild turkey, and various songbirds.

Top two photos: Poorjoe
Bottom two photos: Virginia buttonweed

441

Catchweed bedstraw
Cleavers
Galium aparine

Description and ecology
Spreading annual forb. Stem is square, winged, and covered with fine bristles that give it a rough feel. Leaves are whorled (occur all the way around stem). Plant may be prostrate or erect. Often mat-forming. Flowers March – May. Catchweed bedstraw may be superficially similar to carpetweed, but stem of carpetweed is round and smooth.

Origin and distribution
Some sources say catchweed bedstraw is native to Europe; others say native to Europe and North America; found throughout U.S.

Considerations as a weed and for wildlife
Bedstraw does not usually outcompete planted crops, but it can become widespread and relatively dense. Although seldom necessary, it can be controlled with several preemergence herbicides for various warm-season grain plantings. Various postemergence herbicides, such as glyphosate, carfentrazone-ethyl, and sulfosulfuron, are effective. Dicamba and 2,4-D provide only partial control. Bedstraw is a medium- to high-preference deer forage and may be eaten by eastern cottontail and groundhog.

Florida pusley
Richardia scabra

Description and ecology
Weakly erect or spreading annual, warm-season forb. Stems are hairy. Leaves are opposite with relatively prominent venation. Flowers May – November. Flowers are white with 6 parts.

Origin and distribution
Native to tropical America; naturalized northward along the Atlantic Coastal Plain

Considerations as a weed and for wildlife
Florida pusley is commonly found in warm-season food plots, particularly those managed with conventional tillage (thereby encouraged by disking in early successional areas). It does not usually grow dense enough to compete with planted forages, but may. Florida pusley is a valuable plant for wildlife. It is a high-preference forage for white-tailed deer and rabbits, and the seeds are eaten by northern bobwhite and a variety of songbirds. Pusley can be controlled with various preemergence herbicides, such as S-metolachlor and pendimethalin, and postemergence with selective herbicides, such as dicamba or carfentrazone-ethyl. However, there is rarely a need to control Florida pusley. It is encouraged with annual or biennial disking during winter.

Bottom photo: Florida pusley and common ragweed growing along the edge of a clover/chicory forage food plot provide outstanding forage, seed, and cover value for numerous wildlife species.

443

Balloonvine
Cardiospermum halicacabum

Description and ecology
Climbing or trailing annual vine found most often in moist areas. Stems grow about 6 feet long with tendrils that are green while growing (bottom picture shows plant soon after first frost). Leaves are alternate, divided or dissected, and coarsely lobed. Flower is white with 4 petals. Seeds are black and encased in a membranous, fleshy, capsule.

Origin and distribution
Native; throughout eastern U.S. from southern NY to KS and south.

Considerations as a weed and for wildlife
Balloonvine can spread across forage plots and shade-out planted forages, especially perennial clovers. It can be controlled with imazethapyr or 2,4-DB in perennial clover and alfalfa plots. Forb-selective herbicides will control balloonvine. Various butterflies may feed on the nectar, but there is no reported use of balloonvine by birds or mammals.

Smooth purple gerardia
Purple false foxglove
Agalinis purpurea

Slender gerardia
A. tenuifolia

Description and ecology
Erect, annual, warm-season forbs typically 2 – 4 feet tall. Stem is slender and smooth or slightly hairy. Leaves are opposite, narrow, linear, and only about ½ to 1 inch long; often turn purplish late in season. *A. purpurea* leaves are in tufts, whereas those of *A. fasciculata* occur singly. Flowers appear August – November and are pink-purplish and tubular with 5 spreading lobes.

Origin and distribution
Native; thoughout eastern U.S. and into Great Plains

Considerations as a weed and for wildlife
There are several species of *Agalinis* and they can be difficult to distinguish. False foxglove often appears in fallow plots where it may be parasitic on the roots of native grasses. It can provide fair cover for northern bobwhite and wild turkey broods as well as ground-feeding songbirds. It is a moderate-preference forage for white-tailed deer. The flowers attract bees, especially bumblebees. It rarely presents a competition problem with planted forages, but can be controlled with imazethapyr, glyphosate, and forb-selective herbicides if needed.

Top two photos: Smooth purple gerardia
Bottom photo: Slender gerardia

Old-field toadflax
Blue toadflax
Nuttallanthus canadensis or *Linaria canadensis*

Description and ecology
Slender, erect, annual or biennial forb about 12–18 inches tall. Solitary smooth stem arises from basal rosette. Leaves are alternate and occur primarily on lower portion of stem. Upper leaves are relatively small and linear-shaped. Flowers are irregular in shape with one petal (spur) curving downward; pale purple and occur on upper portion of stem. Typically found on relatively dry, sandy sites, especially newly cleared areas, but may occur on more mesic soils. Spreads by seed.

Origin and distribution
Native to eastern North America; found throughout eastern U.S. and naturalized along Pacific coast.

Considerations as a weed and for wildlife
Old-field toadflax may be widespread in perennial forage plots that are relatively thin in cover and in fallow fields, but it is rarely competitive with other plants. White-tailed deer and eastern cottontail eat the plant sparingly. Bumblebees and other long-tongued bees frequent the plant for nectar. Old-field toadflax is controlled if the plot is sprayed with broadleaf-selective herbicides to control other plants, or by mowing when it begins to flower. However, old-field toadflax is not a problem.

Moth mullein
Verbascum blattaria

Description and ecology
Warm-season biennial forb that forms a basal rosette in first
year, then produces stout central stem with leaves and flowers.
Leaves on the flowering stem are alternate, toothed, and up
to 5 inches long and 2 inches wide near bottom of stem, and
smaller up the stem. Lower leaves clasp the stem. Leaves
are lanceolate (wider at base, pointed at tip) and somewhat
crinkled appearing. Flowers are yellow or white, about 1 inch
wide, with 5 petals and a purple center. The fruit is a round
capsule that contains many tiny seeds that may remain
dormant in the soil for many years. Moth mullein may be found
on recently disturbed sites and tolerates poor soil conditions.

Origin and distribution
Nonnative (Europe); throughout U.S.

Considerations as a weed and for wildlife
Moth mullein can be competitive in perennial food plots, and
may be found in fallow fields and other early successional
plant communities. Bumblebees often pollinate the flowers.
Otherwise, it has no wildlife value. As a biennial, moth mullein
can be suppressed by mowing before capsules form or disking.
It can be controlled with glyphosate, triclopyr, aminopyralid,
and metsulfuron-methyl.

Common mullein
Great mullein
Verbascum thapsus

Description and ecology
Upright, biennial, warm-season forb. Plant very fuzzy hairy. Basal rosette of large leaves forms first year. Tall flowering stem produced second year. Stem is stout and unbranched. Leaves are densely hairy and soft to the touch, almost as toilet tissue, but do not be fooled—the fuzzy soft feel actually is of minutely small bristles that may irritate the skin. Flowers are yellow and produced on a terminal spike.

Origin and distribution
Nonnative (Eurasia); throughout U.S.

Considerations as a weed and for wildlife
Common mullein is often found in warm-season plots planted via conventional tillage and sometimes in perennial forage plots. Mullein is fairly easily controlled with a number of preemergence and broadleaf-selective postemergence herbicides, as well as glyphosate in Roundup-Ready crops. If mullein gets tall in perennial forage crops, it should be spot-sprayed, mowed, or cut down before it produces seed. Various bees, especially bumblebees, visit mullein flowers for nectar. Otherwise, common mullein has no value for wildlife.

Speedwell
Field speedwell
Veronica agrestis, V. polita

Corn speedwell
V. arvensis

Ivyleaf speedwell
V. hederifolia

Bird's eye speedwell
Persian speedwell
V. persica

Description and ecology
Cool-season forbs. Some are annual; some are perennial. Lower leaves are often opposite and upper leaves alternate. Plant may be erect or ascending. Most species will flower spring through early to late summer. Some of the speedwells may look similar to ground ivy (see page 368), but ground ivy has a square stem, whereas speedwells have round stems. Refer to *Weeds of the Northeast* for comparison of species.

Origin and distribution
Nonnative (Asia); various species found throughout U.S.

Considerations as a weed and for wildlife
Speedwells can be competitive in cool-season plots. Reduced competition may be achieved if they are sprayed when young with imazethapyr or 2,4-DB. Glyphosate and carfentrazone-ethyl are effective in the appropriate crops. Speedwells can be controlled in established perennial clover plots and alfalfa by applying imazethapyr in fall before speedwells germinate. If not controlled, speedwells can become very dense and outcompete planted forages. They often grow in association with henbit, purple deadnettle, and chickweeds. They are low-preference forages for white-tailed deer and may be eaten sparingly by rabbits.

From top: Field speedwell, ivyleaf speedwell, bird's eye speedwell, corn speedwell

Jimsonweed
Datura stramonium

Description and ecology
Erect, annual, warm-season forb with branching stems. Leaves are alternate and coarsely toothed. Flowers are white or purple and funnel-shaped. Seed are produced in an egg-shaped pod that is covered in spiny prickles. Plant has strong unpleasant odor if crushed. Jimsonweed is toxic to livestock and humans.

Origin and distribution
Native; throughout eastern U.S. and west of the Rocky Mountains.

Considerations as a weed and for wildlife
Jimsonweed is common in food plots planted via conventional tillage. Jimsonweed can be controlled with several preemergence herbicides, such as imazethapyr, S-metolachlor, atrazine, and postemergence with imazethapyr, dicamba, carfentrazone-ethyl, bentazon, 2,4-D, and glyphosate. If jimsonweed gets large and near flowering, mowing is the best control method. Jimsonweed provides no food value to wildlife. However, it can provide good cover for some species, such as rabbits, wild turkey, and northern bobwhite, with an open structure under a canopy of cover.

Second from bottom photo: Showing seed pod with spiny prickles. Bottom photo: Structure under jimsonweed and pigweed is good for bobwhite, wild turkey poults, and rabbits.

Cutleaf groundcherry
Physalis angulata

Clammy groundcherry
P. heterophylla

Smooth groundcherry
P. subglabrata

Description and ecology
Upright warm-season forbs. Cutleaf groundcherry is annual. Clammy and smooth groundcherry are perennial. Stems are upright, branched, and 3 – 4 feet tall. Stem of clammy groundcherry is pubescent, whereas stem of cutleaf and smooth groundcherry is smooth. Leaves are alternate, pointed, and smooth or coarsely toothed on smooth groundcherry. Flowers are yellow with purple centers. The fruit is a papery calyx (somewhat resembling a lantern) and encloses the berries inside, which are usually orange or tan when mature. Calyx of cutleaf is often purple-veined.

Origin and distribution
Native; cutleaf groundcherry across southern U.S.; clammy and smooth groundcherry from Rocky Mountains east.

Considerations as a weed and for wildlife
Groundcherry may be found in warm-season food plots, especially those prepared via conventional tillage. It can become widespread and relatively dense and can compete with planted crops. Groundcherry can be controlled preemergence with S-metolachlor, imazethapyr, and postemergence with forb-selective herbicides or glyphosate in Roundup-Ready crops. Groundcherry fruits may be eaten sparingly by wild turkey, ring-necked pheasant, northern bobwhite, sharp-tailed grouse, striped skunk, opossum, white-footed mouse, and eastern box turtle. The foliage is a low-preference forage for white-tailed deer, perhaps because of solanum, which is toxic.

Top photo: Cutleaf groundcherry;
Middle photo: Smooth groundcherry,
Bottom photo: Clammy groundcherry

Nightshades

Solanum americanum (American black nightshade)
S. nigrum (European black nightshade)
S. ptycanthum (Eastern or West Indian black nightshade)
S. physalifolium (Hairy or hoe black nightshade)

Description and ecology

There are several species of nightshades. These 4, along with horsenettle, are most common across the U.S. These 4 nightshades are warm-season annual or short-lived perennial forbs. Stems are upright or spreading, growing to about 3 feet tall. Leaves are alternate. Leaf margins are smooth or with shallow, rounded teeth. Flowers are star-shaped with white petals and yellow anthers. Fruits are purple or black (green in hairy nightshade) when mature and in clusters.

Origin and distribution

American and eastern black nightshade are native to the U.S. American black nightshade is common across portions of southern, midwestern, and western U.S. Eastern black nightshade is common from Rocky Mountains east. Hairy black nightshade is nonnative (South America) and occurs throughout U.S. except portions of southeastern U.S. European black nightshade is nonnative (Eurasia) and occurs in scattered locations in eastern and western U.S.

Considerations as a weed and for wildlife

Black nightshades are sometimes found in warm-season food plots, especially those prepared via conventional tillage. They can become widespread, but do not usually become so dense that they compete with planted crops. Black nightshades can be controlled preemergence with S-metolachlor, imazethapyr, and postemergence with forb-selective herbicides or glyphosate in Roundup-Ready crops. All *Solanum* species contain alkaloids, which may be toxic at various levels in various plant parts to many species of animals. The mature fruit of black nightshades are eaten by several bird species, including northern cardinal, northern mockingbird, gray catbird, and wild turkey. The foliage of eastern black nightshade is a moderately selected forage by white-tailed deer.

Photos: Eastern black nightshade

Horsenettle
Solanum carolinense

Description and ecology
Perennial forb that may be erect or spreading. Stems have sharp prickles. Leaves are alternate and have sharp prickles on margin as well as along midvein on underside of leaf. Flowers have white or pale violet petals with yellow anthers. Fruit is a round, green berry that turns yellow upon maturity. Spreads by seed and rhizomes.

Origin and distribution
Native to southeastern U.S.; has spread into northeastern and north-central U.S. as well as several states west of the Rocky Mountains.

Considerations as a weed and for wildlife
Horsenettle can be very problematic in food plots, especially perennial forage plots. It provides little wildlife value. Horsenettle is very resilient to many selective herbicides. It can be spot-sprayed with glyphosate or aminopyralid, but if horsenettle becomes dense in a perennial clover/chicory plot, it is usually best to spray the entire plot with strong rate of glyphosate or triclopyr and rotate the plot to a grass crop so the horsenettle can be selectively sprayed for a couple years and eradicate it from the field. Imazethapyr and S-metolachlor will provide some control when applied preemergence in various plots. Several birds, including wild turkey, northern bobwhite, ring-necked pheasant, northern mockingbird, northern cardinal, gray catbird, eastern meadowlark, fox sparrow, white-crowned sparrow, and song sparrow, may eat the fruit of horsenettle. It is not eaten by white-tailed deer.

Middle photo: Horsenettle in annual clover plot;
Bottom photo: Horsenettle and hedge bindweed (arrowhead-shaped leaf)

Beaked cornsalad
Valerianella radiata

Description and ecology
Cool-season annual forb that forms a rosette in fall, then bolts to flower in spring. May reach 18 inches in height. Stem relatively stout, 4-sided, and often branched in two about midway up stem. Leaves on stem are opposite, 1 – 3 inches long, clasp stem, and have coarse teeth near base of leaf. Clusters of small white flowers at ends of stems. Relatively common in disturbed areas and fallow fields.

Origin and distribution
Native; PA to KS and south

Considerations as a weed and for wildlife
Beaked cornsalad can become relatively dense in forage food plots in spring. It is a low-preference forage of white-tailed deer and eastern cottontail. It has little cover value as it is declining when broods and other young appear in late spring. Various bees obtain nectar from flowers. Beaked cornsalad can be controlled with imazethapyr, 2,4-DB when young, dicamba, and metsulfuron-methyl.

454

Purpletop vervain
Verbena bonariensis

Brazilian vervain
Verbena brasiliensis

Rigid vervain
V. rigida

Hoary vervain
Verbena stricta

Description and ecology
Upright perennial forbs. Stems are green, relatively rough, square, and often branched about midway up stem. Leaves are opposite, relatively narrow, and widely spaced with toothed margins. Leaves and stem of hoary vervain exceptionally hairy, and leaves more oval in shape. Flowers are purple and on spikes at the tip of slender stalks, except purpletop, which has rounded cluster of flowers. These vervains tolerate relatively poor soil conditions.

Origin and distribution
Purpletop, Brazilian, and rigid vervain nonnative (South America) and found throughout southeastern U.S.; hoary vervain native and found throughout most of U.S.

Considerations as a weed and for wildlife
These vervains may be found in perennial forage plots and old-field sites. They can become dense, but rarely outcompete planted forages. These vervains can be controlled with several preemergence herbicides in warm-season crops and with imazethapyr or 2,4-DB if sprayed when young in perennial forage plots. Brazilian vervain is a moderately preferred forage plant for white-tailed deer. Rigid, hoary, and purpletop vervain are low-preference forages for deer. Seeds of each may be eaten by some songbirds, including northern cardinal and various sparrows.

Top photo: Hoary vervain; Second from top: Brazilian vervain; Third from top: rigid vervain; Bottom photo: purpletop vervain

455

White vervain
Nettle-leaf vervain
Verbena urticifolia

Description and ecology

Upright perennial forb (may be annual in some areas). Reproduces by seed. The stem is green or chalky-green (may turn red at maturity), somewhat rough, square, and sparsely hairy. Leaves are opposite, toothed, often somewhat shiny on top with a wrinkled appearance, and may be slightly hairy. Flowers are in a panicle of floral spikes. Flowers are small and white and are in bloom from mid-summer through early fall.

Origin and distribution

Native; throughout much of Midwest and eastern U.S.

Considerations as a weed and for wildlife

White vervain may occur in perennial food plots, but is not usually dense. It is not a selected forage by white-tailed deer. Several bees use the plant for nectar, and several species of sparrows and juncos may eat the small seed. Relatively sparse white vervain can enhance structure of clover plot for wild turkey broods. It can be controlled preemergence with imazethapyr, and postemergence with dicamba, 2,4D, triclopyr, or glyphosate.

European field pansy
Field violet
Viola arvensis

Description and ecology
Erect or spreading annual cool-season forb. Plant arises from a basal rosette. Stems may be branched or unbranched. Leaves are alternate with rounded toothed margins. Flowers have 5 petals and are pale yellow-white with or without purple.

Origin and distribution
Nonnative (Europe); throughout most of eastern U.S., central Great Plains, and portions of western U.S.

Considerations as a weed and for wildlife
European field pansy can be problematic in warm-season plots established via conventional tillage and may occur in perennial plots. It can be tough to control. Imazethapyr may provide some control preemergence. Postemergence applications are usually most effective in fall. Metsulfuron-methyl and triclopyr have performed better than dicamba and glyphosate, but repeat applications may be necessary. There is no reported use of European field pansy by wildlife.

Field pansy
Johnny-jump-up
Viola bicolor or *V. rafinesqueii*

Description and ecology
Erect, annual, cool-season forb, usually about
3 – 5 inches tall. Stem is slender and branched.
Leaves are alternate and linear, but wider toward
tip than base. Flowers appear February – May
and generally are whitish with purple veins and a
yellow center; the 5 petals sometimes are tinged
with pink toward tip.

Origin and distribution
Native; throughout most of eastern U.S. except
upper New England and upper Midwest.

Considerations as a weed and for wildlife
Field pansy is commonly found growing with
speedwells, chickweeds, and geraniums. It
is common in perennial forage plots and in
fallow plots. It rarely competes with planted
forages, but may be widespread. It can be
controlled if necessary with imazethapyr or
forb-selective herbicides, such 2,4-DB, dicamba,
and thifensulfuron-methyl+tribenuron-methyl.
Field pansy may be eaten sparingly by rabbits,
groundhog, white-tailed deer, wild turkey, and
ruffed grouse.

Top and bottom photos show field pansy in clover food plot.

Common blue violet
Viola sororia

Description and ecology
Perennial cool-season forb that occurs in colonies. Lacks stem. Leaves are basal and arise from the crown of a stout rhizome (underground stem). Leaves are generally somewhat heart-shaped. Flowers are blue-violet with 5 petals and appear April – June.

Origin and distribution
Native; throughout eastern U.S.

Considerations as a weed and for wildlife
Common blue violet often occurs in older perennial forage food plots, and particularly those with partial shade. It is a moderate- to high-preference forage for white-tailed deer and eastern cottontails The small black seed of common blue violet are eaten by several birds, including mourning dove, ground dove, ruffed grouse, northern bobwhite, wild turkey, and dark-eyed junco. Common blue violet can become widespread in perennial clover plots. However, this is usually a result of the planted forage thinning and the violet taking advantage of available space. If necessary, it can be

controlled in perennial clovers with imazethapyr and 2,4-DB if sprayed when young. Otherwise, most forb-selective herbicides, including triclopyr and dicamba, will kill common blue violet. Mowing does not get rid of common blue violet. Most often, blue violet is a beneficial plant growing along the edge of forage plots and adjacent woods.

Middle photo by J. Brooke; Bottom right photo: Common blue violet showing crown of rhizome; Bottom left photo: Common blue violet grazed by white-tailed deer adjacent to lush clover food plot

459

Index

Acknowledgements

I have synthesized a lot of information in this book, representing the research and work of many people over many years. However, of particular interest and appreciation to me is the information generated by several of my graduate students who worked with food plots as part of their graduate project or a side project: Ryan Basinger (2001 – 03), John Gruchy (2003 – 07), Chris Shaw (2004 – 08), Marcus Lashley (2007 – 09), Michael McCord (2006 – 11), Jessie Birckhead (2009 – 12), Seth Basinger (2010 – 13), Ashley Unger (2010 – 14), Wade GeFellers (2016 – 19), and Katie Harris (2016 – 19). They spent countless hours helping with soil testing, burning, plowing, disking, rotovating, liming, fertilizing, planting, spraying, collecting coverage, biomass, and nutritional data, grinding samples, and running nutritional assays. Jarred Brooke (2012 – 15) and Jordan Nanney (2014 – 17) helped generate information provided in the Plant Identification Guide. Their work has benefitted many thousands of people. Results of their work have been published in professional journals, conference proceedings, Extension publications, and provided in countless Extension presentations. They should be very proud of their accomplishments; I know I am proud of them.

I sincerely appreciate and acknowledge the experiences and opportunities that I have benefitted from at the University of Tennessee through UT Extension and the Department of Forestry, Wildlife, and Fisheries. The freedom and ability to work with natural resources professionals and private landowners on land management issues related to wildlife is a blessing and something I do not take for granted.

I also appreciate the landowners and managers I have worked with over the years. There are many, but landowners and managers such as Jim Phillips, Bennie Riddle, David Hopkins, Bill Smith (Tennessee Wildlife Resources Agency), Arthur Dick, Stu Lewis, Barry Smith, Jason Hewett, and Bobby Watkins (BASF, retired) have been tremendous to work with and have helped me in numerous ways. In particular, they try new management approaches and share their experiences with me, and several have provided land to monitor management strategies, which has led to a better understanding of various land management practices and strategies for wildlife over the years. Always remember: if you keep an open mind, you never stop learning.

Kip Adams (Director of Education and Outreach, QDMA), Dr. Don Ball (Extension Forage Crops Agronomist, Auburn University), Marion Barnes (Extension Agent, Clemson Cooperative Extension), Dr. Gary Bates (Extension Forage Specialist, UT Extension), Dr. Billy Higginbotham

(Extension Wildlife Specialist, Texas A&M University Cooperative Extension), Dr. Karl Miller (Professor of Wildlife Ecology and Management, University of Georgia), and Dr. Neil Rhodes (Extension Weed Specialist, UT Extension) reviewed the entire manuscript or sections of the book and have helped me in numerous other ways. Dr. Fred Allen (Retired Professor of Plant and Soil Sciences, University of Tennessee), Ryan Basinger (Wildlife Biologist, Westervelt Wildlife Services), Dr. Blake Brown (Director, UT Research and Education Center at Milan), Donnie Buckland (Director of Upland Programs, National Wild Turkey Federation), Dr. Debbie Joines (Retired Soils Lab Director, UT Extension), Dr. Heather Kelly (Associate Professor of Entomology and Plant Pathology, University of Tennessee), Dr. Tom Mueller (Professor of Plant and Soil Sciences, University of Tennessee), Gary Pogue (Waterfowl Biologist, US Fish and Wildlife Service), Dr. Hugh Savoy (Retired Soils Specialist, UT Extension), Dr. Scott Stewart (Professor and Integrated Pest Management Coordinator, University of Tennessee), Dr. Bronson Strickland (Professor of Wildlife Management, Mississippi State University), Carl Wirwa (Retired Waterfowl Biologist, Tennessee Wildlife Resources Agency), Dr. John Waller (Retired Professor of Animal Sciences, UT), Tim White (Waterfowl Biologist, Tennessee Wildlife Resources Agency), and Dr. Grant Woods (Wildlife Biologist, Woods and Associates, Inc.) also reviewed specific sections of this book.

I thank Kristy Keel-Blackmon for reviewing manuscripts, providing a great-looking layout, and being patient working with me through this project. I thank Stephanie McManus (AgCentral Farmers Co-op) for perennially helping me with herbicide availability. All of these folks' help certainly led to a better publication.

I sincerely appreciate Dr. Dwayne Estes (Professor of Biology, Austin Peay State University and Botanical Research Institute of Texas) and Dr. Joseph Neal (Professor of Weed Science, NC State University) for reviewing the Plant Identification Guide. Dr. Shawn Askew (Associate Professor of Weed Science, Virginia Tech), Dr. Geir Friisoe (Director of Plant Protection, and Dr. Joe DiTomaso (Professor of Weed Science, University of California—Davis) also provided photos for the Plant Identification Guide.

I also appreciate the support and friendship of many who have spurred and encouraged me to write this book. Kip Adams (QDMA), Clint Borum (TWRA), Dr. David Buehler (University of Tennessee), Aubrey Deck (TWRA), Dr. Dwayne Elmore (OK State University), Joe Hamilton (QDMA), Dr. Chris Moorman (NC State University), Brian Murphy (QDMA), Matt Ross (QDMA), and Dr. Bronson Strickland (MS State University) have been consistent sources of support and encouragement over the years. There are others; you know who you are, and I thank each of you.

Finally, I want to thank my wife, Theresa, and daughters, Ashley, Rachel, and Shelby, for their continued support and patience, again. How many times did I hear "You're still working on that food plot book?!?" Their love and support are boundless. God has blessed me richly. Proverbs 31:10-31.

C.A.H.

Notes

Notes

Notes

Notes

About the Author

Craig A. Harper is a Professor of Wildlife Management and the Extension Wildlife Specialist at the University of Tennessee. His primary responsibility is assisting natural resource professionals and landowners with issues concerning wildlife management.

Dr. Harper's Extension programming and research efforts have concentrated on applied wildlife management, especially as related to forest management, early succession management, prescribed fire effects, herbicide applications, quality deer management, and food plot management. He has conducted research on food plots, wildlife openings, and their use by wildlife since 1994.

Craig earned an A.A.S. in Fish and Wildlife Management from Haywood Community College (1988), a B.S. in Natural Resources Management from Western Carolina University (1990), an M.S. in Biology from the University of North Carolina at Wilmington (1993), and a Ph.D. in Forest Resources from Clemson University (1998). Between degrees, he worked as a wildlife technician with the North Carolina Wildlife Resources Commission where managing food plots and wildlife openings was one of his duties. Craig is a Certified Wildlife Biologist® and Certified Prescribed Fire Manager.

Craig is a 10th generation North Carolinian. His interest in food plots came naturally as he grew up hunting on his family's farm in the Piedmont. Craig and his family live in Blount County, Tennessee.